Polymer Biocatalysis and Biomaterials II

ACS SYMPOSIUM SERIES **999**

Polymer Biocatalysis and Biomaterials II

H. N. Cheng, Editor
Hercules Incorporated Research Center

Richard A. Gross, Editor
Polytechnic University

Sponsored by the
ACS Division of Polymer Chemistry, Inc.

American Chemical Society, Washington, DC

ISBN 978–0–8412–6970–5

The paper used in this publication meets the minimum requirements of American National Standard for Information Sciences—Permanence of Paper for Printed Library Materials, ANSI Z39.48–1984.

Distributed by Oxford University Press

PRINTED IN THE UNITED STATES OF AMERICA

Foreword

The ACS Symposium Series was first published in 1974 to provide a mechanism for publishing symposia quickly in book form. The purpose of the series is to publish timely, comprehensive books developed from ACS sponsored symposia based on current scientific research. Occasionally, books are developed from symposia sponsored by other organizations when the topic is of keen interest to the chemistry audience.

Before agreeing to publish a book, the proposed table of contents is reviewed for appropriate and comprehensive coverage and for interest to the audience. Some papers may be excluded to better focus the book; others may be added to provide comprehensiveness. When appropriate, overview or introductory chapters are added. Drafts of chapters are peer-reviewed prior to final acceptance or rejection, and manuscripts are prepared in camera-ready format.

As a rule, only original research papers and original review papers are included in the volumes. Verbatim reproductions of previously published papers are not accepted.

ACS Books Department

Contents

Enzyme Immobilization and Assembly

New Synthetic Approaches

Polyesters and Polyamides

Polysaccharides, Glycopolymers, and Sugars

Silicone-Containing Materials

Indexes

Preface

Biocatalysis and biomaterials are dynamic areas of research that have continued to attract a lot of attention. Developments in these areas are largely fueled by demands for sustainable technologies, a desire to decrease our dependence on petroleum, and commercial opportunities to develop 'green' products. Publications and patents in these fields continue to grow as more people are involved in research and commercial activities.

The purpose of this book is to bring together leading researchers from different labs and different countries and publish their latest findings. Because of the multidisciplinary nature of these fields, publications tend to be spread out over different journals. There is great value in publishing a book with all relevant papers in one volume.

This book is based on an international symposium held at the American Chemical Society (ACS) National Meeting in Boston, Massachusetts on September 11–14, 2006. Many exciting new techniques and findings were reported at that symposium. In addition, we have also invited several leaders of the fields to write special reviews of their ongoing work. It is our hope that this book will provide a good representation of what is happening in the forefront of research in polymer biocatalysis and biomaterials.

The book is targeted for scientists and engineers (chemists, biochemists, chemical engineers, biochemical engineers, microbiologists, molecular biologists, and enzymologists) as well as for graduate students who are engaged in research and developments in polymer biocatalysis and biorelated materials. It can also be a useful reference book for people who are interested in these topics.

We thank the authors for their contributions and their patience while the manuscripts were being reviewed and revised. Thanks are also due to the ACS Division of Polymer Chemistry, Inc. for sponsoring the symposium. We also acknowledge the generous funding from the ACS Division of Polymer Chemistry, Inc. and the Petroleum Research Fund.

H. N. Cheng
Hercules Incorporated Research Center
500 Hercules Road
Wilmington, DE 19808–1599

Richard A. Gross
NSF I/UCRC for Biocatalysis and Bioprocessing of Macromolecules
Polytechnic University
6 Metrotech Center
Brooklyn, NY 11201

Polymer Biocatalysis and Biomaterials II

Chapter 1

Polymer Biocatalysis and Biomaterials: Current Trends and Developments

H. N. Cheng[1] and Richard A. Gross[2]

[1]Hercules Incorporated Research Center, 500 Hercules Road,
Wilmington, DE 19808–1599
[2]NSF, Center for Biocatalysis and Bioprocessing of Macromolecules,
Polytechnic University, Six Metrotech Center, Brooklyn, NY 11201

This paper reviews current trends and developments in polymer biocatalysis and biomaterials. Developments in biocatalysis and biomaterials are largely fueled by demands for sustainable chemical technologies, a desire to decrease our dependence on foreign petroleum, and real commercial opportunities to develop 'green' products. Commercial activities continue to increase as is evident from a review of the patent literature. The following topics are covered in this review: 1) novel biomaterials, 2) new and improved biocatalysts, 3) new biocatalytic methodologies, 4) polyesters and polyurethanes, 5) polyamides and polypeptides, 6) polysaccharides, 7) silicon-containing materials, 8) biocatalytic redox polymerizations forming C-C bonds, 9) enzymatic hydrolysis and degradation of natural polymers, 10) biocatalytic routes to monomers. Examples featured are mostly taken from contributions to the Symposium on Biocatalysis in Polymer Science held during the ACS National Meeting in San Francisco in September 2006.

Biocatalysis involves the use of enzymes, microbes, and higher organisms to carry out chemical reactions. Because the reaction conditions are often mild, water-compatible, and environmentally friendly, they offer excellent examples of "green chemistry". In view of the wealth of enzymes and whole-cell approaches available and the quality of researchers, this field continues to be very dynamic and productive. Biocatalysis in polymeric materials has been reviewed periodically (1-2). It is clear from the reviews that there is no shortage of activity and creativity in this field.

Biomaterials constitute an equally exciting field of research that finds numerous applications in medical and industrial areas (3,4). In fact, much commonality is found in biomaterial and biocatalytic research. For example, many researchers in biomaterials work with enzymes, and, often, investigators in biocatalysis are producing natural polymers or polymers that are closely related but are are modified by chemical or enzymatic methods to improve their physical properties.

This paper does not attempt to provide a comprehensive review of research in biocatalysis or biomaterials. Instead, it highlights major themes and developments in these fields, using selected literature and emphasizing research presented during the ACS National Meeting in San Francisco in September 2006 as published in *ACS Polymer Preprints* (31-57), and papers published in expanded form in this book (5-30). This paper has been divided into the following ten sections: 1) novel biomaterials, 2) new and improved biocatalysts, 3) new biocatalytic methodologies, 4) polyesters and polyurethanes, 5) polyamides and polypeptides, 6) polysaccharides, 7) silicon-containing materials, 8) biocatalytic redox polymerizations forming C-C bonds, 9) enzymatic hydrolysis and degradation of natural polymers, and 10) biocatalytic routes to monomers.

1. Novel Biomaterials

As noted earlier, biomaterials (and other bio-related materials) comprise one of most active research areas today. Major research themes in biomaterials include tissue engineering (58), molecular imprinting strategies (59), biosensors (60), stimuli responsive materials (61), biodegradable polymers (62), and smart biomaterials (63).

In this book, a large number of biomaterials are reported. These include polypeptides/proteins, carbohydrates, lipids/triglycerides and synthetic polymers. In addition, it is understood that many of the polymeric materials involved in all the chapters can potentially be used as biomaterials although they may not be specified as such.

Polypeptides/proteins.

A major research trend is to tailor the structure of proteins and polypeptides so they can function for a wide range of advanced material applications. In their paper, Kiick et al (5) described their work in using biosynthetic routes to produce polypeptides with non-natural amino acids that have desired conformation and side-chain placement. Montclare et al (6) described novel research on elastins where the aim is to generate new biomaterials that have the desired biological activity, optimal function in delivery of therapeutics, and more applications.

Ito et al (16) used combinatorial bioengineering methods to produce new biomaterials based on amino acids, nucleic acid, and non-natural components. In a different way, Silvestri et al (7) produced biomaterials by combining enzymes with synthetic polymers; some examples were combinations of α-amylase with poly(vinyl alcohol), poly(ethylene glycol), and poly(hydroxyethyl methacrylate). Jong (9) used soy protein as a reinforcement material in elastomers and observed an increase in the rubber modulus.

Polysaccharides

Sophorolipids are extracellular glycolipids that can be produced in high yields from yeast fermentation. Waller et al (57) carried out metathesis reactions to produce high molecular weight polymers from sophorolipids. Another example of the use of polysaccharide is the paper by Jong (9), where carbohydrates are incorporated in polymer composites.

Lipids and triglycerides

Biswas et al (8) modified soybean oil and fatty acids under mild conditions to introduce nitrogen-containing functionalities. Further derivatives could be made via enzymatic or chemical transformations. Another example is the polymerization of sophorolipids, as mentioned above (57).

Synthetic polymers

Albertsson and Srivastava (53) produced copolyesters from 1,5-dioxepan-2-one and ε-caprolactone via lipase catalysis; the resulting material was porous and potentially suited for tissue engineering. As noted earlier, Silvestri, et al (7) developed "bioartificial polymeric materials" with the aim of producing biomaterials that combine features of synthetic polymers (good mechanical properties, easy processability, lower costs) with specific tissue- and cell-compatibility found in natural materials such as extracellular matrix glycoproteins.

2. New or Improved Biocatalysts

Since biocatalysts are the prerequisites for biocatalysis, it is not surprising that much effort has been expended on improving existing biocatalysts or devising new biocatalysts.

New or Improved Enzymes

Nolte et al (46) produced an artificial enzyme based on the T4 replisome and applied it to the epoxidation of double bonds in synthetic polymers. Smith et al (51) reported that horseradish peroxidase catalyzes the oxidative polymerization of glucuronic acid. In recent literature, many biomimetic macromolecules with enzyme-like structures or functions have been reported including those that are dendrimers (64-66), those that have specified three-dimensional structures or recognition elements created by molecular imprinting (67), and other enzyme mimics (68).

The genetic modification of existing enzymes to improve their properties is now well established and often practiced. Many methods are available, including directed evolution (69) and gene, pathway and genome shuffling (70). Minshull and coworkers (45) at DNA 2.0 developed a new, engineering-based method using computational tools. This allows the design and synthesis of small, information-rich libraries of protein variants (50 to 100). Information gained from evaluation of these variants is then fed back to learning algorithms that generate improved sets of protein variants. This iterative process leads to large improvements in enzyme activities using relatively small total numbers of protein variants. A key advantage of this method is that, with small protein variant libraries, evaluations of proteins can be carried out with sufficient number of replicants under conditions that closely resemble that in the target application. Mang et al (47) described research where they used *Candida antarctica* Lipase B (CALB) variants synthesized by the DNA 2.0 team. The research goal was to engineer CALB that has hyperactivity for catalysis of polyester synthesis using long chain (C14-C18) ω-hydroxyfatty acid monomers.

DeAngelis (50) transferred the glycosaminoglycan (GAG) synthase genes into *Bacillus subtilis* to produce GAG polymers. Well-defined GAG oligosaccharides were also produced using engineered variants of these enzymes.

Furthermore, developments by the DeAngelis group were the bases of a new commercial process launched by Novozymes for the fermentative production of hyaluronic acid (see Brown et al [49]) using recombinant *B. subtilis* strains.

Enzyme Immobilization

Enzyme immobilization has several benefits, including enzyme recovery and reuse. The abilitiy to recycle enzyme-catalysts so they may be reused hundreds of times is critical to achieve economically viable biocatalytic processes. Furthermore, by their immobilization, enzyme thermal and chemical stability as well as catalytic activity can be greatly improved. Numerous enzymes are available commercially in their immobilized form (e.g. Novozym® 435, immobilized CALB). Not surprisingly, many immobilization methods have been devised (*71-74*). The following account provides several good examples; these are also summarized in Table 1.

Gitsov et al (*10*) produced a "nanoreactor" from laccase and linear poly(ethylene oxide)-dendritic poly(benzyl ether) diblock copolymers. A notable feature of this system is the presence of hydrophobic dendritic pockets that increase the local concentration of water-insoluble organic compounds near the active site where they are oxidized. Li and Hsieh (*11*) employed a hydrogel fiber membrane, with a large surface area and improved organic solvent solubility in order to facilitate lipase reactions. He (*12*) grafted lipase onto silica particles, which permitted him to carry out synthesis of polyesters, polycarbonates, polyphosphates, and their copolymers at temperatures up to 150°C.

In order to explore how resin parameters, such as particle and pore size, affect CALB catalytic activity, Chen et al (*13,14*) immobilized CALB on a series of macroporous resins consisting of poly(methyl methacrylate), PMMA, and polystyrene, PS. Surprisingly, systematic variations in resin particle size had very different effects on CALB catalytic activity for polyester synthesis for resins consisting of PMMA and PS. These effects of surface chemistry and enzyme distribution throughout resins are discussed in detail (see chapters 10 & 11). Vaidya et al (*52*) explored covalent strategies for CALB immobilization in order to decrease CALB leaching from resins and, thereby, improve catalyst recyclability. Since methods are not readily available to visualize chemically immobilized enzymes within macroporous resins, Chen et al. (*48*) immobilized CALB on a series of flat films. Chemical immobilization was achieved through reactions with epoxy pendant groups of glycidyl acrylate units. Variations in film hydrophobicity was achieved by changing the ratio of hydroxyethylacrylate and butyl acrylate repeat units in copolymers. AFM studies were performed to determine changes in CALB volume and extent of aggregation as a function of film hydrophobicity. These physical phenomena were correlated with CALB

6

hydrolytic activity. Even though enzyme immobilization on macroporous supports provides many benefits, resins used are often costly and, therefore, must provide sufficient benefits in enzyme performance in order to recap the increased cost incurred. An alternative, explored by Vaidya et al. (*35*), is to develop immobilized recyclable enzyme systems that do not require an additional resin component. This was achieved by forming self-crosslinked enzyme aggregates (CLEA) of CALB that have sufficient size that allow their recovery by filtration. A problem with this technology is the small pore size of the CALB CLEAs that results in poor substrate diffusivity during polyester synthesis reactions.

To facilitate lipase-reactions in supercritical CO_2, Bruns et al (*32*) impregnated a lipase within amphiphilic co-networks that comprised a fluorophilic phase. This resulted in substantially higher enzyme activity relative to the enzyme powder. It may be noted that the bio-artificial materials reported by Silvestri et al (*7*) can also be considered a form of enzyme immobilization.

Table 1. Examples of Immobilized Enzymes

Author(s)	Enzyme[a]	Immobilization method[a]
Gitsov, et al (*10*)	laccase	Nano-reactor (PEO-dendritic PBE)
Li and Hsieh (*11*)	C. rugosa lipase	hydrogel fibrous membranes (PAA, PVA electrospinning)
He (*12*)	porcine pancreatic lipase	silica particles
Chen et al (*13*)	CALB	PMMA macroporous resins
Chen et al (*14*)	CALB	polystyrene macroporous resins
Bruns et al (*30*)	R. miehei lipase	amphiphilic co-networks
Vaidya et al (*35*)	CALB	self-crosslinked networks
Chen et al (*48*)	CALB	polymethacrylate films
Vaidya et al (*52*)	CALB	macroporous resins
Silvestri et al (*7*)	amylase	PVA hydrogel, HEMA template polymer, EVAL-dextran sponge

[a]PEO = poly(ethylene oxide), PBE = poly(benzyl ether), CALB = *Candida antarctica* lipase B, PVA = poly(vinyl alcohol), HEMA = hydroxyethyl methacrylate, EVAL = ethylene-vinyl alcohol copolymer, PAA = poly(acrylic acid), PMMA = poly(methyl methacrylate)

Enzyme Assemblies

By strategic assembly of enzymes it may be possible to achieve enhanced enzyme activities or efficient multiple-step biotransformations. Two prominent literature examples that describe how enzyme assemblies can be exploited are the "superbeads" technology developed by Wang et al (*75*) where several enzymes were immobilized along the biosynthetic pathway onto beads, and the "enzyme reactor" of DeAngelis (*76*) where two single-action enzymes were immobilized for stepwise oligosaccharide synthesis.

Another concept, introduced by van Hest et al (*39*), used enzymes non-covalently assembled in polymeric capsules forming a metabolic pathway; thus, the product of one catalytic step could be directly used as substrate for the next. Lambermon et al (*40*) developed scaffolds to immobilize enzymes to produce synthetic metabolons. The scaffolds were based on self-assembling organized systems, such as virus particles or aggregates, formed in water by an ammonium terminated surfactant.

Smart Biocatalysts

Alternatively, researchers have been developing "smart" enzyme-polymer conjugates (see, for example, Refs 61b and 63c). An example is to create enzyme-polymer conjugates that are reversibly soluble or insoluble in a solvent medium. Other examples are enzyme-polymer conjugates that are stimuli-responsive, changing their solubility as a function of changes in pH, temperature, ionic strength, or metal ion(s). As a different approach that mimics natural systems, enzyme-activity can be regulated through their functional connection or via molecular recognition processes on lipid bilayer membranes (*77*).

3. New or Improved Biocatalytic Methodologies

Below are examples, taken primarily from the ACS Symposium on Biocatalysis in Polymer Science at the National Meeting in San Francisco (September 2006), that highlight new biocatalytic methodologies. Indeed, throughout this book, many newly developed methodologies are reported that enable enzymes to do new or improved biocatalytic conversions.

Combinatorial technology

In their chapter, Ito et al (*16*) used combinatorial bioengineering, in which target molecules were selected from a random molecular library, to prepare functional polymers that had molecular recognition and catalysis capability. Other recent examples of where combinatorial methods have been applied to solve problems are given in Ref. 78.

Biocatalysis on surfaces

Although most organic reactions can be performed on surfaces, steric hindrance and diffusion barriers can hamper the yield or rate of reactions. Mahapatro (15) reviewed this field and noted there are few reports on biocatalytic methodologies for surface modification reactions. He believed that because of the mild reaction conditions and enforced positioning of functional groups in self-assembled monolayers, biocatalysts on surfaces could have great potential for selective rate enhancement and inhibition (15,79). Cummins et al (56) reported the use of enzymes to catalyze hydroxy-functionalized surface reactions, e.g., introduction of hydroxy group to silicon wafers, and (subsequent) enzymatic grafting of polymers.

Chemoenzymatic synthesis of block copolymers

Heise, Palmans, de Geus, Villarroya and their collaborators (17,41,42) have been working on a chemoenzymatic cascade synthesis to prepare block copolymers. They combine enzymatic ring-opening polymerization (eROP) and atom transfer radical polymerization (ATRP). The synthesis of block copolymers was successful in two consecutive steps, i.e., eROP followed by ATRP. In the one-pot approach, block copolymers could be obtained by sequential addition of the ATRP catalyst, but side reactions were observed when all components were present from at the onset of reactions. A successful one-pot synthesis was achieved by conducting the reaction in supercritical carbon dioxide.

Dynamic kinetic resolution polymerization

Hilker et al (44) combined dynamic kinetic resolution with enzymatic polycondensation reactions to synthesize chiral polyesters from dimethyl adipate and racemic secondary diols. The concept offered an efficient route for the one-pot synthesis of chiral polymers from racemic monomers. Palmans at al (18,43) generalized the approach to *Iterative Tandem Catalysis* (ITC), in which chain growth during polymerization was effected by two or more intrinsically different catalytic processes that were compatible and complementary.

Sono-enzymatic polymerization of catechol

Fernandes et al (31) used laccase in combination with ultrasound to improve coloration of wool by "in situ" radical polymerization of catechol. The hydroxyl radicals produced by ultrasound react with intermediate molecules produced by the enzyme, enhancing the enzymatic catechol polymerization. Moreover, ultrasound improved the diffusion processes and might also have a positive effect on laccase activity.

Different Reaction Media

Biocatalysis in supercritical fluids, fluorous solvents, and under solvent-free conditions was recently reviewed (*80*). In this book, de Geus et al (*17*), Villarroya (*41*) and Bruns et al (*32*) all provide important examples of how supercritical CO_2 can be used for enzyme-catalyzed reactions. Furthermore, Srienc et al (*38*) used ionic liquid media for enzyme-catalyzed polymerizations of β-butyrolactone in order to prepare poly(hydroxyalkanoic acids), PHA. The role of ionic liquids was to both maintain enzyme activity and propagating chain solubility so that high molecular weight products could be obtained in monophasic media.

4. Polyesters and Polyurethanes

The use of whole cell catalysts for the preparation of PHAs has been an area of intense interest by many researchers (*81*). Investigations in this area have led to discoveries of biocatalytic methodologies to prepare a wide range of PHAs with highly diverse structures (See Ref. 82 for recent reviews). This is largely due to the promiscuity of enzymes involved in PHA synthesis. Examples of repeat unit types incorporated in PHAs include those with aromatic, fluorinated, terminal alkene, alkyne and much more. β-Ester links have been replaced by thioester moieties as well as wide range of ester links with hydroxyl groups that are beyond the β-position. Maybe the most important accomplishment has been the ability to achieve high PHA volumetric yields and produce copolymers, such as those with combinations of C4, C6 and C8 repeat units that have desirable properties and can be thermally processed forming coatings on paper, bottles, handles for razor blades and much more (*81d*). A recent collaboration between ADM and Metabolix seeks to further reduce costs, improve product performance and increase the volume of PHAs that are used in commodity applications.

Important progress has been made on the use of immobilized enzyme catalyzed polycondensation reactions to prepare a wide range of polyesters (*2,83*). In this book, Hunsen et al (*20*) describe how an immobilized catalyst from Humilica insolens has excellent activity for polycondensation reactions between diols and diacids. The natural role of cutinases is hydrolysis of the outer 'cutin' polyester layer on plant leaves. Cutin is made of long chain hydroxyl fatty acids that are crosslinked at various positions. Also, Hu et al (21) have a chapter in this book describing the effects of different alditol substrates on polycondensation reactions with adipic acid. These natural polyols differ in chain length and stereochemistry. Azim et al (22) used the immobilized lipase catalyst Novozym 435 to prepare polyesters from relatively short chain building blocks (e.g. butanediol and succinate). Hilker et al (44) use dyamic kinetic resolution to prepare chiral polyesters from polycondensation reactions catalyzed by Novozym 435 between a racemic diol and dimethyl adipate. Another contribution from Hu at al (34) describes the selectivity of CALB for polycondensation reactions between glycerol, adipic acid and octanediol. Glycerol polyesters are particularly interesting given the high abundance of glycerol as a by-product of biodiesel production. In an attempt to increase branching during lipase-catalyzed polycondensation reactions, Kulshrestha et al (54) explored the preparation of hyperbranched aliphatic polyesters by introducing trimethlyol propane (TMP) as a polyol building block. TMP differs from glycerol in that glycerol has two primary and one secondary hydroxyl whereas all three hydroxyl moieties of TMP are primary.

As an alternative route to polyesters, considerable progress has also been made in using enzymes (most often CALB) to catalyze lactone ring-opening polymerization reactions (2,84). This book also provides a number of examples of recent advances in this research area. For example, as above, Hunsen et al (20) explored the kinetics and mechanism by which the cutinase from *Humilica insolens* catalyzes lactone (e.g., ε-caprolactone) ring-opening polymerization reactions. Martinez-Richa et al (33) explore that activity of *Yarrowia lipolytica* lipase for ε-caprolactone ring-opening polymerization reactions. Furthermore, in an effort to increase the hydrolytic degradation rate of poly(ε-caprolactone), Albertsson and Srivastava (53) used Novozym 435 to copolymerize ε-caprolactone and 1,5-dioxepan-2-one. Moreover, Duxbury et al (55) explored the activity of immobilized CALB for reactions in which ε-caprolactone is grafted from poly(styrene-co-4-vinyl benzyl alcohol) copolymers. The question addressed in their research is the extent that CALB can access internal hydroxyl groups along polyvinyl chains that can either initiate chain growth or react with a propagating poly(ε-caprolactone) chain via either a condensation reaction with a terminal carboxyl moiety or by transesterification reactions at an internal site along preformed chains.

Research by Matsumura et al (19) describes a different concept whereby lipase-catalysis is used to first degrade polyesters into cylic oligomers. Subsequently, cyclic oligomers formed are then polymerized using the same enzyme-catalyst to prepare high molecular weight polyesters. If such a process could be efficient performed with commercial polyesters, it could function as a method to recycle existing materials into new polyesters.

Polyurethanes can also be prepared making use of precursor molecules first prepared by enzyme-catalysis. Some examples of polyurethane synthesis that involved enzyme-catalysis are given in Ref. 85.

5. Polyamides and Polypeptides

In contrast to polyesters, very few publications deal with enzymatic synthesis of polyamides. Gu et al (24) discovered a process of using lipase at high temperatures in a single-step reaction to produce high molecular weight polyamides. A large number of polyamides were made enzymatically, including a few that could not be synthesized via chemical means.

In a separate contribution to this book, Li et al (23) used papain to prepare oligo(γ-ethyl-L-glutamate) in a phosphate buffer at pH 7. Their synthesis was rapid, giving yields of about 80% within 10 to 15 minutes. Key to driving the process is the use of diethyl esters that increase the kinetics of reactions. Furthermore, reactions are thermodynamically driven to product by the precipitations of oligomers once they reach chain lengths of about eight units. Other protease-mediated amide and oligopeptide syntheses have been reported elsewhere (86).

6. Polysaccharides

Biocatalysis is a key route to both natural and non-natural polysaccharide structures. Research in this area is particularly rich and generally involves at least one of the following three synthetic approaches: 1) isolated enzyme, 2) whole-cell, and 3) some combination of chemical and enzymatic catalysts (i.e. chemoenzymatic methods) (87-90). Two elegant examples that used cell-free enzymatic catalysts were described by Makino and Kobayashi (25) and van der Vlist and Loos (27). Indeed, for many years, Kobayashi has pioneered the use of glycosidic hydrolases as catalysts for polymerizations to prepare polysaccharides (88,91). In their paper, Makino and Kobayashi (25) made new monomers and synthesized unnatural hybrid polysaccharides with regio- and stereochemical-control. Van der Vlist and Loos (27) made use of tandem reactions catalyzed by two different enzymes in order to prepare branched amylose. One enzyme catalyzed the synthesis of linear structures (amylose) where the second enzyme introduced branches. In this way, artificial starch can be prepared with controlled quantities of branched regions.

An excellent example of whole-cell polysaccharide synthesis is provided by important work in the biosynthesis of glycosaminoglycans (GAGs). For example, Brown et al (*49*) reported the use of a recombinant *Bacillus subtilis* strain as the host for hyaluronic acid (HA) production. This work has been developed into a commercial scale process by Novozymes for non-animal derived HA. Fermentatively produced HA is used in numerous cosmetic products, as therapeutic polymers to treat arthritic conditions and as scaffolds for tissue engineering. DeAngelis (*50*) also reported novel chemoenzymatic methods with synchronized reactions that allow the preparation of well-defined oligomeric and polymeric GAGs.

For many years Wang and his group (*26*) have carried out extensive research on glycopolymers (e.g., glycosylated linear polymers or hydrogels). In their review article, they described some of the characteristics of these polymers, particularly their use as anti-adhesion drugs.

An entirely different approach to sugar-based polymers involves the use of selective enzymatic catalysts to prepare vinyl sugar monomers that are subsequently polymerized via chemical catalysts. Tokiwa and Kitagawa (*28*) published extensively on this subject, and their contribution within this book describes a wide range of sugar monomer structures.

7. Silicon-Containing Materials

Silicon-containing materials represent an exciting new area of research. In their paper, Clarson and coworkers (*29*) studied the silica-forming ability of three peptides derived from diatoms, including kinetics and morphological characterization. McAuliffe et al (*30*) described a number of enzymatic and whole-cell reactions to oxidize aryl silanes. The resulting silicon-containing chiral cis-diols and catechols might be used in a number of applications.

8. Biocatalytic Redox Polymerizatons forming C-C Bonds

It is now well-established that some enzyme families, including various peroxidases and laccases, catalyze the polymerization of vinyl monomers and other redox active species such as phenol-type structures. Vinyl polymerization by these redox catalysts has recently been reviewed (*93*). These catalysts have been used to prepare polyanilines (*94*) and polyphenols (*95,96*). A few examples of related research are included in this book. For example, Smith et al (*51*) described a novel reaction catalyzed by horseradish peroxidase (HRP). In the presence of HRP and oxygen, D-glucuronic acid was polymerized to a high molecular weight (60,000) polyether. However, the authors have not yet illucidated the polyether structure. Two other oxidative biotransformations were discussed above: i) the sono-enzymatic polymerization of catechol via laccase (*31*), and ii) the oxidation of aryl silanes via aromatic dioxygenases (*30*).

9. Enzymatic Hydrolysis and Degradation of Natural Polymers

Studies on hydrolases as catalysts for polymer degradation have been an area of fundamental research for many years. Early research focused on degradation pathways of natural polymers such as lignin, cellulose and starch. The breakdown of various biomass constituents for conversions to biofuels is currently under intense study. Improved enzymes are being development that can function together with other enzymes for biomass-to-fuel conversions. Major technical hurdles include the development of enzymes that are not inhibited by byproducts formed during these complex multi-enzyme multi-substrates processes. Although many reviews are currently available on these subjects, readers may wish to go through the following publications that include a general review (*97*), focus on cellulose and hemicellulose hydrolysis (*98*), lignin bioconversions (*99,100*), and lignin-containing effluent treatment (*101*).

10. Biocatalytic Routes to Monomers

Biocatalytic routes to monomers have resulted in breakthrough new technologies. Prominent examples include biocatalytic routes to 1,3-propane-diol, lactic acid, and acrylamide. Contributions towards biocatalytic routes to monomer synthesis, presented at the Symposium on Biocatalysis in Polymer Science held during the ACS National Meeting in San Francisco (September 2006) include Haering (*36*) who described a series of specialty acrylic acid esters that were prepared by lipase-catalyzed transesterification of inexpensive bulk alkyl acrylates. Synthesized monomers had hydroxy, carbamate, sugar, benzophenone, amide and acetal functional groups. Haynie et al (*37*) documented accomplishments by DuPont scientists where enzymes were used to prepare the following monomers: para-hydroxycinnamic acid (pHCA) (from tyrosine), p-hydroxystyrene (from pHCA), and 1,3-propanediol (from glucose).

References

1. Some recent books on biocatalysis include: (a) *Polymer Biocatalysis and Biomaterials*; Cheng, H.N.; Gross, R. A., Eds.; Amer. Chem. Soc., Washington, DC, 2005. (b) *Biocatalysis in Polymer Science*; Gross, R. A.; Cheng, H. N., Eds.; Amer. Chem. Soc., Washington, DC, 2003. (a) *Biocatalysis and Biodegradation*; Wackett, L. P.; Hershberger, C. D.; ASM Press, Washington, DC, 2001. (c) *Enzymes in Polymer Synthesis*; Gross, R. A.; Kaplan, D. L.; Swift, G., Eds.; Amer. Chem. Soc., Washington, DC, 1998.

14

2. Some reviews on polymer biocatalysis include: (a) Cheng, H. N.; Gross, R. A. *ACS Symp. Ser.* **2005**, *900*, 1. (b) Cheng, H. N.; Gross, R. A. *ACS Symp. Ser.* **2002**, *840*, 1. (c) Gross, R. A.; Kumar, A.; Kalra, B. *Chem. Rev.* **2001**, *101*, 2097. (d) Kobayashi, S.; Uyama, H.; Kimura, S. *Chem. Rev.* **2001**, *101*, 3793.

3. Some books on biomaterials include: (a) *Biomaterials.* 2nd Ed.; Bhat, S.V.; Alpha Science, 2005. (b) *Biomaterials Science: An Introduction to Materials in Medicine.* 2nd *Ed.*; Ratner, B. D.; Academic Press, 2004. (c) *Biomaterials: From Molecules to Engineered Tissues*; Hasirci, N.; Hasirci, V.; Kluwer Academic, 2004. (d) *Biorelated Polymers: Sustainable Polymer Science and Technology*; Chiellini, E.; Springer, 2001.

4. Some reviews on biomaterials include: (a) Roach, P.; Eglin, D.; Rohde, K.; Perry, C.C. *J. Materials Sci.: Materials in Medicine* **2007**, *18*, 1263. (b) Stupp, S.I. (Editor), *Annual Review of Materials Research*, Vol. 31, August 2001. (c) Stikeman, A. *MIT Technology Rev.* November 2002.

5. Charati, M.; Kas, O.; Galvin, M.E.; Kiick, K. L. *ACS Symposium Ser.* (this volume), Chapter 2.

6. Baker, P. J.; Haghpannah J. S.; Montclare, J. K. *ACS Symposium Ser.* (this volume), Chapter 3.

7. Silvestri, D.; Cristallini, C.; Barbani, N. *ACS Symposium Ser.* (this volume), Chapter 4.

8. Biswas, A.; Shogren, R. L.; Willett, J. L.; Erhan, A. Z.; Cheng, H. N. *ACS Symposium Ser.* (this volume), Chapter 5.

9. Jong, L. *ACS Symposium Ser.* (this volume), Chapter 6.

10. Gitsov, I.; Simonyan, A.; Krastanov, A.; Tanenbaum, S. *ACS Symposium Ser.* (this volume), Chapter 7.

11. Li, L.; Hsieh, Y.-L. *ACS Symposium Ser.* (this volume), Chapter 8.

12. He, F. *ACS Symposium Ser.* (this volume), Chapter 9.

13. Chen, B.; Miller, E. M.; Miller, L.; Maikner, J.J.; Gross, R.A. *ACS Symposium Ser.* (this volume), Chapter 10.

14. Chen, B.; Miller, E.M.; Gross, R.A. *ACS Symposium Ser.* (this volume), Chapter 11.

15. Mahapatro, A. *ACS Symposium Ser.* (this volume), Chapter 12.

16. Ito, Y.; Abe, H.; Wada, A.; Liu, M. *ACS Symposium Ser.* (this volume), Chapter 13.

17. de Geus, M.; Palmans, A.R.A.; Duxbury, C. J.; Villarroya, S.; Howdle, S. M.; Heise, A. *ACS Symposium Ser.* (this volume), Chapter 14.

18. Palmans, A.; van As, B.; van Buijtenen, J.; Meijer, E. W. *ACS Symposium Ser.* (this volume), Chapter 15.

19. Kondo, A.; Sugihara, S.; Okamoto, K.; Tsuneizumi, Y.; Toshima, K.; Matsumura, S. *ACS Symposium Ser.* (this volume), Chapter 16.

20. Hunsen, M.; Azim, A.; Mang, H.; Wallner, S. R.; Ronkvist, A.; Xie, W.; Gross, R.A. *ACS Symposium Ser.* (this volume), Chapter 17.

15

21. Hu, J.; Gao, W.; Kulshrestha, A.; Gross, R. A. *ACS Symposium Ser.* (this volume), Chapter 18.
22. Azim, H.; Dekhterman, A.; Jiang, Z.; Gross, R. A. *ACS Symposium Ser.* (this volume), Chapter 19.
23. Li, G.; Vaidya, A.; Xie, W.; Gao, W.; Gross, R.A. *ACS Symposium Ser.* (this volume), Chapter 20.
24. Gu, Q.-M.; Maslanka, W. W.; Cheng, H. N. *ACS Symposium Ser.* (this volume), Chapter 21.
25. Makino, A.; Kobayashi, S. *ACS Symposium Ser.* (this volume), Chapter 22.
26. Zhang, Y.; Wang, J.; Xia, C.; Wang, P. G. *ACS Symposium Ser.* (this volume). Chapter 23.
27. van der Vlist, J.; Loos, K. *ACS Symposium Ser.* (this volume). Chapter 24.
28. Tokiwa, Y.; Kitagawa, M. *ACS Symposium Ser.* (this volume). Chapter 25.
29. Whitlock, P. W.; Patwardhan, S. V.; Stone, M. O.; Clarson, S. J. *ACS Symposium Ser.* (this volume). Chapter 26.
30. Smith, W. C.; Whited, G.M.; Lane, T. H.; Sanford, K.; McAuliffe, J. C. *ACS Symposium Ser.* (this volume). Chapter 27.
31. Fernandes, M.; Basto, C.; Zille, A.; Munteanu, F.; Guebitz, G. M.; Cavaco-Paulo, A. "Sono-enzymatic polymerization of catechol" *ACS Polymer Preprints*, **2006**, *47*(2), 273.
32. Bruns, N.; Tiller, J.C. "Amphiphilic conetworks comprising a fluorophilic phase" *ACS Polymer Preprints*, **2006**, *47*(2), 205.
33. Martinez-Richa, A.; Barrera-Rivera, K. A.; Flores-Carreon, A. "Synthesis and characterization of poly(ε-caprolactone) obtained by enzymatic polymerization with Yarrowia lipolytica lipase" *ACS Polymer Preprints*, **2006**, *47*(2), 277.
34. Hu, J.; Gao, W.; Kulshrestha, A.S.; Gross, R. A. "Synthesis of glycerol-based oligomers and polymer: Comparison between lipase and dibutyltin oxide catalyzed polymerization" *ACS Polymer Preprints*, **2006**, *47*(2), 279.
35. Vaidya, A.; Xie, W.; Gao, W.; Bohling, J.C.; Miller, M.E.; Gross, R. A. "Enzyme immobilization without a support: Candida antartica lipase B (CALB) Self-crosslinked aggregates" *ACS Polymer Preprints*, **2006**, *47*(2), 236.
36. Häring, D. "Industrial applications of enzymes in polymer science" *ACS Polymer Preprints*, **2006**, *47*(2), 249, and references therein.
37. Haynie, S. L.; Nakamura, C.E.; Sariaslani, S. "Fermentative and biocatalytic processes to prepare 1,3-propanediol, p-hydroxycinnamic acid and p-hydroxystyrene" *ACS Polymer Preprints*, **2006**, *47*(2), 241.
38. Gorke, J.; Kazlauskas, R. J.; Srienc, F. "Enzymatic synthesis of PHAs using ionic liquids as solvents" *ACS Polymer Preprints*, **2006**, *47*(2), 233.
39. van Hest, J.C.M.; Vriezema, D.M.; Garcia, P.M.L.; Rowan, A.E.; Nolte, R.J.M.; Cornelissen, J. "Enzyme positional assembly in polymeric capsules" *ACS Polymer Preprints*, **2006**, *47*(2), 238.

40. Lambermon, M.; Hendriks, L.J.A.; Schoffelen, S.; Vos, M.; Hatzakis, N.S.; Carette, N.; Rowan, A.E.; Cornelissen, J.; Michon, T.; Sommerdijk, N.A.J.M.; van Hest, J.C.M. "Assembly strategies for enzyme immobilization" *ACS Polymer Preprints*, **2006**, *47*(2), 209.

41. Villarroya, S.; Zhou, J.; Thurecht, K. J.; Howdle, S. M. "Enzymatic polymerization in supercritical carbon dioxide" *ACS Polymer Preprints*, **2006**, *47*(2), 231.

42. Heise, A.; Peeters, J.; Xiao, Y.; Palmans, A.R.A.; Koning, C. "Cross-linked polymers by chemoenzymatic polymerization" *ACS Polymer Preprints*, **2006**, *47*(2), 224.

43. Palmans, A.R.A.; Van As, B.A.C.; van Buijtenen, J.; van der Mee, L.; Meijer, E.W. "Iterative tandem catalysis: A novel tool for the synthesis of chiral polymers" *ACS Polymer Preprints*, **2006**, *47*(2), 251.

44. Hilker, I.; Verzijl, G.K.M.; Palmans, A.R.A.; Heise, A. "Chiral polyesters by dynamic kinetic resolution polymerization" *ACS Polymer Preprints*, **2006**, *47*(2), 222.

45. Minshull, J.; Ness, J.; Mang, H.; Yang, J.; Wallner, S.R.; Gao, W.; Lu, W.; Martin, A.; Shah, V.; Gross, R.A.; Gustafsson, C.; Govindarajan, S. "Protein engineering for polymer biocatalysis" *ACS Polymer Preprints*, **2006**, *47*(2), 267.

46. Nolte, R.C.M.; Rowan, A.E.; Benkovic, S.J.; Clerx, J.; Cornelissen, J.J.L.M.; Spiering, M.M.; Zhuang, Z. "Artificial processive enzymes based on the T4 replisome" *ACS Polymer Preprints*, **2006**, *47*(2), 267.

47. Mang, H.; Yang, J.; Wallner, S.R.; Gao, W.; Minshull, J.; Ness, J.; Govindarajan, S.; Gross, R.A. "Protein engineering of Lipase B from Candida antarctica for polyester synthesis" *ACS Polymer Preprints*, **2006**, *47*(2), 269.

48. Chen, B.; Gross, R.A.; Raflailovich, M.H. "Protein immobilization on epoxy activated polymer films: The effect of surface wettability" *ACS Polymer Preprints*, **2006**, *47*(2), 219.

49. Widner, W.; Behr, R.; Sloma, A.; DeAngelis, P. L.; Weigel, P.; Guillaumie, F.; Brown, S. "Hyaluronic acid production by recombinant Bacillus subtilis strains" *ACS Polymer Preprints*, **2006**, *47*(2), 213.

50. DeAngelis, P. L. "Glycosaminoglycan production systems" *ACS Polymer Preprints*, **2006**, *47*(2), 244.

51. Smith, B.T.L.; Omrane, K.; Mandalaywala, M.; James, K.; Balantrapu, K.; Mueller, A. "Novel horseradish peroxidase activity" *ACS Polymer Preprints*, **2006**, *47*(2), 261.

52. Vaidya, A.; Bohling, J.C.; Miller, M.E.; Gross, R.A. "Immobilization of Candida antarctica Lipase B on macroporous resins: Effects of resin chemistry, reaction conditions and resin hydrophobicity" *ACS Polymer Preprints*, **2006**, *47*(2), 247.

53. Albertsson, A.-C.; Srivastava, R.K. "High molecular weight aliphatic polyesters from enzyme-catalyzed ring-opening polymerization and their use" *ACS Polymer Preprints*, **2006**, *47*(2), 245.

54. Kulshrestha, A.; Gao, W.; Fu, H.; Gross, R.A. "Lipase catalyzed route to hyperbranched polymers with dendritic trimethylolpropane units" *ACS Polymer Preprints*, **2006**, *47*(2), 254.

55. Duxbury, C.J.; Heise, A. "Selective enzymatic grafting by steric control" *ACS Polymer Preprints*, **2006**, *47*(2), 271.

56. Cummins, D.M.; Heise, A.; Koning, C.E. "Biocatalytic modification of functionalized surfaces" *ACS Polymer Preprints*, **2006**, *47*(2), 211.

57. Wallner, S.R.; Gao, W.; Hagver, R.; Shah, V.; Xie, W.; Mang, H.; Ilker, M. F.; Bell, C. M.; Burke, K.A.; Coughlin, E.B.; Gross, R.A. "Metathesis polymerization of natural glycolipids" *ACS Polymer Preprints*, **2006**, *47*(2), 258.

58. Some recent papers in tissue engineering include a) Correlo, V. M.; Gomes, M. E.; Tuzlakoglu, K.; Oliveira, J. M.; Malafaya, P. B.; Mano, J. F.; Neves, N. M.; Reis, R. L. *Biomedical Polymers*, **2007**, 197. b) Yoon, D. M.; Fisher, J. P. *Tissue Engineering* **2007**, 8/1-8/18. c) Silva, G. A.; Ducheyne, P.; Reis, R. L. *Journal of Tissue Engineering and Regenerative Medicine* **2007**, *1*(1), 4. d) Hubbell, J. A. *Macromolecular Engineering* **2007**, *4*, 2719. e) Mooney, D. J.; Silva, E. A. *Nature Materials* **2007**, *6*(5), 327.

59. Some recent papers in molecularly imprinted polymers are: a) Ulubayram, K. *Adv. Exp. Med. Biol.(Biomaterials)* **2004**, *553*, 123. b) Lavignac, N.; Allender, C. J.; Brain, K. R. *Anal. Chim. Acta* **2004**, *510*(2), 139. c) Hillberg, A. L.; Brain, K. R.; Allender, C. J. *Adv. Drug Delivery Rev.* **2005**, *57*(12), 1875. d) Mahony, J. O.; Nolan, K.; Smyth, M. R.; Mizaikoff, B. *Anal. Chim. Acta* **2005**, *534*(1), 31.

60. Some recent papers on polymer/biocatalytic biosensors include: a) Ramanavicius, A.; Malinauskas, A.; Ramanaviciene, A. *NATO Sci. Ser., II* (Advanced Biomaterials for Medical Applications) **2004**, *180*, 93. b) Karan, H. I. *Compr. Anal. Chem.* (Ed. Gorton, Lo), **2005**, *44*, 131. c) Miao, Y.; Chen, J.; Wu, X. *Trends Biotechnol.* **2004**, *22* (5*)*, 227. d) Xu, H.; Wu, H.; Fan, C.; Li, W.; Zhang, Z.; He, L. *Chinese Science Bulletin* **2004**, *49*(21), 2227.

61. For example, a) Hoffman, A.S.; Stayton, P.S. *Prog. Polym. Sci.* **2007**, *32*, 922. b) Rapoport, N. *Prog. Polym. Sci.* **2007**, *32*, 962. c) Hoshino, K.; Taniguchi, M, *Smart Polym. Biosep. Bioprocess* (Eds. Galaev, I. Y.; Mattiasson, B. **2004**, pp. 257-283. d) Ulijn, R. V. *J. Materials Chem.* **2006**, *16*(23), 2217.

62. For a recent review, see Nair, L.S.; Laurencin, C.T. *Prog. Polym. Sci.* **2007**, *32*, 762.

63. For example, a) Patil, N. V. *BioProcess Int.* **2006**, *4*(8). 42. b) Pennadam, S. S.; Firman, K.; Alexander, C.; Gorecki, D. C. *J. Nanobiotechnology*

18

2004, *2*, pp. (c) Roy, I.; Sharma, S.; Gupta, M. N. *Adv. Biochem. Eng./Biotechnol.* (New Trends and Developments in Biochemical Engineering), **2004**, *86,* 159. (d) Chaterji, S.; Kwon, I.K.; Park, K. *Prog. Polym. Sci.* **2007**, *32*, 1083.

64. Kofoed, J.; Reymond, J.-L. *Curr. Opin. Chem. Biol.* **2005**, *9*(6), 656.
65. Delort, E.; Darbre, T.; Reymond, J.-L. *Chimia* **2005**, *59*(3), 77.
66. Breslow, R.; Wei, S.; Kenesky, C. *Tetrahedron* **2007**, *63,* 6317.
67. Severin, K. *Mol. Imprinted Mater.* (Eds. Yan, M.; Ramstroem, O.), **2005**, pp. 619-640.
68. Enzyme mimicry is a popular topic as shown by these examples from 2007: a) Hessenauer-Ilicheva, N.; Franke, A.; Meyer, D.; Woggon, W.-D.; van Eldik, R. *J. Amer. Chem. Soc.* **2007**, *129*(41), 12473. b) Kuwabara, J.; Stern, C. L.; Mirkin, C. A. *J. Amer. Chem. Soc.* **2007**, *129*(33), 10074. c) Wang, Q.; Yang, Z.; Zhang, X.; Xiao, X.; Chang, C. K.; Xu, B. *Angew. Chem. Intl. Ed.* **2007**, *46*(23), 4285.
69. Some recent papers on directed evolution include: a) Farinas, E. T. *Combinatorial Chemistry & High Throughput Screening* **2006**, *9*(4), 321. b) Bloom, J. D.; Meyer, M. M.; Meinhold, P.; Otey, C. R.; MacMillan, D.; Arnold, F. H. *Current Opinion in Structural Biology* **2005**, *15*(4), 447. c) Eijsink, V. G. H.; Gaseidnes, S.; Borchert, T. V.; van den Burg, B. *Biomolecular Engineering* **2005**, *22*(1-3), 21.
70. For example: a) Bergquist, P. L.; Gibbs, M. D. *Methods in Molecular Biology* (Protein Engineering Protocols), **2007**, *352*, pp. 191-204. b) el Cardayre, S. B. *Natural Products* **2005**, 107. c) Gomez, A.; Galic, T.; Mariet, J.-F.; Matic, I.; Radman, M.; Petit, M.-A. *Appl. Environmental Microbiology* **2005**, *71*(11), 7607.
71. Basso, A.; Braiuca, P.; Ebert, C.; Gardossi, L.; Linda, P. *J. Chem. Technol. Biotechnol.* **2006**, *81*(10), 1626.
72. Knezevic, Z. D.; Siler-Marinkovic, S. S.; Mojovic, L. V. *Acta Period. Technol.* **2004**, *35*, 151.
73. Costa, S. A.; Azevedo, H. S.; Reis, R. L. *Biodegradable Systems in Tissue Engineering and Regenerative Medicine* **2005**, 301.
74. Romaskevic, T.; Pielichowski, K.; Budriene, S.; Pielichowski, J. *Modern Polymeric Materials for Environmental Applications*, International Seminar, 1st, Krakow, Poland, Dec. 16-18, 2004, **2004**, *1*, pp. 129-132.
75. Li, H.; Zhang, H.; Yi, W.; Shao, J.; Wang, P. G. *ACS Symp. Ser.* (Polymer Biocatalysis and Biomaterials), **2005**, *900*, 192, and references therein.
76. DeAngelis, P. L. *ACS Symp. Ser.* (Polymer Biocatalysis and Biomaterials), **2005**, *900*, 232, and references therein.
77. For example, Tian, W.-J.; Sasaki, Y.; Fan, S.-D.; Kikuchi, J.-I. *Supramol. Chem.* **2005**, *17*(1-2), 113.
78. For example, a) Dolle, R. E. *J. Combinatorial Chem.* **2005**, *7*(6), 739. b) Meldal, M. *QSAR & Combinatorial Science* **2005**, *24*(10), 1141. c) Fumoto,

M.; Hinou, H.; Ohta, T.; Ito, T.; Yamada, K.; Takimoto, A.; Kondo, H.; Shimizu, H.; Inazu, T.; Nakahara, Y.; Nishimura, S. *J. Amer. Chem. Soc.* **2005**, *127*(33), 11804.

79. Fischer-Colbrie, G.; Heumann, S.; Guebitz, G. *Modif. Fibers Med. Spec. Appl.* (Ed. Edwards, J. V.; Buschle-Diller, G.; Goheen, S. C.), **2006**, pp. 181-189.

80. Hobbs, H. R.; Thomas, N. R. *Chemical Reviews* **2007**, *107*(6), 2786.

81. For example, a) Stubbe, J.; Tian, J.; He, A.; Sinskey, A. J.; Lawrence, A. G.; Liu, P. *Annu. Rev. Biochem.* **2005**, *74*, 433. b) Foster, L. J. R. *Appl. Microbiology & Biotechnology* **2007**, *75*(6), 1241. c) Dias, J. M. L.; Lemos, P. C.; Serafim, L. S.; Oliveira, C.; Eiroa, M.; Albuquerque, M. G. E.; Ramos, A. M.; Oliveira, R.; Reis, M. A. M. *Macromolecular Bioscience* **2006**, *6*(11), 885. d) Noda, I.; Bond, E.B.; Green, P.R.; Melik, D.H.; Narasimhan, K.; Schechtman, L.; Satkowski, M. *ACS Symp. Ser.* (Polymer Biocatalysis and Biomaterials) **2005**, *900*, 280.

82. Some recent reviews include: a) Suriyamongkol, P.; Weselake, R.; Narine, S.; Moloney, M.; Shah, S. *Biotechnology Advances* **2007**, *25*(2), 148. b) Tan, I. K. P. *Concise Encyclopedia of Bioresource Technology* **2004**, pp. 653-662. c) Steinbuchel, A. *ACS Symp. Ser.* (Biocatalysis in Polymer Science) **2003**, *840*, 120.

83. For example, Uyama, H.; Kobayashi, S. *Adv. Polym. Sci.* (Enzyme-Catalyzed Synthesis of Polymers) **2006**, *194*, 133.

84. For example, Matsumura, S. *Adv. Polym. Sci.* (Enzyme-Catalyzed Synthesis of Polymers) **2006**, *194*, 95.

85. For example, a) Matsumura, S.; Soeda, Y.; Toshima, K. *Appl. Microbiol. Biotechnol.* **2006**, *70*(1), 12. b) Soeda, Y.; Toshima, K.; Matsumura, S. *Macromolecular Bioscience* **2005**, *5*(4), 277. c) McCabe, R. W.; Taylor, A. *Green Chemistry* **2004**, *6*(3), 151.

86. For example, a) Fan, Y.; Chen, G.; Tanaka, J.; Tateishi, T. *Key Engineering Materials* (Advanced Biomaterials VI) **2005**, *288-289*, pp. 469-472. b) Belyaeva, A. V.; Bacheva, A. V.; Oksenoit, E. S.; Lysogorskaya, E. N.; Lozinskii, V. I.; Filippova, I. Yu. *Russian J. Bioorganic Chem.* **2005**, *31*(6), 529. c) Cheng, H. N.; Gu, Q.-M. *ACS Polymer Preprints* **2000**, *41*(2), 1873. d) Gu, Q.-M.; Nickol, R. G.; Cheng, H. N. *ACS Polymer Preprints* **2003**, *44*(2), 608.

87. Ohmae, M.; Fujikawa, S.-I.; Ochiai, H.; Kobayashi, S. *J. Polym. Sci., Part A: Polym. Chem.* **2006**, *44*(17), 5014.

88. Kobayashi, S.; Ohmae, M. *Adv. Polym. Sci.* (Enzyme-Catalyzed Synthesis of Polymers), **2006**, *194*, 159.

89. Faijes, M.; Planas, A. *Carbohydrate Research* **2007**, *342*(12-13), 1581.

90. Yanase, M.; Takaha, T.; Kuriki, T. *J. Science of Food and Agriculture* **2006**, *86*(11), 1631.

20

91. Kobayashi, S. *J. Polym. Sci., Part A: Polym. Chem.* **2005**, *43*(4), 693.
92. For example, a) Derango, R. A.; Chiang, L.C.; Dowbenko, R.; Lasch, J. G. *Biotechnology Techniques* **1992**, *6*(6), 523. b) Ikeda, R.; Tanaka, H.; Uyama, H.; Kobayashi, S. *Macromolecular Rapid Commun.* **1998**, *19*(8), 423. c) Durand, A.; Lalot, T.; Brigodiot, M.; Marechal, E. *Polymer* **2001**, *42*(13), 5515. d) Singh, A.; Roy, S.; Samuelson, L.; Bruno, F.; Nagarajan, R.; Kumar, J.; John, V.; Kaplan, D. L. *J. Macromol. Sci., Pure Appl. Chem.* **2001**, *A38*(12), 1219. e) Kalra, B.; Gross, R. A. *Green Chemistry* **2002**, *4*(2), 174.
93. Singh, A.; Kaplan, D. L. *Adv. Polym. Sci.* (Enzyme-Catalyzed Synthesis of Polymers) **2006**, *194*, 211.
94. Xu, P.; Singh, A.; Kaplan, D. L. *Adv. Polym. Sci.* (Enzyme-Catalyzed Synthesis of Polymers) **2006**, *194*, 69.
95. Uyama, H.; Kobayashi, S. *Adv. Polym. Sci.* (Enzyme-Catalyzed Synthesis of Polymers) **2006**, *194*, 51.
96. Reihmann, M.; Ritter, H. *Adv. Polym. Sci.* (Enzyme-Catalyzed Synthesis of Polymers) **2006**, *194*, 1.
97. Madras, G. *Biodegrad. Polym. Ind. Appl.* (Ed. Smith, R.), **2005**, pp. 411-433.
98. Wyman, C. E.; Decker, S. R.; Himmel, M. E.; Brady, J. W.; Skopec, C. E.; Viikari, L. *Polysaccharides,* 2nd Ed. (Ed. Dumitriu, S.), **2005**, pp. 995-1033.
99. Saha, B. C. *Handb. Ind. Biocatal.* (Ed. Hou, C. T.), **2005**, pp. 24/1-24/12.
100. Martinez, A. T.; Speranza, M.; Ruiz-Duenas, F. J.; Ferreira, P.; Camarero, S.; Guillen, F.; Martinez, M. J.; Gutierrez, A.; del Rio, Jose C. *Int. Microbiol.* **2005**, *8*(3), 195.
101. Guerra, A.; Ferraz, A. *Wastewater Treat. Using Enzymes* (Ed. Sakurai, A.), **2003**, pp.73-91.
102. Matsumura, S.; Osanai, Y.; Soeda, Y.; Suzuki, Y.; Toshima, K. *Handbook of Biodegradable Polymeric Materials and Their Applications* **2006**, *1*, pp. 239-275.

Novel Biorelated Materials

Chapter 2

Chemically Reactive Peptides for the Production of Electroactive Conjugates of Specified Conformation and Side-Chain Placement

Manoj Charati[1,2], Onur Kas[1], Mary E. Galvin[3],
and Kristi L. Kiick[1,2,*]

[1]Department of Materials Science and Engineering, University of Delaware,
201 DuPont Hall, Newark, DE 19716
[2]Delaware Biotechnology Institute, University of Delaware, 15 Innovation
Way, Newark, DE 19711
[3]Air Products and Chemicals, 7201 Hamilton Boulevard,
Allentown, PA 18195

Biosynthetic routes to polymeric materials provide important opportunities for the production of well-defined macromolecular templates. The sequence control in the synthesis of both proteins and peptides has been employed to produce macromolecules that can be useful in toxin neutralization, electroactive device applications, and integration with inorganic materials. Alanine-rich polypeptides of both random coil and helical conformations have been produced via biosynthetic strategies. Spectroscopic characterization of the molecules indicates that they adopt conformations that are retained upon either the incorporation of non-natural amino acids or upon chemical modification with multiple biological ligands or electroactive groups. The functional behavior of these molecules has been characterized via biochemical and physical methods; results indicate control of properties (immunochemical or luminescence) as a direct function of macromolecular structure. These results point to the exciting opportunities for the biosynthetic production and chemical elaboration of an increasing array of advanced multifunctional materials.

Introduction

Peptides designed with known secondary structure have been shown to be excellent model systems to study side chain interactions, such as aromatic-aromatic (*1-4*), cationic-π (*5-7*) and charge-dipole (*8, 9*) interactions. In order to understand the factors that govern the secondary structures of these peptides, a variety of non-natural amino acids with well defined stereo-chemical and functional properties have been studied. In nature, non-natural amino acids are found in peptides and serve both functional and structural purposes. For example, α-aminoisobutyric acid (Aib) has been shown to be present in chlamydocin, an antimicrobial antibiotic that is used as a cytostatic agent by *Diheterospora chlamydosporia* (*10*). Aib is also present in other naturally occurring antimicrobial peptides, such as alamethicin, that produce voltage gated ion channels in lipid membranes (*11, 12*). α-ethylallanine (isovaline) has also been found to be present in the antimicrobial antibiotic emermicin (*13*).

In experimental studies, several non-proteinogenic amino acids have been used in peptides to stabilize their secondary structures; Aib has been used most frequently to stabilize α-helical peptides, owing to the steric restrictions of the methyl group on the C^α atom, which restricts the possible rotations around the N-C^α and the C^α-C' bond (*12*). Aib homologues such as dipropylglycine (Dpg) and dibutylglycine (Dbg) have also been observed to induce α- and 3_{10} helical conformations in 5-10 residue peptides (*14*). Cornish *et al* incorporated non-natural amino acids containing hydrophobic side chains such as α-aminobutyric acid (Abu), norvaline (Nva), norleucine (Nle) and tert-leucine (Tle) in helical domains of T4 lysozyme to study the stabilizing effect β-branched amino acids on α-helicies (*15*). Abu, Nva and Nle were shown to be as helix stabilizing as alanine; however, Tle was found to be more helix destabilizing due to the presence of a β-branched methyl group. In another report, incorporation of ε-(3,5-dinitrobenzoyl)-lysine at *i, i+4* position was shown to reinforce α-helices (*16*). The helical content of the peptide was observed to increase with the percentage of water in the solution, leading to π-π stacking of the nitrobenzoyl groups.

In addition to imparting structural stability to peptides, a variety of non-natural amino acids have been incorporated into peptides in order to study side chain-side chain interactions. Waters and coworkers have reported the use of phenylalanine (Phe) analogues to investigate the aromatic interactions in model peptide systems. The interaction of Phe in the *i* position of an α-helix with a natural and several non-natural amino acids in the *i + 4* position has been explored; the amino acids included Phe, homophenylalanine (hPhe), biphenylalanine (bPhe), and pentafluorophenylalanine (f5Phe) (*1*). It was observed that the naturally occurring *i, i + 4* Phe-Phe interaction is the most efficient aromatic side-chain interaction for inducing α-helical structure with a stronger interaction at the C-terminus than at the center of an α-helix. In another report, halogen-aromatic interaction was investigated in context of a β-hairpin peptide in aqueous solution and it was observed that the interaction of Phe with 3,5-diiodophenylalanine imparts remarkable stability to the β-hairpin structure (17). Over the years, introduction of non-natural amino acids in peptides has expanded the scope of peptide engineering and facilitated *de novo* design of novel peptide biomolecules for a variety of applications.

Non-natural amino acids can be incorporated into peptides and polypeptides via several different methodologies. Solid-phase peptide synthesis (SPPS) is a straightforward method for incorporation of non-natural amino acids and allows the incorporation of essentially any amino acid but is limited by the size of the peptides produced (18). Suppression-based strategies, both *in vitro* and *in vivo*, have been developed for site specific incorporation of diverse set non-natural amino acids into natural and synthetic polypeptides (19). Alternatively, auxotrophic expression hosts have been used for multisite incorporation of non-natural amino acid in protein polymers, where multiple natural amino acids of one type can be replaced with non-natural analogues during protein biosynthesis (20, 21). Multisite incorporation of non-natural amino acids in the synthesis of protein polymeric materials facilitates chemical modification at multiple sites and can modulate the physical properties of the protein polymers (22).

Accordingly, our research efforts include the design of chemically reactive, recombinant polypeptides onto which additional functional groups can be appended. Main aims include the control of functional group presentation, and the correlation of this presentation to specific macromolecular properties. The incorporation of non-natural amino acids into these polypeptides plays a central role in the expansion of their chemical versatility; thus, details of the impact of this incorporation on the conformational properties of the polypeptides are of interest. Toward these ends, we report here the design and synthesis of a series of alanine rich, α-helical peptides with different Phe analogues at i, $i+4$ and i, $i+5$ positions and evaluation of their contribution towards the stability of α-helical peptides. The use of chemically reactive Phe analogues such as *p*-bromophenylalanine (BrPhe), *p*-iodophenylalanine (IPhe) and *p*-cyanophenylalanine (CNPhe) allows subsequent modification of the side chain via various chemistries for grafting functional hybrid molecules. In addition to characterizing the impact of these non-natural amino acids on the secondary structure for protein polymer design, we also synthesized alanine rich, Aib-stabilized α-helical peptide templates containing non-natural amino acid, BrPhe, at tunable desired spacing and orientations. The BrPhe side chains were modified via Heck reaction with α-methyl styrene to afford electroactive methylstilbene side chains. The control over the placement of methylstilbene side chains on the peptide backbone permits the manipulation in the electronic behavior of these conjugates. These peptide templates have been utilized to manipulate the association of electroactive molecules and study their photophysical properties as a function of molecular orientation and spacing (23).

Table 1. Peptide sequences employed in these studies.

PEPTIDE SEQUENCE		
Ac-KAAAKAAAAXAAAXAAAAKAAAKGGY-NH$_2$	**X-D**[a]	i,i+4
Ac-KAAAKAAAAXAAAAXAAAKAAAKGGY-NH$_2$	**X-O**[a]	i,i+5
Ac-KAA-Aib-KAA-Aib-AYA-Aib-AYA-Aib-AAK-Aib-AAKGGY-NH$_2$ **Aib-Y-D**[b]		i,i+4
Ac-KAA-Aib-KAA-Aib-AYA-Aib-AAY-Aib-AAK-Aib-AAKGGY-NH$_2$ **Aib-Y-O**[b]		i,i+5

[a] X in the peptide sequence (Table 1) may represent Phe, p-BrPhe, p-IPhe and p-CNPhe.
[b] Y in the peptide sequence represents either p-BrPhe or p-methylstyryl phenylalanine.
The **X-D** peptide containing Phe will be represented throughout the text as **Phe-D** and the **X-O** peptide containing Phe will be represented throughout the text as **Phe-O**; similar notation will be used to represent the other peptides.

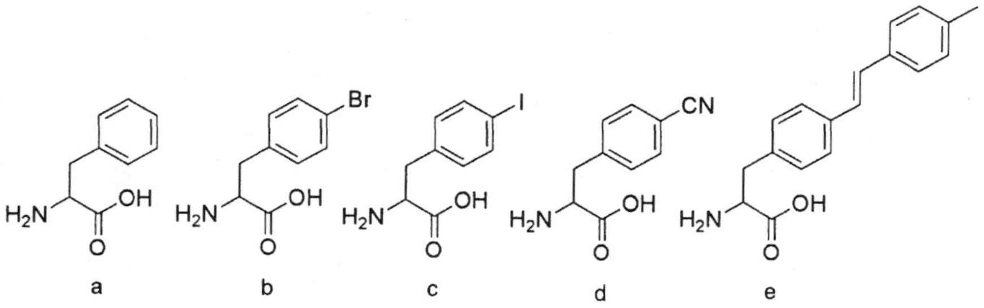

Figure 1. Structures of the natural and non-natural amino acids.
(a) phenylalanine; (b) p-bromophenylalanine (BrPhe); (c) p-iodophenylalanine (IPhe); (d) p-cyanophenylalanine (CNPhe); (e) p-methylstyryl phenylalanine.

Results and Discussion

Peptide Design

The peptide sequences employed for this study are shown in Table 1. Each peptide is based on a general AAAX repeating sequence, where the high helical propensity of alanine facilitates the formation of an α-helix (24, 25). Lysine residues were introduced to impart water solubility and prevent peptide aggregation. Tyrosine was placed at the C-terminus of the peptide to allow concentration determination. The placement of Phe and Phe analogues at i, i+4 arrangements permits the positioning of the side chains on the same side of the helix and enables monitoring of possible side chain-side chain interactions. The corresponding i, $i + 5$ control peptides were also synthesized wherein the side chains are located on opposite sides of the helix, minimizing any possible side chain interactions.

Circular Dichroic Spectroscopy

The helical content of the peptides was assessed via circular dichroic (CD) spectroscopy under aqueous conditions (Figure 2). CD spectroscopy was performed on the peptides, **X-D** and **X-O**, to assess the contribution of the non-natural amino acids towards the α-helix stability. All of the peptides display a characteristic α-helical CD spectrum with double minima near 208 nm and 222 nm and a single maximum near 192 nm. All of i, $i + 4$ peptides incorporating Phe and Phe analogues showed similar helical contents at 5 °C. The helical content of all these peptides were comparable to i, $i + 5$ control peptides at 5 °C (Figure 2a), indicating the absence of significant stabilizing side chain interactions between the Phe analogues compared to Phe.

Thermal denaturation experiments were performed on the peptides, **X-D** and **X-O** (Table 1), to estimate the folded fraction of each peptide as a function of temperature (Figure 2b). Comparisons of full wavelength spectra at various temperatures indicate that all of the peptides undergo a reversible conversion between the α-helix to random-coil conformations (data not shown), and plots of the mean residue ellipticity at 222nm (θ_{MRE222}) as a function of temperature indicate similar temperatures at which the α-helix unfolds into a non-helical conformation (Figure 2b). The temperature dependence of θ_{MRE222} was also studied between 10 μM and 100 μM of peptide solution, to assess any potential aggregation of the peptides; the lack of concentration dependence of the melting behavior indicate that the peptides were monomeric (data not shown). The similarity in the helical content and the thermal unfolding behavior for the peptides containing Phe and Phe analogues indicate that incorporation of this set of Phe analogues does not have any substantial positive or negative effect on the α-helix stability (Figure 2), in contrast to some previous reports. In particular, Waters and coworkers reported an increase in helical content of the peptide when the aromatic side chains are placed on a peptide in a i, $i + 4$ arrangement, as compared to the i, $i + 5$ arrangement (1). Despite the apparent lack of any helix stabilization upon analogue incorporation in these experiments, the results reported here demonstrate that a variety of phenylalanine analogues can be included in peptides without causing any significant alteration of peptide conformation. These results therefore suggest the potential versatility of incorporating specific chemically reactive amino acids into peptide backbone of desired conformations to create functional materials.

NMR Experiments

^{1}H NMR experiments were performed on all the **X-D** and **X-O** peptides to probe aromatic proton shifts with alterations in peptide architecture. ^{1}H NMR experiments were performed at low concentrations (1 mg/ml) to prevent intermolecular interactions; the sharp resonances of the aromatic peaks suggest that the peptides do not aggregate under these experimental conditions. In addition, the aromatic resonances of the tyrosine, δ (ppm) 6.65-6.725, overlap in spectra of the peptides, allowing direct comparison of the chemical shifts of other aromatic protons in this region. Figure 3 presents the NMR data in the aromatic region for the peptides **BrF-D** and **BrF-O**. As shown in Figure 3, the peak positions of the phenylalanyl aromatic protons in **BrF-D** are very similar to those in **BrF-O**, although small splitting of the resonances in **BrF-D** is observed that is absent in the spectrum of **BrF-O,** which may reflect the proximity of the sidechains in the **BrF-D**. Nevertheless, upfield shifts of at least 0.1 ppm are indicative of significant $\pi-\pi$ overlap (25). Thus, the similarity in the positions of the aromatic resonances in these spectra suggests a lack in π-π overlap of the sidechains in the two peptides; this is consistent with the lack of sidechain-induced helix stabilization observed in the CD experiments. Taken together, the results indicate that the incorporation of these Phe analogues does not substantially alter sidechain interactions, and therefore does not affect the helical content of the peptides.

The confirmation that Phe analogue incorporation is not likely to exert undesired changes in polypeptide conformation suggests the potential for these types of substitutions in protein polymer design. Toward ultimate goals of producing chemically modified recombinant polypeptides equipped with conjugated sidechains, we first synthesized an additional series of α-helical peptide templates (**Aib-Y-D, Aib-Y-O**), as shown in Table 1, to test chemical modification strategies and the impact of controlling orientation of the sidechains. These peptides are decorated with BrPhe residues at desired spacing and orientation and are enriched with Aib residues to increase the helical content of the short sequences (26). It is anticipated that increased lengths of recombinant constructs will provide sufficient helix stability in future experiments. The peptide templates were modified via Heck reaction to create methylstilbene side chain moieties (Figure 4).

Circular Dichroic Spectroscopy

The CD spectra of these peptides as a function of increasing temperature were collected for the peptide templates containing BrPhe and methylstilbene as side chains to understand the effect of bulkier side group on the helical content of the peptide. The temperature dependence of the θ_{MRE222} as a function of

28

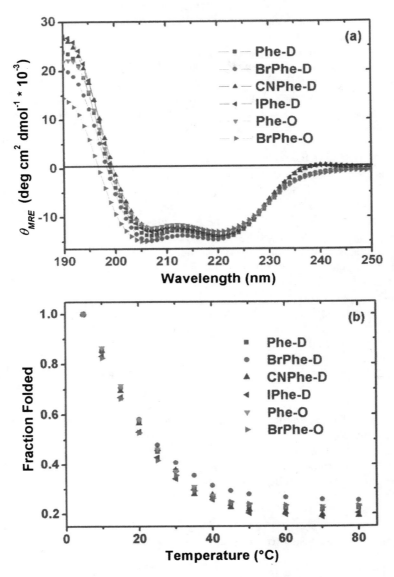

*Figure 2. Circular dichroic spectra for the **X-D** and **X-O** peptides. (a) CD scans reported at 5 °C in water. (b) Thermal unfolding scans of peptides in water.*

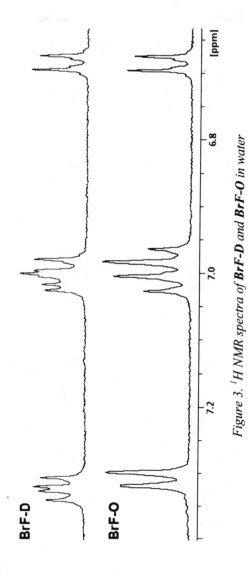

*Figure 3. ¹H NMR spectra of **BrF-D** and **BrF-O** in water*

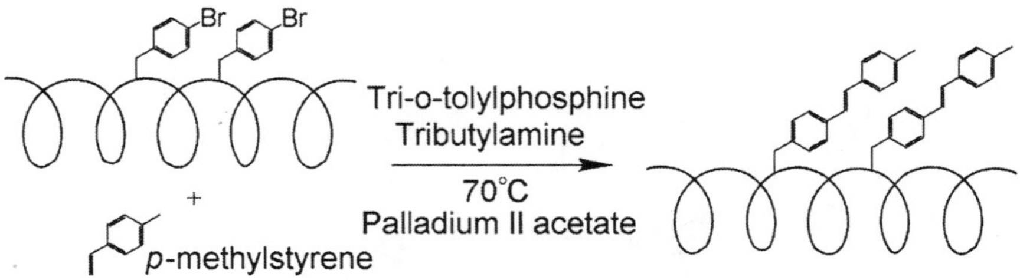

*Figure 4. Schematic of the Heck reaction between Aib-BrPhe-D peptide
and p-methylstyrene in DMF.*

temperature over a range of concentrations from 10 µM to 100 µM was monitored in TFE. The melting behavior was essentially identical for the peptides, revealing that even the modified peptides do not aggregate in solution. As shown in Figure 5, all peptides exhibit very similar CD spectra with peak minima and maxima as expected for an α-helix (Figure 5a), with similar thermal denaturation profiles (Figure 5b). The CD spectra at elevated temperatures show a reduction in MRE values for the helical conformation and an increase in peak intensity near 200nm, indicative of the random coil conformation (data not shown). The thermal unfolding scans indicate that all these peptides undergo a reversible helical to non-helical confirmation (not shown), but are quite stable, retaining more than 50% of their original helicity at elevated temperatures (80 °C) (Figure 5b). The higher helical content and thermal stability in the **Aib-Y-D** and **Aib-Y-O** peptides compared to the **X-D** and **X-O** peptides is mainly due to the presence of Aib, as expected (*26*). The similarity between the CD spectra and the thermal denaturation curves between **Aib-BrPhe-D, Aib-BrPhe-O, Aib-Stb-D** and **Aib-Stb-O** peptides suggest that even the bulkier side groups do not have a considerable effect on the helical content and thermal stability of the peptide (Figure 5); such helical stability is also indicated for PPV-based side chains of greater length and steric bulk (in preparation). These results confirm the broad utility of such *de novo* designed α-helical peptide templates for desired presentation of electroactive species. We then further investigated the effect of conformation/orientation on the photo-physical behavior of electroactive methylstilbene side chains using the above mentioned peptide templates.

Photoluminescence Experiments

Photoluminescence (PL) experiments were performed on methylstilbene-equipped peptides (**Aib-Stb-D** and **Aib-Stb-O**) in TFE to assess the effect of the relative orientation of electroactive side chains on their electronic properties. **Aib-Stb-D** and **Aib-Stb-O** show identical absorption spectra (data not shown) and display monomeric methylstilbene emission with a peak maximum at 357 nm (Figure 6); however, they exhibit notably different PL intensities. **Aib-Stb-O**

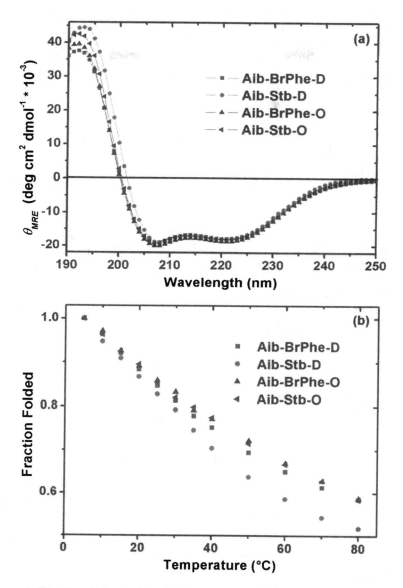

*Figure 5. Circular dichroic data for the peptides. (a) CD scans reported at 5 °C for the **Aib-Y-D** and **Aib-Y-O** peptides in TFE. (b) Thermal unfolding scans of the peptides in TFE.*

displays only monomeric stilbene emission with a peak intensity and fluorescence quantum yield that is comparable to isolated stilbene molecules in dilute solution. Thus, it can be inferred from these results that the methylstilbene side chains do not interact with each other. On the other hand, **Aib-Stb-D** displays more than 50% reduction in PL intensity, along with a tail in the low energy region of the PL spectrum. This transition in the low energy region of the spectrum is attributed to the presence of an excimer (*27, 28*). The formation of an excimer and the reduction in PL intensity, in these very dilute peptide solutions, suggests that the methyl stilbene side chains in **Aib-Stb-D** are interacting with each other and this controlled proximity is mediated by the α-helical peptide. The PL results clearly demonstrate that the interaction of conjugated side chains can be precisely prescribed via presentation of these side chains on an appropriate peptide template.

Exciton-coupled circular dichroism experiments

Exciton-coupled circular dichroism (ECCD) spectroscopy was performed on the **Aib-Stb-D** peptide to further substantiate the interaction between the methyl stilbene side-chains, which was observed via PL experiments. The interaction between the excited state of chromophores in a chiral environment causes split Cotton effects upon absorption of circularly polarized light by the chromophores (*29, 30*). It can be observed from Figure 7 that **Aib-Stb-D** exhibits a split CD Cotton effect presumably because of the chiral presentation of the methyl stilbene molecules on the same side of the α-helical peptide; the asymmetric nature of the split observed may be due to some other electronic transitions or of additional background ellipticity, as has been observed in other systems (*29, 31*). The ECCD results confirm the close proximity of the methyl stilbene side chains mediated by the peptide backbone, which permits interaction between the side-chains in the excited state.

Conclusions

The impact on the helicity of alanine-rich peptides by several Phe analogues was assessed via circular dichroic spectroscopy. The results of these studies demonstrate that the Phe analogues under investigation do not disrupt the expected secondary structure of the peptide. Introduction of Aib residues in these peptide sequences increased the helical content and the helix stability of the short sequences and permitted investigation of the effect of controlled presentation of methylstilbene chromophores on photoluminescence behavior. A strong interaction between the methylstilbenes was observed when the molecules are located on the same side of the helix; these interactions are not present when the molecules are located on opposite sides of the helix. These results suggest the potential use of *de novo* designed peptide templates to arrange functional

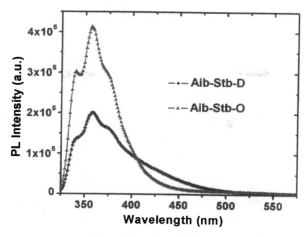

*Figure 6. PL spectra of **Aib-Stb-D** and **Aib-Stb-O** after excitation at 313 nm in TFE.*

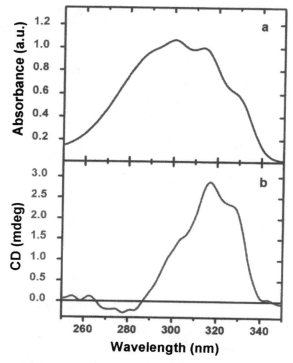

*Figure 7. (a) Absorption spectrum of **Aib-Stb-D** in TFE. (b) Exciton coupled circular dichoric spectrum of **Aib-Stb-D** peptide in TFE.*

molecules where intermolecular spacing on the angstrom lengthscale is critical. Furthermore, the lack of secondary structure change upon incorporation of chemically and structurally diverse Phe analogues suggests the potential for such incorporation in the production of chemically reactive, recombinant polypeptides of well-defined conformation.

Experimental Section

Peptide Synthesis and Purification

Peptides were synthesized on a rink amide MBHA resin (EMD Biosciences, CA) via automated solid-phase peptide synthesis (PS3, Protein Technologies, Inc, Tucson, AZ) using standard Fmoc protected amino acids. Non-natural amino acids (bromophenylalanine, iodophenylalanine and cyanophenylalanine) were purchased from Chem-Impex International Inc (Wood Dale, IL). The amino acid residues were activated for coupling with HBTU (O-benzotriazole-N,N,N',N',-tetramethyluronium hexafluorophosphate) in the presence of 0.4M methyl morpholine in DMF. Deprotections were carried out in 20% piperidine in DMF (N,N-dimethyl formamide) for approximately 30 min. Standard coupling cycles were used for the first 9-12 couplings (60 min) and extended coupling cycles (2 h) were used to complete the sequence. The N-terminus was acetylated with 5% acetic anhydride in DMF for 30 min. Cleavage of the peptide from the resin was performed in 95:2.5:2.5 Trifluoroacetic acid (TFA):Triisopropylsilane (TIPS):water for 6-8 h. TFA was evaporated and cleavage products were dissolved in ether. The water-soluble peptides were extracted with water and lyophilized. Peptides were purified via reverse-phase HPLC (Waters, MA, USA), using a Symmetry C-18 column. The identity of each peptide was confirmed via electrospray ionization mass spectrometry (ESI-MS) (AutospecQ, VG Analytical, Manchester, UK).

CD Measurements

Stock solutions of the purified peptides were prepared by dissolving in water. The concentration of each peptide was then determined using the absorbance of the tyrosine residue at 274 nm ($\varepsilon = 1408$ M^{-1}cm^{-1}). Three solutions were prepared for each peptide to permit error determination for fractional helicities caused by probable errors in the concentration measurements. CD spectra were acquired on a Jasco-810 spectropolarimeter (Jasco, Inc., Easton, MD, USA) using a 0.1 cm path length cell and a scan rate of 50 nm/min. CD spectra were recorded for each peptide as a function of increasing temperature from 190 to 250 nm to determine helicity. Data points for the wavelength dependent CD spectra were recorded with a 1 nm bandwidth. Data points for the thermal denaturation scans were recorded at 222 nm, at every 1 °C intervals with an equilibration time of 1 minute. The mean residue ellipticity, θ_{MRE}, was calculated using the molecular weight of the peptide and cell path length employed for CD experiments. ECCD experiments were performed in very dilute solution of peptide in TFE from 250 to 350 nm, using a 1 cm path length cell.

Heck Reaction

p-Methylstyrene was purified via distillation over CaH_2 to remove the inhibitors. Heck coupling between *p*-methyl styrene and peptide was performed in an inert atmosphere with anhydrous DMF. To a solution of 25 mmol p-methylstyrene and 20 mmol tributylamine 2 ml of DMF, 10 mmol peptide template (for a peptide with two Br functionalities), 4 mmol tri-o-tolylphosphine and 0.8 mmol palladium II acetate mixture was added in the reaction flask. Reaction temperature was monitored at 70°C in an oil bath. The progress of the reaction was monitored via ESI-MS and when necessary, initial amounts of the reactant, ligand, and catalyst were added again. The final reaction mixture was precipitated in ether to extract the water-soluble peptide. The reacted peptides were purified via reverse phase HPLC as mentioned above. The identity of each peptide was confirmed via ESI-MS.

PL experiments

PL experiments were carried out on SPEX Fluoromax 3 spectrofluorimeter using Datamax Software. PL experiments were performed on dilute solutions of peptides with absorption maximum of ~0.1. Peptides were dissolved in TFE, under conditions similar to those employed in CD studies.

^{1}H NMR experiments

NMR spectra were performed on Bruker DRX 400 MHz spectrometer at room temperature. 1 mg/mL solution of the peptide in D_2O was prepared and data was collected for 512 scans.

References

1. S. M. Butterfield, P. R. Patel, M. L. Waters, *J. Am. Chem. Soc.* **2002**, *124*, 9751.
2. C. D. Tatko, M. L. Waters, *J. Am. Chem. Soc.* **2002**, *124*, 9372.
3. S. Aravinda, N. Shamala, R. Rajkishore, H. N. Gopi, P. Balaram, *Angew. Chem. Int. Ed.* **2002**, *41*, 3863.
4. S. Aravinda *et al.*, *J. Am. Chem. Soc.* **2003**, *125*, 5308.

36

5. C. A. Olson, Z. S. Shi, N. R. Kallenbach, *J. Am. Chem. Soc.* **2001**, *123*, 6451.
6. Z. S. Shi, C. A. Olson, N. R. Kallenbach, *J. Am. Chem. Soc.* **2002**, *124*, 3284.
7. C. D. Andrew *et al.*, *J. Am. Chem. Soc.* **2002**, *124*, 12706.
8. H. Nicholson, W. J. Becktel, B. W. Matthews, *Nature* **1988**, *336*, 651.
9. J. Sancho, L. Serrano, A. R. Fersht, *Biochemistry* **1992**, *31*, 2253.
10. A. Closse, R. Huguenin, *Helvetica Chimica Acta* **1974**, *57*, 533.
11. R. O. Fox, F. M. Richards, *Nature* **1982**, *300*, 325.
12. I. L. Karle, P. Balaram, *Biochemistry* **1990**, *29*, 6747.
13. R. C. Pandey, J. C. Cook, K. L. Rinehart, *J. Am. Chem. Soc.* **1977**, *99*, 5205.
14. I. L. Karle, R. B. Rao, S. Prasad, R. Kaul, P. Balaram, *J. Am. Chem. Soc.* **1994**, *116*, 10355.
15. V. W. Cornish, M. I. Kaplan, D. L. Veenstra, P. A. Kollman, P. G. Schultz, *Biochemistry* **1994**, *33*, 12022.
16. J. S. Albert, A. D. Hamilton, *Biochemistry* **1995**, *34*, 984.
17. C. D. Tatko, M. L. Waters, *Org. Lett.* **2004**, *6*, 3969.
18. W. C. Chan, P. D. White, *Fmoc Solid Phase Peptide Synthesis: A Practical Approach.* Oxford University Press: Oxford, **2000**.
19. L. Wang, P. G. Schultz, *Angew. Chem. Int. Ed.* **2004**, *44*, 34.
20. K. L. Kiick, D. A. Tirrell, *Tetrahedron* **2000**, *56*, 9487.
21. K. Kirshenbaum, I. S. Carrico, D. A. Tirrell, *ChemBioChem* **2002**, *3*, 235.
22. R. F. Farmer, M. B. Charati, K. L. Kiick, In *Macromolecular Engineering: From Precise Macromolecular Synthesis to Macroscopic Materials Properties and Applications*, K. Matyjaszewski, Y. Gnanou, L. Leibler, Eds. WILEY-VCH Verlag GmbH & Co.: Weinheim, 2007; Vol. 1.
23. O. Y. Kas, M. B. Charati, K. L. Kiick, M. E. Galvin, *Chem. Mater.* **2006**, *18*, 4238.
24. A. Chakrabartty, T. Kortemme, R. L. Baldwin, *Protein Sci.* **1994**, *3*, 843.
25. S. Padmanabhan, E. J. York, J. M. Stewart, R. L. Baldwin, *J. Mol. Biol.* **1996**, *257*, 726.
26. V. De Filippis, F. De Antoni, M. Frigo, P. P. de Laureto, A. Fontana, *Biochemistry* **1998**, *37*, 1686.
27. I. D. W. Samuel, G. Rumbles, C. J. Collison, *Phys. Rev. B* **1995**, *52*, 11573.
28. M. Pope, C. E. Swenberg, *Electronic Processes in Organic Crystals and Polymers.* Oxford University Press: New York, **1999**.
29. N. Harda, K. Nakanishi, *Circular Dichroic Spectroscopy-Exciton Coupling in Organic Stereochemistry.* University Scienec Books: Mill Calley, CA, **1983**.
30. K. Nakanishi, N. Berova, In *Circular Dichroism, Principles and Applications*, K. Nakanishi, N. Berova, R. Woody, Eds. VCH Publishers, Inc.: New York, 1994; pp 337-395.
31. S. Matile, N. Berova, K. Nakanishi, J. Fleischhauer, R. W. Woody, *J. Am. Chem. Soc.* **1996**, *118*, 5198.

Chapter 3

Elastin-Based Protein Polymers

Peter James Baker[1], Jennifer S. Haghpanah[1], and Jin Kim Montclare[1,2]

[1]NSF, Department of Chemical and Biological Sciences, Polytechnic University, Six Metrotech Center, Brooklyn, NY 11201
[2]Department of Biochemistry, State University of New York, Downstate Medical Center, Brooklyn, NY 11203

The synthesis of well-defined macromolecular structures with controlled properties is critical for the production of advanced materials for biological and industrial applications. Inspired by nature's ability to create proteins with exquisite control, we focus on the applications of elastin and elastin-derived polymers for materials design. The elucidation of elastin biochemical, conformational and physical properties offers insight into the fabrication of novel biomaterials. As part of this review, we highlight some of the recent advances that permit the generation of customized elastin-based polymers. These developments provide an added level of control vital to the future construction of tailor-made supramolecular structures with emergent physical, mechanical and biological properties.

Introduction

The ability to tailor functional polymeric structures is highly dependent on the level of control in their synthesis. Although tremendous progress has been made in the chemical synthesis of polymers, the degree of control and diversity of monomers afforded by protein polymers remains unsurpassed (1). Natural proteins possess precise molecular weights and high sequence specificity that permits the formation of critical supramolecular structures. These molecular architectures can fold and self-assemble, often translating into unique physical, mechanical and biological properties. In particular, elastin is the major protein component of the extracellular matrix (ECM) providing the structural and elastic integrity for tissue function, the smooth movement of joints and limbs, and the elasticity of large blood vessels. The stability of elastin polymers provides the unique mechanical behavior of elastin where it can undergo a life-time of stretching and recoiling cycles without structural deformation (2). Recent advances in recombinant DNA technology have assisted researchers in understanding these structures. By employing these techniques, it is possible to modify elastin-derived polymers with specified functional groups and designated physical properties. In this review, we demonstrate that exploiting the structural understanding of elastin polymers provides a tool-box to develop new biomaterials, three-dimensional (3D) tissue engineering scaffolds, site-specific drug delivery vehicles and protein purification as well as immobilization schemes.

Nature's Polymers: Tropoelastin and Elastin

Tropoelastin is the water-soluble precursor protein for the insoluble elastin fiber. The tropoelastin gene is encoded by 34-36 exons depending on the species, and is expressed as ≈70 kDa monomeric protein (3). The protein is hallmarked by repeating hydrophobic and cross-linking domains. Hydrophobic domains are composed of mostly repeating non-polar amino acids, while the cross-linking domains are an alanine-rich sequence punctuated with lysine residues, which are extracellularly deaminated to form cross-links. These sequences and their arrangements allow tropoelastin to exhibit rubber-like extensible properties as well as maintain distinct morphological networks in different tissues.

Elastogenesis, the formation of elastin fibers, is a multi-step process where the individual tropoelastin monomers undergo a series of post-translational modifications to form the fibrous polymer network. To prevent intracellular aggregation and ensure proper release from the cell surface, tropoelastin is modified with a 67 kDa elastin-binding glycoprotein (EBP), which binds through the hydrophobic domain (4). At the cell membrane, two essential membrane proteins interact with the tropoelastin/ EBP complex to form a transmembrane link. At the cell surface, lectins in the extracellular matrix (ECM) bind to the EBP, inducing the release of tropoelastin into the ECM where the tropoelastin is used as a substrate to generate the elastin polymer (5). The molecular mechanisms involved in the transportation of the tropoelastin monomer to the growing elastin polymer are not fully elucidated. Recent investigations into the roles of fibrillin and micro-fibril associated glycoprotein (MAGP) suggests that the alignment of tropoelastin may be more tightly regulated than originally thought (6).

The cross-linking of tropoelastin is proceeded by the association and alignment of tropoelastin molecules via coacervation. Coacervation is a thermodynamically controlled and fully reversible process influenced by hydrogen bonding. The hydrogen bonding of water molecules to the hydrophobic regions of the protein keep it unfolded. As the temperature increases, there is disruption of hydrogen bonding and aggregation of the tropoelastin molecules between the hydrophobic domains (7). This aggregation mechanism can be influenced by ionic strength, pH, protein concentration and proportionality of hydrophobic residues. When released from the cell membrane, tropoelastin will experience high concentrations of tropoelastin monomers as well as the growing elastin fiber, promoting a favorable condition for coacervation (7).

Coacervated tropoelastin undergoes further extracellular post-translational modifications, where the primary amines of lysine residues (in the cross-linking domains) undergo enzymatic oxidative deamination via lysyl-oxidase (8). The resultant aldehyde is able to spontaneously cross link with nearby lysines and aldehydes, exhibiting self-assembly like properties to form lysinonorleucine and allysine aldol, respectively. Further condensation of these molecules results in the formation of desmosine and isodesmosine irreversible cross-links which maintain structural integrity (Figure 1) (9). These intermolecular cross-links are unique to the elastin fiber and contribute to the tensile strength and elasticity of native elastin.

Recent studies have shown that natural elastin can undergo approximately a billion stretching and relaxation cycles with minimal deformations (10). This ability supplied by tropoelastin bestows elastin with its outstanding durability. Upon stretching and relaxing, the elastin networks will exhibit unique properties. In the stretched state, elastin will stiffen and the system will be of lower entropy. Under this condition, elastin molecules will have high frequency vibrations called solitions. During relaxation, fibers will recoil back to their original unstructured conformation of higher entropy. This relaxation mechanism entails the movement of molecules in motions of low frequencies called chaos (10). Once these motions are complete, the energy expended in the deformation of the elastin will be recovered and exhibit minimal energy loss, preserving the original structure (11).

Elastin is simply a network of tropoelastin monomers with uniform chain extensions arising from the cross-linked structure. The advantage of this network is that it prevents slippage when elastic fibers are subject to external forces. Elastin's high tensile strength comes from its cross-linked network (12,13). Studies on electrospun protein fibers contradict this idea by revealing that the tensile strength for elastin is 1.6 MPa and the ultimate elongation is 0.01, while tropoelastin is 13 MPa and 0.15, respectively (14). Although tropoelastin is mechanically favorable over elastin for biological functions, it is feasible to mimic the desired properties of tropoelastin by structurally modifying elastic fibers.

40

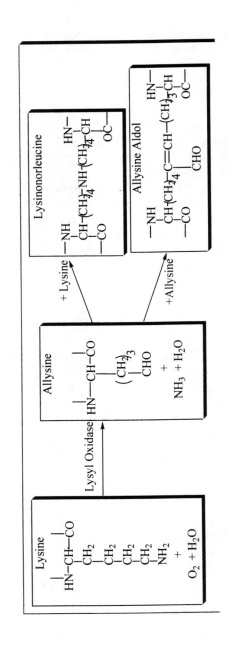

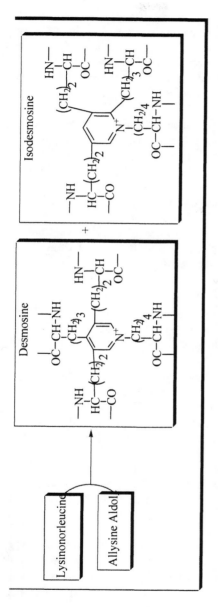

Figure 1. Elastin cross-links formation. Lysine is converted into allysine via lysyl oxidase, which leads to the formation of lysinonorleucine and allysine aldol (blue). The spontaneous condensation between lysine and allysine will generate tetra-substituted desmosine and isodemosine cross-link structures (red). All of the boxes on the top are blue and all of the boxes on the bottom are red. (8).

Structural Understanding of Elastin

The precise understanding of the structural nature of elastin remains elusive because the unusual physical properties of native elastin have prevented the use of traditional structural biology techniques. The hydrophobic domains are dominated by a pentapeptide repeat sequence of (Val-Pro-Gly-Val-Gly). This repeat sequence is believed to endow elastin its remarkable elastic properties, allowing continual expansion and contractions without loss of structural integrity *(2)*.

Pioneering work by the Alix laboratory on the secondary structure of human elastin and the solubilized κ-elastin, estimated the molecule to be composed of 10% α-helices, 35% β-strands and 55% undefined conformation. These estimations were based on Fourier transform infrared (FTIR), near infrared Fourier transform Raman spectroscopy and circular dichroism (CD) *(15)*. To further investigate the nature of the elasticity, polypeptides of hydrophobic sequences containing exons 3, 7, and 30 of human elastin were analyzed by CD and Classic Raman spectroscopy, revealing polyproline II (PPII) helix secondary structures in both the aqueous and solid phase. Further analysis of exon 30 by FTIR spectroscopy determined that this sequence was characterized by both PPII as well as β-sheets structures *(15)*. The presence of these structures were dependent on temperature, concentration and / or time, where lower temperatures and concentrations favored the PPII structure and higher temperatures and concentrations favored β-sheets *(16)*.

PPII helices are comprised of a left-handed, helical structure characterized by ϕ and ψ torsion angles of -75° and 145°, respectively (Figure 2). Unlike α-helices and β-sheets, these structures lack main chain hydrogen bonding but are sterically possible for hydrogen bond formation between side chains and nonprolyl amides *(17)*. The lack of main chain hydrogen bonding creates an inherent flexibility which may mislead traditional methods of secondary structure spectroscopy into classifying these structures as undefined conformations. Surveys of structural databases suggest that Gly and aromatic residues are disfavored in PPII helices *(17,18)*. Notably, peptides containing PPII helices exhibit a thermal transition from PPII helices to β-structures as a function of increasing temperature *(19)*. PPII helices are associated with the unfolded or denatured conformations of proteins, thereby reducing the initial amount of conformational entropy of the unfolded state *(20,21)*.

Seminal work by Urry on cyclic analogs of elastin revealed a Pro-Gly type II β-turn by NMR and crystallography, which served as the basis for the β-spiral model *(22,23,24)*. The model described a structure in which consecutive β-turns formed a helical arrangement. In this structure, there was one type II β-turn per pentameric unit of elastin, which served as spacers between the spiral turns. Further evidence for this β-spiral structure was compiled by molecular dynamic simulations detailing the type II β-turn at the Pro-Gly diad *(25)*. Although alternatives to this model have been proposed, the β-spiral model is more commonly accepted *(26)*.

β-structures are a ubiquitously found secondary structure of proteins, forming either a parallel or anti-parallel hydrogen bonding conformation *(27,28)*. A common structural motif of β-structures is a β-hairpin turn, which essentially results in the reversal of direction of the polypeptide backbone. A typical β-turn is the result of a four residue sequence in an *i* to *i*+ 3 formation where there is hydrogen bond formation between C=O (*i*) and the N-H (*i*+3) (Figure 2) *(29)*. Statistical analysis of Protein Data Bank structures suggests type II β-turns energetically favor Pro in position *i*+1 and Gly in position *i*+2. This compliments structural analysis of elastin where the type II β-turns occur predominantly between the *i*+1 and *i*+2 positions *(30)*. Pro and Gly are unique in their conformational flexibility; the side chain of Pro allows very little conformational freedom where the *cis* form is found at higher frequencies over the *trans* form. In the case of Gly, the absence of a side chain allows for the most conformational flexibility *(31)*. Recent investigations into the stereoelectronic effect of proline analogs in type II β-turns have revealed that self-assembly properties of these molecules can be manipulated *(32,33)*.

Tissue-Engineering

Researchers have long been inspired by exploiting nature's use of self-assembly to fabricate hosts of novel biomaterials *(34)*. Biomaterials from collagen stimulated researchers to develop materials from proteins with similar self-assembling properties. The potential of elastin, having similar non-covalent molecular interactions (hydrogen bonding, hydrophobic interactions, van der Waals forces and electrostatic interactions) as collagen, was naturally explored (Figure 3A).

Simulating the appropriate size, geometry and architecture of natural extracellular matrix is of critical importance for tissue engineering and three-dimensional (3D) tissue culture. Essential parameters for tissue scaffolds are microstructures, porosity, pore size, surface area / surface chemistry and mechanical properties *(35)*. With these properties in mind, several iterations of scaffolds have been produced and evaluated.

Initially, natural decellularized arterial (ECM) was previously used as a scaffold however, this material was found to form matrices too tight for cellular migration when seeded with cells *(36)*. To create more porous scaffolds, pure elastin and pure collagen scaffolds derived from arterial ECM were generated. The elastin scaffolds exhibited 120 μm infiltration of fibroblast cells *in vitro* and *in vivo* models showed improved cell infiltration and repopulation of the scaffold

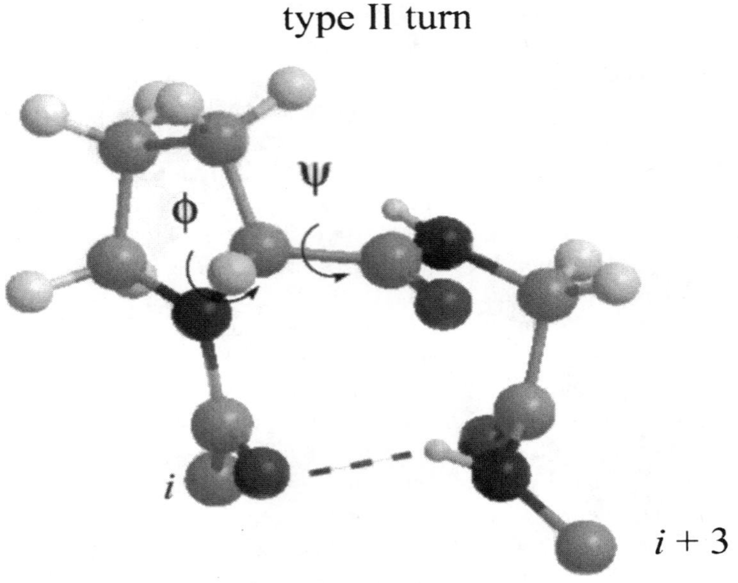

type II turn

Figure 2. Representation of a type II β-turn. The φ and ψ bonds are highlighted. Hydrogen bonding between the i and i+3 carbonyl oxygen and amide hydrogen is represented as a dashed line.

presumeably due to the presences of chemotactic factors *(37,38)*. Further, electrospun scaffolds of both α-elastin and recombinant tropoelastin seeded with human embryonic palatal mesenchymal (HEPM) cells exhibited a higher rate of cell proliferation over other scaffolds such as collagen and gelatin *(14)*. The α-elastin / tropoelastin scaffolds retained their elastic and wavy properties after electrospinning where the other scaffolds were mostly straight, suggesting the scaffold topology may play a role in the rates of cell proliferation *(14)*. Stable 3D scaffolds for smooth muscle cells were also generated by electrospinning elastin, in which the diameters of the electrospun scaffolds was manipulated by electrospinning different proportions of collage or elastin *(39)*. Manipulating the diameter of elastin scaffolds could provide better control over the scaffold topology where it may be possible to fabricate scaffolds to parameters favored by individual cell types may be possible. Elastin / collagen / poly (D,L-lactide –*co*-glycolide) PLGA blends electrospun into fibers exhibited a burst pressure range similar to that of native blood vessels, cell attachment ability as well as *in vivo* biocompatibility, suggesting these materials could be useful in generating vascular grafts *(40)*. Moreover, electrospun polydioxanone (PDO) elastin-blends demonstrated potential benefits for developing vascular grafts *(41)* (Figure 3A).

Despite the success of these scaffolds, calcification remained problematic after implantation. To address this limitation, the rate of calcification could be altered depending on the mode of elastin purification and formation (38).

Elastin-like polypeptides (ELP) consisting of the sequence Val-Pro-Gly-Xaa-Gly (where Xaa can be any amino acid except for Pro) exhibit physical properties as the native elastin. Under changes in temperature, pH and ionic strength, they are able to undergo phase transition from a water soluble polypeptide to an aggregated insoluble form (42). This phase transition is fully reversible resulting in ELP 'smart biomaterial' properties to assemble and / or disassemble in response to environmental conditions. In addition, ELPs have a certain tunibility that is achieved by length manipulation of the repeat sequence and the 'ghost' amino acid residing in the Xaa position (35).

Synthetic ELP hydrogels have been shown to exhibit characteristics similar to those of native elastin in response towards temperature and salt concentrations. Subjecting to mechanical tests, the recombinant human tropoelastin isoform SHELΔ26A (synthetic human elastin without domain 26A) is extensible up to four times the natural elastin. These hydrogels are able to support cellular growth both *in vivo* and *in vitro* (43) (Figure 3A).

Hydrogels of ELP possess potential for 3D tissue engineering. Research on understanding the molecular mechanism that regulate cell proliferation and differentiation has been limited because of the poor topologies of two-dimensional tissue culture. The 3D architecture of the extracellular matrix (composed mainly of collagen and elastin) has been shown to provide an adequate environment to reproduce the physiological surroundings necessary for proliferation and differentiation studies (44). Human adipose derived stem cells (hADAS) incubated with coacervated three-dimensional ELP demonstrated histological and immunohistological characteristics of differentiated chondrocytes (45). Although this differentiation occurs on a gelatin matrix as well, it is necessary to supplement the media with growth factors to induce differentiation (46). ELPs have also been genetically engineered for *in vivo* cross-linking where tissue transglutaminase (tTG) catalyses the introduction of covalent bonds, resulting in the synthesis of a stabilized 3D cartilage matrix (Figure 3A) (47).

Elastin Fusion Proteins

Recombinant DNA technology not only allows researchers the ability to manipulate individual genes but to fuse genes encoding individual proteins together to develop chimeric or fusion proteins. These fusion proteins may have the function of one of the proteins, exhibit properties of both proteins, or acquire a new function altogether. The ELP sequence derived from elastin has been used for a variety of functions. We will evaluate ELP fusion proteins used for drug delivery, protein purification and developing artificial polypeptide scaffolds.

The control of inverse transition temperatures by sequence manipulation and biocompatibility of ELPs make them useful polymers for drug delivery. Cultured cancer cells and solid tumors in animal models uptake fluorescently labeled ELPs in a thermally responsive manner *(48,49)*. Two major limitations in cancer therapy have been the inability of therapeutic molecules to cross the cell membrane and the target-specificity of the compounds. To overcome these limitations cell-penetrating, peptides (CPP) have been fused with ELPs (CPP-ELP) to develop thermally responsive therapeutics with the ability to translocate the cell membrane (Figure 3B). CPPs can assist in the transportation of hydrophilic compounds (small molecules, oglionucleotides and peptides) across the cell membrane *(50)*. Fusing ELPs to a variety of CPPs have revealed that the peptide sequence of penetratin demonstrates the most efficient cellular uptake *(51)*. Further, these CPP-ELPs have been fused to a c-Myc inhibitory peptide known to target and inhibit cancer. As proof of principle, these fusion proteins inhibits proliferation of cultured cancer cell lines in a thermally responsive manner *(52)*.

Protein purification is a technique required for the characterization of proteins. It is hindered by technical difficulties and high cost. To combat these problems, a variety of fusion protein methodologies have been developed *(53,54)*. The Chilkoti group have designed a novel method of purifying proteins exploiting the thermally-responsive nature of ELPs (Figure 3C) *(54)*. An ELP sequence with a predictable, reversible inverse transition temperature is fused to thioredoxin, a carrier protein, and tendamistat, an inhibitory protein of α-amylase. Subjecting *Escherichia coli* lysates expressing this fusion protein to increasing temperatures reveals an increase in turbidity due to aggregation of the ELP sequence. This aggregate can be isolated by centrifugation, resuspended and cleaved with thrombin resulting in a purified active enzyme. This methodology has been used in the purification of gp130 family of cytokines as well as carbohydrate binding proteins *(55,56)*. Developing this method further, a self-cleaving intein has been introduced into the fusion proteins in which a slight reduction in pH results in intein cleavage. This step is followed by another temperature cycling and centrifugation, leaving the purified protein of interest in the supernatant *(57)*.

Protein microarray technology has been limited in its progress because of technical difficulties of immobilizing proteins on a solid support. Two different methods of immobilizing elastin fusion proteins on solid support have been described in the literature (Figure 3D). Chilkoti and coworkers have generated an ELP end-grafted to a thermally evaporated·nanoscale gold surface via its amine groups. The ELP- thioredoxin (Trx-ELP) fusion proteins are captured by exploiting the self-assembling properties of ELP-ELP hydrophobic interaction.

Antibodies specific to Trx are able to bind to the complex, suggesting the Trx moiety of Trx-ELP is preferentially oriented outward from the surface. Decreasing the NaCl concentration results in release of the complex *(58)*. This work demonstrates the capture and release properties of ELP peptides on a nanoscale. Recently, Tirrell *et al.* have developed an artificial polypeptide scaffold bearing an ELP fused to a zipper-based protein capture domain *(59)*. In the 'ghost' position of the ELP sequence (VPGXG), a photoreactive phenylalanine analogue *para*-azidophenylalanine was introduced that enabled cross-linking to the solid support upon UV irradiation. Fused to the ELP sequence is a linker element joined to heterodimeric leucine zipper. Green fluorescent protein and glutathione-S-transferase are expressed with a C-terminal heterodimeric leucine zipper complementary to the zipper of the ELP fusion. By exploiting the high affinity of the heterodimeric leucine zipper-zipper interaction, concentrations as low as 50 nM are detected under a broad range of pH values.

Conclusions

Elastin-derived proteins serve as versatile scaffolds that can be used to generate novel biomaterials for particular biological functions, therapeutic delivery, protein separation and immobilization. This is due in part from the knowledge of the basic structure and mechanical properties learned from studying such proteins. The fact that it is now possible to precisely design polymers based on elastin via recombinant DNA techniques further expands the capabilities and potential applications for such materials. This, in addition to the recent advances in incorporating unnatural amino acids, not only broaden the diversity of monomers, but also can influence the overall behavior and properties of the polymer *(56,59)*. The installation of side chains capable of performing orthogonal chemistries and conjugation provides a handle for adorning elastin-derived proteins with additional components to design complex materials that may not have been possible before. Overall these methods grant the researcher an added level of molecular control with the potential to engineer innovative supramolecular structures endowed with emergent properties.

Acknowledgments

The authors acknowledge funding from Polytechnic University, the Othmer Institute (J.K.M), the Wechler Award (J.K.M.) and the Society for Plastic Engineers (J.S.H.). We also thank Dr. Victor Barinov for assistance on the manuscript. [‡]These authors contributed equally to this document.

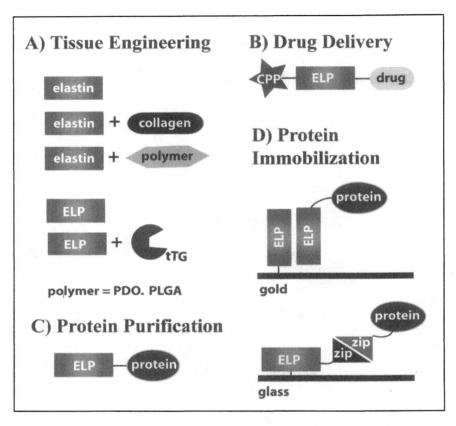

Figure 3. Applications of elastin and elastin-like polypeptides (ELPs). (A)Electrospun elastin, blends of elastin with collagen or polymers andhydrogels of ELP (where the addition of tTG enables crosslinking) used for tissue engineering. (B) ELP-protein fusions employed for purification. Although not shown, a cleavage site could be engineered into the sequences to remove the ELP from the target protein. (C) ELP fused to a cell penetrating peptide (CPP) and drug moiety used for drug-delivery purposes. (D) ELP fusions employed in protein capture and immobilization on chemically coated gold and glass substrates.

References

1. McGrath, K. P.; Tirrell, D. A.; Kawai, M.; Mason, T. L.; Fournier, M. J. *Biotechnol Prog.* **1990**, *6*, 188-192.
2. Rapaka, R. S.; Okamoto, K.; Urry, D. W. *Int J Pept Protein Res.* **1978**, *12*, 81-92.
3. Mecham, R.; Davis, E., *Elastic fiber structure and assembly*; New York: Academic Press: **1994**; pp 281-314.
4. Hinek, A.; Rabinovitch, M. *J Cell Biol.* **1994**, *126*, 563-174.
5. Hinek, A.; Mecham, R. P.; Keeley, F. *J Clin Inves.* **1991**, *88*, 2083-2094.
6. Kielty, C. M.; Sherratt, M. J.; Shuttleworth, C. A. *J Cell Sci.* **2002**, *115*, 2817-2828.
7. Vrhovski, B.; Weiss, A. S. *Eur J Biochem.* **1998**, *258*, 1-18.
8. Reiser, K.; McCormick, R. J.; Rucker, R. B. *Faseb J.* **1992**, *6*, 2439-2449.
9. Mecham, R. P.; Foster, J. A. *Biochem J.* **1978**, *173*, 617-625.
10. Urry, D. W.; Hugel, T.; Seitz, M.; Gaub, H. E.; Sheiba, L.; Dea, J.; Xu, J.; Parker, T. *Philos Trans R Soc Lond B Biol Sci.* **2002**, *357*, 169-184.
11. Debelle, L.; Tamburro, A. M. *Int J Biochem Cell Biol.* **1999**, *31*, 261-272.
12. Tatham, A. S.; Shewry, P. R. *Philos Trans R Soc Lond B Biol Sci.* **2002**, *357*, 229-34.
13. Flory, P. J. *J. Am.Chem. Soc.* **1956**, *78*, 5222-5235.
14. Li, M.; Mondrinos, M. J.; Gandhi, M. R.; Ko, F. K.; Weiss, A. S.; Lelkes, P. I. *Biom.* **2005**, *26*, 5999-6008.
15. Bochicchio, B.; Ait-Ali, A.; Tamburro, A. M.; Alix, A. J. *Biopoly.* **2004**, *73*, 484-493.
16. Tamburro, A. M.; Pepe, A.; Bochicchio, B.; Quaglino, D.; Ronchetti, I. P. *J Biol Chem.* **2005**, *280*, 2682-2690.
17. Stapley, B. J.; Creamer, T. P. *Protein Sci.* **1999**, *8*, 587-595.
18. Cubellis, M. V.; Caillez, F.; Blundell, T. L.; Lovell, S. C. *Prot.* **2005**, *58*, 880-92.
19. Shi, Z.; Olson, C. A.; Rose, G. D.; Baldwin, R. L.; Kallenbach, N. R. *Proc Natl Acad Sci U S A.* **2002**, *99*, 9190-9195.
20. Creamer, T. P.; Campbell, M. N. *Adv Protein Chem.* **2002**, *62*, 263-82.
21. Tiffany, M. L.; Krimm, S. *Biopoly.* **1968**, *6*, 1767-70.
22. Venkatachalam, C. M. K., M. A.; Sugano, H.; Urry, D. W. *J. Am. Chem. Soc.* **1981**, *103*, 2372-2379.
23. Cook, W. J. E., H.; Trapane, Tina L.; Urry, Dan W.; Bugg, Charles E. *J. Am. Chem. Soc.* **1980**, *102*, (17), 5502-5505.
24. Khaled, M. A.; Venkatachalam, C. M.; Sugano, H.; Urry, D. W. *Int J Pept Protein Res.* **1981**, *17*, 23-33.
25. Li, B.; Alonso, D. O.; Bennion, B. J.; Daggett, V. *J Am Chem Soc.* **2001**, *123*, 11991-11998.

50

26. Gross, P. C.; Possart, W.; Zeppezauer, M. *Z Naturforsch [C]*. **2003**, *58*, 873-878.
27. Pauling, L.; Corey, R. B. *Proc Natl Acad Sci U S A*. **1951**, *37*, 251-256.
28. Blake, C. C.; Koenig, D. F.; Mair, G. A.; North, A. C.; Phillips, D. C.; Sarma, V. R. *Natu*. **1965**, *206*, 757-761.
29. Venkatachalam, C. M. *Biochim Biophys Acta*. **1968**, *168*, 411-6.
30. Gunasekaran, K.; Gomathi, L.; Ramakrishnan, C.; Chandrasekhar, J.; Balaram, P. *J Mol Biol*. **1998**, *284*, 1505-1516.
31. Jabs, A.; Weiss, M. S.; Hilgenfeld, R. *J Mol Biol*. **1999**, *286*, 291-304.
32. Kim, W.; McMillan, R. A.; Snyder, J. P.; Conticello, V. P. *J Am Chem Soc*. **2005**, *127*, 18121-18132.
33. Hodges, J. A.; Raines, R. T. *J Am Chem Soc*. **2003**, *125*, 9262-9263.
34. Chiruvolu, S.; Walker, S.; Israelachvili, J.; Schmitt, F. J.; Leckband, D.; Zasadzinski, J. A. *Sci*. **1994**, *264*, 1753-1756.
35. Leong, K. F.; Cheah, C. M.; Chua, C. K. *Biom*. **2003**, *24*, 2363-2378.
36. Walles, T.; Lichtenberg, A.; Puschmann, C.; Leyh, R.; Wilhelmi, M.; Kallenbach, K.; Haverich, A.; Mertsching, H. *Eur J Cardiothorac Surg*. **2003**, *24*, 358-363.
37. Simionescu, D. T.; Lu, Q.; Song, Y.; Lee, J. S.; Rosenbalm, T. N.; Kelley, C.; Vyavahare, N. R. *Biom*. **2006**, *27*, 702-713.
38. Lu, Q.; Ganesan, K.; Simionescu, D. T.; Vyavahare, N. R. *Biom*. **2004**, *25*, 5227-5237.
39. Buttafoco, L.; Kolkman, N. G.; Engbers-Buijtenhuijs, P.; Poot, A. A.; Dijkstra, P. J.; Vermes, I.; Feijen, J. *Biom*. **2006**, *27*, 724-734.
40. Stitzel, J.; Liu, J.; Lee, S. J.; Komura, M.; Berry, J.; Soker, S.; Lim, G.; Van Dyke, M.; Czerw, R.; Yoo, J. J.; Atala, A. *Biom*. **2006**, *27*, 1088-1094.
41. Sell, S.; McClure, M.; Barnes, C.; Knapp, D.; Walpoth, B.; Simpson, D.; Bowlin, G. *Biomedical Mat*. **2006**, *1*, 72-80.
42. Urry, D. W.; Trapane, T. L.; Prasad, K. U. *Biopoly*.**1985**, *24*, 2345-2356.
43. Mithieux, S. M.; Rasko, J. E.; Weiss, A. S. *Biom*. **2004**, *25*, 4921-4927.
44. Schindler, M.; Nur, E. K. A.; Ahmed, I.; Kamal, J.; Liu, H. Y.; Amor, N.; Ponery, A. S.; Crockett, D. P.; Grafe, T. H.; Chung, H. Y.; Weik, T.; Jones, E.; Meiners, S. *Cell Biochem Biophys*. **2006**, *45*, 215-227.
45. Betre, H.; Ong, S. R.; Guilak, F.; Chilkoti, A.; Fermor, B.; Setton, L. A. *Biom*. **2006**, *27*, 91-99.
46. Awad, H. A.; Wickham, M. Q.; Leddy, H. A.; Gimble, J. M.; Guilak, F. *Biom*. **2004**, *25*, 3211-3222.
47. McHale, M. K.; Setton, L. A.; Chilkoti, A. *Tissue Eng*. **2005**, *11*, 1768-1779.
48. Meyer, D. E.; Trabbic-Carlson, K.; Chilkoti, A. *Biotechnol Prog*. **2001**, *17*, 720-728.
49. Raucher, D.; Chilkoti, A. *Cancer Res*. **2001**, *61*, 7163-7170.
50. Temsamani, J.; Vidal, P. *Drug Discov Tod*. **2004**, *9*, 1012-1019.

51. Bidwell, G. L., 3rd; Raucher, D. *Mol Cancer Ther.* **2005**, *4*, 1076-1085.
52. LaVallie, E. R.; DiBlasio, E. A.; Kovacic, S.; Grant, K. L.; Schendel, P. F.; McCoy, J. M. *Biotechnology (N Y).* **1993**, *11*, 187-193.
53. Smith, P. A.; Tripp, B. C.; DiBlasio-Smith, E. A.; Lu, Z.; LaVallie, E. R.; McCoy, J. M. *Nucleic Acids Res.* **1998**, *26*, 1414-1420.
54. Meyer, D. E.; Chilkoti, A. *Nat Biotechnol.* **1999**, *17*, 1112-1115.
55. Lin, M.; Rose-John, S.; Grotzinger, J.; Conrad, U.; Scheller, J. *Biochem J.* **2006**, *398*, 577-83.
56. Sun, X. L.; Haller, C. A.; Wu, X.; Conticello, V. P.; Chaikof, E. L. *J Proteome Res.* **2005**, *4*, 2355-2359.
57. Banki, M. R.; Feng, L.; Wood, D. W. *Nat Meth.* **2005**, *2*, 659-661.
58. Hyun, J.; Lee, W. K.; Nath, N.; Chilkoti, A.; Zauscher, S. *J Am Chem Soc.* **2004**, *126*, (23), 7330-5.
59. Zhang, K.; Diehl, M. R.; Tirrell, D. A. *J Am Chem Soc.* **2005**, *127*, 10136-10137.

Chapter 4

Bioartificial Polymer Matrices as Biomaterials with Enzymatically Controlled Functional Properties

Davide Silvestri[1], Caterina Cristallini[2], and Niccoletta Barbani[1]

[1]Department of Chemical Engineering, Industrial Chemistry and Materials Science, University of Pisa, via Diotisalvi 2, 56126 Pisa, Italy
[2]CNR Institute for Composite and Biomedical Materials IMCB, Pisa c/o Department of Chemical Engineering, Pisa, Italy

A review is given of the authors' work in developing new bio-inspired materials, called "bioartificial materials." The key is to understand the interactions between the synthetic and biological systems and to obtain materials where these interactions are optimized prior to their contact with a living tissue. The work shown in this paper provides encouraging indication for the developments of new biomaterials containing synthetic and natural components with improved performances for new applications.

I. Introduction

During the past few years, there has been great success in attempts to realise a closer relationship between the life science and material science and to use the great potential of biological world for the development of new materials and new industrial processes. The interaction between the natural and synthetic components played a major role, especially in the closely related fields of bioengineering, biomaterials and biotechnologies. Blends of natural and

52

synthetic polymers and, in particular, blends of synthetic macromolecules and biological components (such as proteins and enzymes) were studied in the past decade by our research group with the aim of investigating their functional, chemical and biological properties and their application as biomaterials. The goal of the present paper is to carry out a brief overview of our recent activities involving the use of catalytic molecules for the control of functionality and characteristics (such as chemical properties and transport features) and to describe the development of the study and approach towards these materials from our first attempts to our more recent experiences.

The combination of synthetic polymeric materials with natural macromolecules has gathered growing interest from the scientific community in recent years. Most of these systems have attractive characteristics for the realisation of new materials for bioengineering and biotechnological applications. They have been defined by our group as "bioartificial polymeric materials" (1,2) and were originally conceived with the aim of realising new biomaterials combining the features of synthetic polymers (good mechanical properties, easy processability, low production and transformation costs) with the specific tissue- and cell-compatibility of biopolymers (3,4). The performance of these materials relies on the relationship between the role played by the synthetic-natural polymer interactions and the cell- and tissue-compatibility of the resulting material. In fact, the cell- and tissue-compatibility of a material is determined by the interactions at the molecular level between the material and the constituents of the living tissue of the host body. Through these interactions, the molecules of the living tissues can change their biological functionality, e.g., molecular conformation and/or activity of functional groups bound to or interacting with the synthetic material. The less the biological functionality of the living tissue is affected by the synthetic material, the better its compatibility. The basic idea of bioartificial polymeric materials' development is to reduce the interactions between synthetic and living systems by creating a two-component material, inside which changes at the molecular level have already occurred as a result of the interactions between the synthetic and the biological components, before coming into contact with the living tissue (5). Such a material, with pre-established molecular interactions, should behave better macroscopically than a fully synthetic material, with regard to the biological response of the host. It becomes therefore evident that the study and comprehension of interactions occurring at a molecular level is of crucial importance for a selection at a very early stage of the more promising synthetic polymers in manufacturing new systems with improved cell and tissue compatibility.

Recently, our research attempted to find new materials based on blends of biological macromolecules, such as structural proteins and polysaccharides, and hydrophilic synthetic polymers, such as poly(vinyl alcohol) (PVA), in which the biocompatibility of the former is combined to the mechanical properties of the latter (6).

In particular, the attention of the research was addressed to the use of enzymes (a particular class of proteins) as biological components of the bioartificial material in order to expand the use of enzyme-synthetic polymer blends, for example, in the field of drug delivery systems. In fact, for many significant applications, the drug delivery systems should be able to release *in vivo* high molecular weight drugs as peptides and proteins without any modification in their bioactivity, and the investigation of enzymes, as model molecule to be released, permits the control and the monitoring of biomolecular stability during entrapment and elution phases. Controlled release systems have found wide application in those therapies where a continuous delivery of the active compound (e.g., enzyme and drug) is required (e.g., diabetes, glaucoma, antitumoral or contraceptive therapies) (7).

In our pursuit of bioartificial systems to be delivered or containing an enzyme that plays its specific role of a catalyst for a bioreaction to modify such chemical or functional property of host matrix, our group investigated different possibilities of preparing bio-compatible, bio-functional and bio-inspired materials for advanced applications in the biomaterials field.

II. Enzyme-based Bioartificial Polymeric Materials

One of the first attempts in the realisation of bioartificial matrices, containing as an enzyme as a biological component was carried out in our research group by Cristallini et al. (8). In view of the basic idea of bioartificial materials, the goal of this study was the analysis of interactions and relationship at nanoscale level between a synthetic component and α-amylase enzyme, from chemical and physical points of view.

In particular, the investigation of the molecular level interactions between synthetic and natural materials is important in designing new biomaterials, providing basic information about the biocompatibility characteristics of synthetic component. Common calorimetric or chemical investigation methods are in general useful for investigating these interactions from a thermodynamic point of view; however these approaches are not indicative or sufficient to show changes occurring in the biological activity of the natural component used. It was clear to the authors that a strong reduction of the biological performance would make the biopolymer useless in manufacturing a bioartificial material. For these reasons, the authors decided to study biological-synthetic polymer interactions by means of biological tests based on the use of enzymes, in order to investigate the state of the biological component inside a bioartificial material. The state of a natural component blended with a synthetic material could be studied by considering the biopolymer as a substrate for a specific enzyme, or, alternatively, the state of the enzyme after blending can be investigated by studying its activity against an external substrate. In this specific study, enzyme-synthetic material blend was in itself a bioartificial material, which could be considered either as a bioactive material or a delivery system depending on whether the enzyme was released or not. In this paper, PVA-based hydrogels, containing the amylase molecule, were evaluated as drug (enzyme) delivery systems, paying attention to the effect of the synthetic polymer on thermostability and activity of the enzyme. The cross-linking method employed for the realisation of PVA-amylase hydrogels was in effect selected to maintain unaltered the functionality of the enzyme, performing a freeze-thaw (9) procedure that presented the advantages of obtaining three-dimensional polymer networks without chemical cross-linkers.

A more important findings of this study is the differential scanning calorimetry (DSC) data that clearly detected the effect of the enzyme and showed a certain degree of interaction between α-amylase and PVA, by which α-amylase molecules could influence the crystallinity of PVA. Together with DSC investigation, X-ray diffraction was performed (Figure 1) showing three strong diffraction peaks for α-amylase (14°, 16.8° and 18.7°) and a diffraction maximum at 19° for pure PVA. The α-amylase/PVA blend showed the diffraction peaks at the same angles as the pure components, but the normalized intensity of the PVA peak were higher than that of pure PVA. These results indicated that the crystalline structure of PVA did not change in the blend, and a great number of crystallites were formed. This agreed with DSC, indicating that α-amylase molecules acted as additional nucleation sites for PVA chains.

As previously mentioned, part of our work was dedicated to the study of bio-activity of the enzyme used to prepare bioartificial hydrogels: the enzyme activity was monitored, in a phosphate buffer containing starch substrate (0.2 mg/ml), by measuring the substrate concentration remaining in the batch solution at different times. The results, compared with free α-amylase behaviour, indicated that the catalytic hydrolysis of starch was the same for free α-amylase and α-amylase delivered from bioartificial blend: no deactivation of the enzyme because of the presence of PVA was observed either in solution or after the preparation through casting procedure. On the contrary, α-amylase seemed slightly more stable in the polymer network.

Starting from these first promising results, our studies were geared toward the use of enzymes as biological components of a bioartificial material expanding their potential use for controlled delivery of biologically active substances (10). For many significant clinical applications, drug delivery systems should be able to release high molecular weight drugs such as peptides and proteins, which in-vivo have to maintain their functional properties with preservation of their bioactivity. In fact, the introduction of enzymes in medical therapy has undergone significant developments in the past decade; progress on this road has been made, even if some problems remain, thanks to the new technologies of protein engineering which has led to an increasing number of new therapeutic protein drugs and methods to modify their physico-chemical properties such as stability, activity and specificity (11,12). Physical entrapment

56

or chemical conjugation with biocompatible polymers (such as poly(ethylene glycol) PEG or poly(vinyl alcohol) PVA) was demonstrated to confer important protein drug properties: reduction of the metabolic degradation and immunogenicity, control of the rate and site of delivery. In this context, Cristallini et al. (8) reported the study of the influence of the synthetic polymer at molecular level on a biological component like an enzyme inside a bioartificial polymeric system. Results showed a stable and unaltered biological activity of enzyme, evidencing that neither the tight contact with the PVA nor the casting procedure damage α-amylase and the molecules of biological component seemed to become more stable with respect to the free enzyme.

However, for a potential use of these films as drug delivery systems it is extremely important to limit a rapid film dissolution after contact with aqueous medium, so attempts to induce film stabilization in water have been made by subjecting the samples to a suitable cross-linking method based on dehydro-

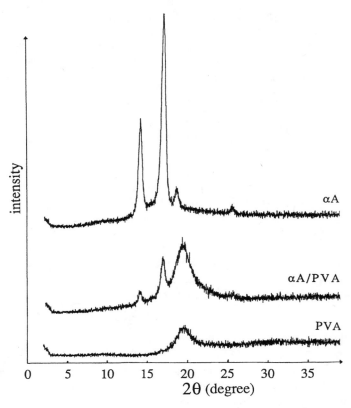

Figure 1. X-ray diffraction scans of pure components and α-amylase/PVA blend

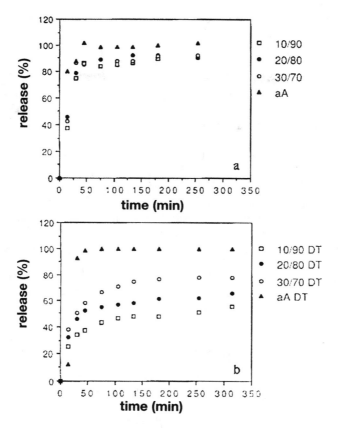

Figure 2. Percentage release of α-amylase from untreated (a) and dehydro-thermal DT treated (b) α-amylase-PVA bioartificial blends. Initial α-amylase-PVA ratio 10/90 (□), 20/80 (●), 30/70 (○) and 100/0 (▲).

thermal treatment. Moreover, after this thermal modification, interactions between the biological and synthetic components have been examined from both physicochemical and biological points of view (*10*). The stability of the enzyme, both pure and in the presence of PVA did not seem to be altered by the dehydro-thermal treatment (no variation of T_d was detected), but these results differed from those obtained for untreated films (*8*) which showed an increase of α-amylase stability attributable to hydrogen bond interactions between the hydrophilic residues of the enzyme and the hydroxyl groups of the synthetic polymer. This behavior variation in terms of thermal stability of the enzyme, together with enthalpy values for treated samples, were associated with the fact that the cross-linked PVA chains developed steric hindrance around the α-amylase molecules and consequently caused a reduction of their interaction with water molecules in the surrounding environment.

The hypothesis that the α-amylase molecules act as additional nucleation sites for PVA chains seemed to be confirmed also for the bioartificial blends subjected to dehydro-thermal treatment.

The PVA and thermal treatment confer improved enzyme stability suggesting the use of this system in protein delivery. Delivery kinetics depend on numerous variables such as chemical nature and processing of the matrix. The release of an enzyme from a matrix with hydrophilic characteristics depends mainly on two simultaneous processes: water migration into the film (with swelling of polymer matrix) and enzyme diffusion through the swollen film. In effect, the delivery results were typical of a release from a film which rapidly reaches swelling equilibrium in an aqueous medium (Figure 2).

This work showed that dehydro-thermal treatment carried out on α-amylase-PVA bioartificial blends was a very promising method for obtaining materials in the biomedical field. The treatment provided stability to the film, ensured that the film did not completely dissolve after contact with water medium, and at the same time seemed to have no significant effect on enzyme activity and delivery. The promising results obtained from this study encouraged us to continue the research of bio-artificial materials based on enzymes.

III. Enzyme-based Bioartificial Polymeric Materials through "Template Polymerisation"

Part of the path towards of the realization of enzyme-based bioartificial materials was the investigation of bioartificial materials deriving from "template polymerisation" procedure. Similar to earlier research, the initial idea of using an enzyme as biological component (and stamp molecule) relied on the possibility to use its functionality as diagnostic tool for evaluating the efficiency of the interaction between the biological and the synthetic component. In addition, in bioartificial polymeric materials study, "Template Polymerisation" should be conceived as a potential procedure to facilitate molecular interactions between the two polymers as the synthetic polymer forms "in situ" taking the "shape" of the enzyme (template).

"Template polymerisation" is usually defined as a reaction in which the growing chains propagate, during most of their lifetime, along macromolecules added to the system as pre-formed polymer (13-18): in this way, there is the formation of a polymer ("daughter polymer") in the presence of the macromolecule ("template" or "parent polymer") added to the reaction mixture (19,20). This particular propagation mode is allowed by specific co-operative interactions (i.e., ionic interactions or hydrogen bonds) between the growing chain units and the polymer (usually defined as template).

The presence of the template may influence the reaction kinetics (i.e., polymerisation rate, reaction order with respect to monomer and initiator, activation energy), the structural characteristics of the produced polymer (i.e., average molecular weight, molecular weight distribution, microstructure and stereoregularity) and the reactivity ratios in the case of co-polymerisations.

Depending on the kind and the strength of the interactions between the monomer and the template monomeric unit, the polymerisation may follow two main classes of mechanisms called *zip* and *pick-up* (Figure 3) (*15,18,21*).

In a *zip* mechanism the affinity of the monomer for the template repeating unit is high (strong ionic interactions may result), the monomer molecules are adsorbed on the template and the adjacent molecules bind themselves to each other (in the case of a step polymerisation) or they add to the active ends of the chain (in the case of a chain polymerisation).

In a *pick-up* mechanism, because of the weaker interactions between the monomer and the template, there is no preferential adsorption and polymerisation starts in the bulk. When the growing oligomer has reached a critical length the co-operative effects allow for the adsorption of the chain on the template and the polymerisation proceeds by adding monomer from solution onto the growing chain adsorbed on the template. Information on the mechanism may be gained from kinetic studies (*15*).

In particular the changes in the polymerisation rate (with respect to blank polymerisation) when the template concentration [T] changes and the monomer concentration [M] is kept constant are studied. The relative initial polymerisation rate $V_R = V/V_B$ (where V is the rate of template polymerisation and V_B is the rate of the blank polymerisation) is reported as a function of the initial template unit to monomer molar ratio ([T]/[M]). If the polymerisation involves a *zip* mechanism the relative rate V_R increases by increasing the [T]/[M] ratio and a maximum is reached for [T]/[M] = 1 (supposing that monomer and template interact in 1:1 molar ratio); beyond this value V_R decreases. However, if the adsorbed monomer molecules are able to move along the template, the polymerisation rate will decrease less rapidly or remain constant.

If a *pick-up* mechanism is involved, the relative rate V_R increases by increasing the [T]/[M] ratio and it reaches a maximum for [T] = [T]* where [T]* is the template concentration for which the template chains are sufficient to complex all the growing radicals. For higher [T]/[M] ratios the excess of template macromolecules does not affect further the reaction rate.

In practice several systems show reaction mechanisms which are intermediate between a *zip* or a *pick-up* mechanism.

The method introduced by our group to monitor and investigate the polymerisation kinetics of acrylic monomers based on the on-line measurement of the reaction mixture conductivity both for strong (*22*) and weak electrolytes (*23*) resulted an interesting way to determine the template polymerisation mechanism. Direct and rapid information on the reaction mechanism can be

60

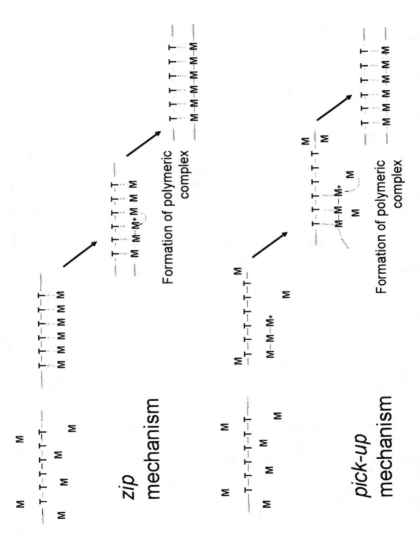

Figure 3. Schematic representation of zip and pick-up mechanisms in the template polymerisation

obtained by the analysis of the conductimetric curve (instead of the more tedious and longer dilatometric measurements).

In the case of a *zip* mechanism a conductivity decrease is observed for monomer addition, due to the monomer adsorption on the template and the consequent mobility decrease of the charge carriers. Also, a conductivity increase is recorded when the initiator is added but no subsequent conductivity change is expected since the polymerisation takes place when the monomer is already pre-adsorbed, with reduced mobility, on the template.

On the other hand, in the case of a *pick-up* mechanism there is no preferential adsorption of monomer onto the template; both during oligomer formation phase and during the subsequent phase in which oligomers are adsorbed onto the template, the chain growth takes place by picking up monomers from the solution as in a blank polymerisation.

A decrease of the solution conductivity is expected as a consequence of the decrease of the charge carriers mobility. Therefore the specific conductivity vs. time curves will show a trend qualitatively similar to the curves for blank polymerisations. In conclusion, it is possible to obtain information on the mechanism of a template polymerisation by recording the changes in conductivity with time.

First attempts at template polymerisation were carried out in the radical polymerisation of acrylic monomers onto synthetic templates in aqueous environment and the development of efficient monitoring systems for the investigation of template influence on the reaction kinetics of homo- and co-polymerisation (*24-30*). After these studies, attention was devoted to transferring the lesson learned by studying fully synthetic systems to the preparation of bioartificial materials by polymerising acrylic monomers onto a template of natural origin.

As mentioned in the Introduction, for several years our research group proposed the production of bioartificial polymeric materials obtained from the blending of pairs of synthetic and biological polymers. The production of these materials is generally performed by blending aqueous solutions of the pre-formed polymers and the subsequent transformation of the blend into a film, hydrogel or sponge, through the use of different techniques such as solvent evaporation, lyophilisation and freeze-thaw cycles. The chemico-physical, mechanical and biological characteristics of these materials strongly depend on the degree of compatibility between the two components, which is a function of the interactions between the functional groups present in the material.

The degree of intimacy that is obtained between the two blended components depends in any case not only on the chemical structure of the polymers used but also on the method of preparation. For example, when polymers with a high molecular weight or noticeable steric hindrance are blended, it is difficult to obtain the maximum degree of interaction between the potentially co-operative groups.

The general idea of the "template polymerisation" strategy focuses on attempts to control the constitution of synthetic-biological complex by specific interactions of the synthetic component with a natural template molecule. So the template polymerisation should offer the opportunity of controlling the formation of a synthetic component on a suitable biological template in order to obtain a bioartificial system in which the forming polymer and the template are intimately bound to give a blend with improved properties with respect to a blend obtained by simple mixing.

The diagnostic methods successfully applied to the kinetic study of the synthetic systems were extended to the polymerisation in the presence of biological templates. For the preparation of new bioartificial systems obtained by template polymerisation, different biological classes of templates such as proteins (gelatine), enzymes (α-amylase) and carbohydrates (dextran and chitosan) were used. Typical functional monomers have been used such as AA, NaMAA, sodium styrenesulfonate (NaSSA) and HEMA.

As discussed previously, blends of PVA and α-amylase were prepared from aqueous solutions of the two polymers, proposed by our group as the first bioartificial materials based on enzyme.

Blends, either as films by casting or in the form of hydrogels by freeze-thaw technique, were prepared. The blends could act as delivery systems or bioactive materials depending on their capability to release the enzyme or to retain it, respectively.

Therefore a new bioartificial system obtained directly by template polymerisation of HEMA in the presence of the enzyme α-amylase was investigated (5). By using a functional protein (an enzyme), the blend obtained by template polymerisation can be differentiated from the one prepared by simple mixing through the observation of the biological activity of the natural component and also through the properties of the synthetic counterpart.

The polymerisation of HEMA in the presence of different amounts of α-amylase was studied. The results of the kinetic study are reported in Figure 4 indicating that for a weight ratio enzyme/monomer = 1/2 the reaction rate becomes higher than the rate of the blank polymerisation, while for the other contents no rate enhancements can be observed.

Evidence of the influence of α-amylase on the complex was also gained by means of DSC analysis. A shift toward higher values of the glass transition temperature (T_g) for PHEMA in the blends with respect to the pure synthetic polymer was found independently from the blend composition.

As in the case of simple blending of α-amylase with PVA polymer (presented before), from the isolated poly-complex (that precipitates from the polymerisation solution as a hydrogel) only a small amount of enzyme is released, and the retained enzyme maintained a considerable activity (Figure 5).

In our opinion, the results described in this paper provide encouraging indication for the realisation of new biomaterials containing a synthetic component and a natural one for new applications or for improved performances.

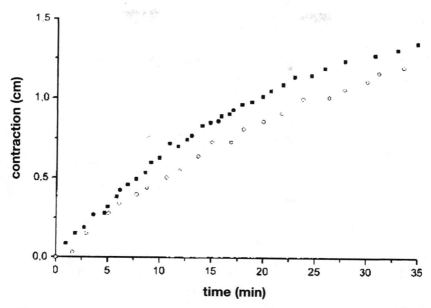

Figure 4. Meniscus displacement vs time curves for the polymerisation of HEMA 0.2 M onto α-amylase at weight ratioT/M = 0 (○) and T/M = 1/2 (■).

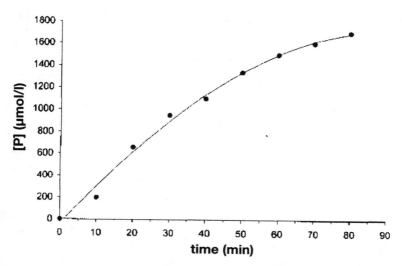

Figure 5. The kinetics of product (maltose) formation from starch hydrolysis catalysed by immobilised α-amylase in a α-amylase/PHEMA complex (weight ratio T/M = 1/2). Production conditions of the complex: 60°C, phosphate buffer (0.02 M, pH = 6.9), reaction time 20 min.

If the idea of anticipating at a molecular level the interactions between the biological and the synthetic systems through the preparation of blends has resulted in successful biomaterials with improved biocompatibility, then the template polymerisation can share the same advantages as those of blends from preformed polymer, or even represent further progress in such a direction.

Finally, the synthetic component, after separation from the biological matrix, can act as a mould for the synthesis of biomolecules mimicking the functional or structural properties of the original template or as a molecular detector having preferential interaction with natural occurring macromolecules in some way similar to the template itself.

These synthetic approaches of bioimprinting aim at reducing the degree of complexity of biological systems in simpler ("miniaturised") models and resulting in several advantages, such as their low production cost, their stability, and simple preparation procedure.

The use of chemically prepared macromolecules with pre-selected specificity lies in a large number of applications starting from the artificial antibodies, to polymers capable of selective synthesis like artificial enzyme systems in catalytic applications and substrate sensitive sensor devices.

In a later paper, in order to improve the engineering of such a bioactive material, the inclusion of poly(N-vinylpyrrolidone) (PVP) in the polymerisation system was also attempted (31). In the evaluated system, the polymers interact by means of hydrogen bonds occurring between hydroxyl groups of PHEMA and functional groups in the biological macromolecule. Further interactions intended for an improved system stabilisation occur with the amide moieties of the PVP repeating unit. The choice of PVP as a stabilising agent for the enzyme/polymer complex was mainly due to the ability of PVP to mimic a natural compound since it contains an amide functional group which is also common to several natural polymers, acting as a "bridge" between the synthetic and natural component. Moreover, PVP showed a stabilising effect on the structure of some proteins (32).

Concerning the investigation of interaction between natural- and synthetic-origin materials and the effect of α-amylase on reaction procedure, the relative initial reaction rate V_R ($V_R = V/V_B$, as described before) as a function of the T/M ratio was plotted (Figure 6): from this graph, a template effect on the polymerisation rate was observed, and the maximum value of the reaction rate was obtained for T/M = 1/2 weight ratio.

This effect was ascribed to the presence of chemical interactions between the functional group of the HEMA monomer and those present in the template molecule, that showed a maximum for the T/M weight ratio indicated. In order to identify the nature and entity of interactions between the monomer or the growing chain and the template, occurring during the polymerisation process, polycomplexes between PHEMA and α-amylase were also prepared by mean of a simple mixing procedure of the preformed components and compared with those obtained by template polymerisation (where PHEMA was obtained by a blank polymerisation of HEMA in the same conditions as for the template polymerisation).

Polycomplexes from both sources were tested for solubility. The two components in the polycomplexes obtained by simple mixing were readily solubilised when treated first with water (solvent for α-amylase) and then with ethanol (solvent for PHEMA). Polycomplexes by template polymerisation showed no detectable solubility in the same solvents. The separation of the two components constituting the polycomplexes failed also by means of hot solvent extraction (Soxhlet), suggesting the presence of stronger interactions in the polycomplexes obtained by template polymerisation. This effect was interpreted as more efficient hydrogen bonding between the two components taking place during template polymerisation (the good solvent for one component was not able to swell sufficiently the other component in order to solubilize the whole material) or the formation of covalent bonds (grafting) between the template and the "daughter" polymer. Another possible interpretation, supported by enzyme release trials, was a self-crosslinking of PHEMA to form a polymeric network entrapping the enzyme. In effect, no significant release of the enzyme (α-amylase) from the polycomplex was revealed by a proper enzymatic assay. A further improved design of the functional material where an enzyme is immobilised in a polymeric matrix by means of template polymerisation was attempted by introducing a third component with a prospective stabilising action.

As a third component, as previously noted, PVP was selected for its chemical structure possessing amide bonds which should show an affinity for the enzyme, helping to maintain its activity and in the same time interacting by hydrogen bonding with the functional groups of PHEMA obtained by template polymerisation (33). In order to verify the effective stabilising action of PVP in this system, a comparison of the biological activity of enzyme-containing polycomplexes prepared by template polymerisation in the presence or without PVP was performed. Results showed higher activity values for polycomplexes containing PVP, confirming the expected stabilising effect. All the results obtained in this study showed that the presence of the α-amylase (biological template) influenced the polymerisation of 2-hydroxyethyl methacrylate. In particular, an increase of the initial reaction rate with the increasing of the enzyme amount was observed (with the maximum rate for the ratio T/M = 1/2). The biological tests indicated that the retained enzyme maintained an adequate enzymatic activity and in addition the presence of PVP played a protective action towards the enzyme immobilised inside the polycomplexes. This property can be used to retain the enzymatic functionality also in other immobilisation systems obtained by means of radical polymerisation.

In effect, the immobilisation of biological molecules on solid polymeric supports is receiving great interest as a way to obtain selectivity, catalytic and recognition properties in a variety of application areas (34-36).

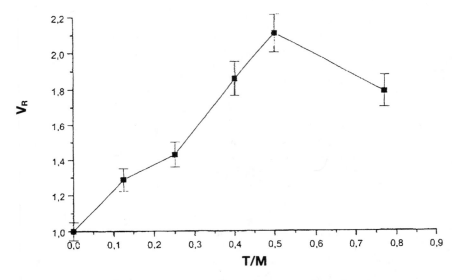

Figure 6. Diagram of V_R as function of T/M for the polymerisation of HEMA in the presence of α-amylase

IV. Enzyme-based Bioartificial Polymeric Materials as Bioactive Immobilisation Systems

Bio-immobilisation has to result in a high number of immobilised biomolecules with sufficient activity, long-term stability and easy accessibility for the substrate. Recently, the interest of scientific researchers has been focused on immobilisation of enzymes in food processing and in medical applications (*37,38*) because of the efficiency and selectivity of the catalytic action that can be obtained using an enzyme in comparison with chemical catalysts. Immobilisation inside a polymer matrix has to guarantee high catalytic activity concentrated in a small volume and higher stability of the enzyme. This last aspect is due to the formation of a micro-environment around the protein that can cause a structural modification resulting in an increase of the active site stability. Also, the entrapment of the catalyst results in a reduction of equipment cost (no need for enzyme recovery) and in the number of processing steps. The introduction of the immobilised enzymes substantially reduces the cost of the biocatalytic processes, thus permitting the use of several fine chemicals through this approach. Enzyme immobilisation finds application in the biomedical field especially in the treatment of metabolic deficiencies via the administration of microencapsulated enzymes (e.g., through liposomes) or in hematic detoxification by means of extracorporeal hemoperfusion (*34-36*). Further applications stem from coupling enzyme immobilisation with analytical electrochemistry for the realisation of (bio)sensors for clinical uses or in developing bioreactors (*39,40*). It is important to consider also that the realisation of a matrix with good entrapping properties towards large molecules presents several technical limitations. The loss of activity of the enzyme entrapped in the support is the major problem that has been observed: in fact, denaturation or inaccessibility to the active site of the immobilised enzyme can occur when a polymer matrix surrounds the biomolecule (*41*).

In this approach, sponges based on dextran (DEX) and poly[ethylene-*co*-vinyl alcohol] (EVAL) (a bioartificial material), containing α-amylase, were produced by phase inversion (*42*). Four aspects seemed to contribute significantly to the efficiency of immobilisation systems: (1) high thickness of the sponges (with respect to the films) was favourable to obtain immobilisation devices for granular enzymes and to realise activity tests for the entrapped biomolecule; (2) pre-freezing of the cast solution reduced the speed of diffusive phenomena during the phase inversion process, thus favouring symmetric porous matrix formation (Figure 7); (3) pre-freezing assured a homogeneous distribution of α-amylase throughout the polymer matrix section; (4) use of a polymer blend containing a synthetic and a natural polymer allowed the design of systems with good interactions towards an immobilised protein and towards body fluids or cells (like in the case of PHEMA-PVP-α-amylase three-component system).

The biocatalysts obtained were evaluated with respect to the composition, morphology, activity and stability of the immobilised enzyme in the starch hydrolysis reaction. In general, two alternative methods can be used, considering the bioartificial matrix as a substrate for the enzyme (this method is used for example to drive drug release into erosion control devices), or alternatively, as in the case of this work, after blending the enzyme with a polymer, and investigating its activity against an external substrate. The apparent kinetic parameters of the reaction catalyzed by the immobilised and native enzymes were determined and compared.

However, the main goal was to study the behavior and performance of a novel bioartificial material (obtained with an innovative method involving pre-freezing and inversion steps) as a suitable matrix for the entrapment of proteins or enzymes in a stable manner. The tests performed (determination of enzyme activity, determination of K_m and V_{max} kinetic parameters, repeatability test) confirmed both that the enzyme immobilised in the bioartificial polymeric matrix maintained its catalytic activity unchanged and that the catalytic reaction rate was comparable with that of the free α-amylase reaction (used as control).

An adequate interaction between the polymeric components and α-amylase and a good residual enzyme activity of the immobilised α-amylase clearly indicate that a porous bioartificial polymer matrix of DEX and EVAL can be a good support for the immobilisation of proteins.

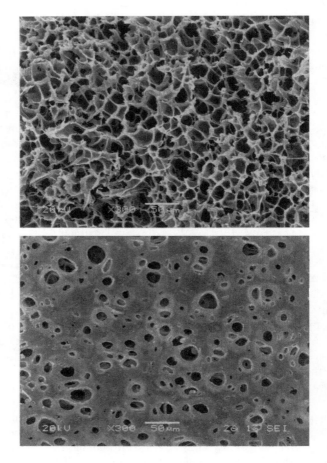

Figure 7. SEM micrographs of the internal structure and surface of a DEX-EVAL-α-amylase system.

V. Enzyme-based Bioartificial Polymeric Materials as Bioerodable Drug Delivery Systems

As indicated, one of the alternatives to preparing a bioactive structure immobilising an enzyme is to use the polymer structure considering the bioartificial matrix as a substrate for the enzyme. In this case, it is possible for example to drive drug release using erosion control devices, where the enzyme plays the catalytic role, and one of the components of the matrix is digested (in a controlled way) by the catalysed reaction.

In this context, a degradable controlled release system is based on a homogenous erosion of a polymer matrix through a physical or chemical mechanism (e.g., ionization, protonation, solubilization, hydrolytic or enzymatic degradation of the polymer). In the literature the erosion of biodegradable polysaccharide matrices has already been investigated; in particular, starch and starch-based blends containing α-amylase were prepared to obtain an accelerated drug release, useful, for example, for drugs with limited solubility or for drugs whose solubility can be influenced by the variation of gastro-intestinal pH. The introduction of matrices based on biological polymers is interesting since the degradation of natural products occurs naturally in the human body, producing no harmful metabolites; in fact a potential disadvantage of the biodegradable polymers is the eventual toxicity of the degradation products (43-47). Coluccio et al. (48) studied the preparation and characterization of bio-artificial polymeric materials through blending of synthetic and natural polymers, in order to exploit the properties of α-amylase-controlled starch-based matrices. The idea was to obtain a material in which the porosity and, accordingly, the drug release were dependent on the amount of enzyme and its activity. In fact, the main release mechanism of a controlled drug delivery system is the drug diffusion through the matrix that is usually described by Fick's law since, as the drug diffuses out, its concentration decreases gradually with time. In the presence of α-amylase, both diffusion and erosion can operate simultaneously with a higher control of the release (49). In most controlled drug release applications a zero-order kinetics is desirable and the appropriate combination of diffusion and erosion permits it. In particular the zero-order drug release is obtained with a surface-erosion controlled system, where the drug release rate is equal to the erosion rate of the delivery system, whilst the drug release rate from the delivery system undergoing bulk degradation is difficult to control because the release rate may change as the polymer degrades (50). However, the bulk erosion of a matrix can be useful, for example, to accelerate the drug release. The enzymatic erosion of starch/PVA/α-amylase films was investigated together with their drugs (model drugs used were theophylline and caffeine) release properties (48). As verified also for an other bioartificial material (51), in the presence of α-amylase both diffusion and erosion mechanisms could operate simultaneously to control the drug release, even if, as the results obtained suggested, at low α-amylase concentrations the release of theophylline and caffeine was mainly controlled by diffusion, while at high α-amylase concentrations erosion became more important with a consequent increased rate of release. In effect, several parameters influenced the behavior of enzymatically eroding systems. Crystallinity and surface area are considered the most important parameters affecting the kinetic of starch hydrolysis: α-amylase can more easily degrade the amorphous regions than the crystalline ones; therefore the drug delivery is lower for a matrix with a higher crystallinity. Another important factor affecting the kinetic of starch hydrolysis is the nature of interaction between the enzyme and the polymeric chains of the substrate. Hydrolysis of insoluble starch substrate differs from that of soluble one and takes place in three steps: 1) diffusion/adsorption of the enzyme from the aqueous phase (in the pores of the polymer network) to the contact with starch substrate, 2) hydrolysis of the α-(1,4)-glycosidic bonds, 3) diffusion of the soluble degradation products from the solid substrate into the aqueous phase.

The properties of the films depend mostly on the amount of enzyme and on the characteristics of the drug. The increase of α-amylase content in the films resulted in a higher porosity of the matrix and consequently both in an increase of drug release in a phosphate buffer solution and in an increase of the transport parameters. Drug release, both for theophylline and caffeine, was almost instantaneous due to starch hydrolysis that occurs during the preparation of the films. This event was in part a consequence of the activity of the enzyme and in part due to the thermal treatment necessary for the crosslinking of the matrix. This study evidenced the possibility to control the transport properties and the film morphology with the amount of enzyme contained in the preparation: the release increased with increasing α-amylase content and the release mechanism was governed by the combined action of drug diffusion and polymer erosion. In particular, the kinetic behavior of drug delivery bioartificial films studied is affected first by the matrix degradation, measured in terms of maltose production, and then by the erosion-porosity increase (Figure 8).

These characteristics suggested the potential application of these materials as fast drug release systems. However, by modulating and controlling the action of the enzyme, they could also lead to formulations to be used where a prolonged release is required. In particular, a natural development of these systems was the production of the same films with the inactivated enzyme (avoiding the starch degradation during preparation procedure): the inactivation of the enzyme, and the following reactivation guaranteed a better control of the functionality of the enzyme and of the composition of the bioartificial matrix, together with a higher control of the erosion kinetic (after the implant).

VI. Conclusions and Future Prospects

In the past decade, the authors' group investigated the properties and the potential of new bio-inspired materials, named "bioartificial materials", in which

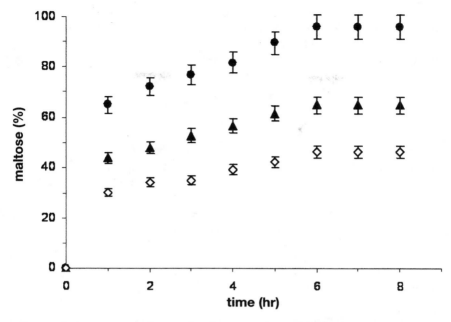

Figure 8. Percentage release of maltose (product of the reaction catalysed by α-amylase) originating from starch in the bioartificial films: film without α-amylase (◇) and with low (▲) or high (●) enzyme content.

the starting idea was to smooth away the interactions between the synthetic and biological systems for obtaining a material where changes at a molecular level, due to synthetic-biological polymer interactions, have been already accomplished before getting in touch with the living tissue. Such a material, with already established molecular interactions, should behave macroscopically better than a fully synthetic material as far as the biological response of the host is concerned.

In our opinion, the efforts described in this paper provide encouraging indication for the realisation of new biomaterials containing a synthetic component and a natural one for new applications or improved performances. Since the idea of anticipating at a molecular level the interactions between the biological and synthetic systems by the preparation of blends has resulted in successful biomaterials with improved biocompatibility, the use of enzymes as biological component can represent an interesting approach for preparing added-value systems showing controlled drug delivery characteristics or bio-catalytic properties. Additionally, the "template polymerisation" procedure can act as a further important straightforward progress in this direction. By starting from the investigation of fully synthetic materials, efficient diagnostic methods for evaluating the presence of template effects on the polymerisation mechanisms and fundamental characterisation methods for the determination of material properties were perfected. The general idea focused on the attempt to transfer these experiences to bioartificial systems (containing a natural-origin component as template polymer), in order to design the constitution of the synthetic-biological complex by specific interactions between the monomeric units of the synthetic component and natural template molecule. Thus, template polymerisation seemed to offer the opportunity of controlling the formation of a synthetic component on a suitable biological template in order to obtain a bioartificial system in which the forming polymer and the template were intimately bound to give a blend with improved properties with respect to a blend obtained by simple mixing.

Moreover, the synthetic component, after separation from the biological matrix, can act as a mould for the synthesis of biomolecules mimicking the functional or structural properties of the original template or as a molecular detector, having preferential interaction with natural occurring polymers in some way similar to the template itself. These synthetic approaches of bio-imprinting aimed at reducing the degree of complexity of biological systems in simpler ("nanoscale level") models resulting in several advantages for their low cost of production, ready preparation, and their stability. The use of chemically prepared macromolecules with pre-selected specificity lies in a large number of applications from the artificial antibodies to artificial enzymes and substrate sensitive sensor devices.

Finally, this study evidenced the possibility to control the transport properties of a bioartificial material, introducing the α-amylase enzyme as third component that play a catalytic functional role, affecting on the morphology of the bioartificial drug delivery system (depending on the amount of enzyme contained in the preparation mixture). The release increases with the increasing level of α-amylase loaded, and the release mechanism is determined by the combined action of drug diffusion and polymer erosion. In particular, the delivery kinetics of these bioartificial films are shown to be related to the matrix degradation, measured by the determination of maltose production and then with the erosion-porosity increase. It will be also possible, by modulating and controlling the action of the enzyme contained into the material (by studying and regulating the parameters of the transport properties, such as Darcy Permeability and Effective Diffusivity), to achieve polymeric formulations to be used where a controlled, prolonged and local release is required.

Further work is necessary to realise materials with such precisely tailored characteristics, but the possibility to better control the structure of the product is one of the most stimulating challenges for the polymer chemist and material scientist of this century.

One current goal of our research group is to attempt to overcome some of the problems associated with traditional chemical treatments; in fact, recent research has focused on enzyme-catalysed reactions as a tool to modify commercially useful protein matrices (52). In this work, gelatin was crosslinked via an enzymatic treatment using tissue transglutaminase (tTGase) and microbial transglutaminase (mTGase), which catalyse the formation of isopeptide bonds between the γ-carbonyl group of a glutamine residue and the ε-amino group of a lysine residue. The reaction resulted in an interesting alternative to the traditional glutaraldehyde (GTA) crosslinking, which in general has several drawbacks (e.g., in medical application) due to the toxicity of the chemical reagent.

Acknowledgements

The authors are sincerely grateful to all the people that worked in the last decade on Bioartificial Polymeric Materials with Enzymatically Controlled Functional Properties.

References

1. P. Giusti, L. Lazzeri, L. Lelli, *Trends Polym. Sci.* **1993**, *1*, 261.
2. L. Lazzeri, *Trends Polym. Sci.* **1996**, *4*, 249.
3. P. Giusti, L. Lazzeri, N. Barbani, L. Lelli, M. G. Cascone, S. De Petris, *Makromol. Chem., Macromol. Symp.* **1994**, *78*, 285.
4. P. Giusti, N. Barbani, L. Lazzeri, L. Lelli, *J. Macromol. Sci., Pure Appl. Chem., Macromol. Rep.* **1994**, *A31* (6-7), 839.
5. C. Cristallini, N. Barbani, P. Giusti, L. Lazzeri, M.G. Cascone, G. Ciardelli, *Macromol. Chem. Phys.* **2001**, *202*, 2104.
6. P. Giusti, L. Lazzeri, M.G. Cascone, N. Barbani, C. Cristallini in *Polymers and Other Advanced Materials: Emerging Technologies and Business Opportunities.* Edited by P.N. Prasad et al., Plenum Press, New York, **1995**, 563.
7. A.F. Kydonieus, *Controlled release technologies: Methods, Theory and Applications*, CRC Press, Boca Raton, FL **1979**, 1.
8. C. Cristallini, L. Lazzeri, M.C. Cascone, G. Polacco, D. Lupinacci, N. Barbani, *Polymer International* **1997**, *44*, 510.
9. M. Watase, k. Nishinari, *Makromol. Chem.* **1988**, *189*, 871.
10. C. Cristallini and N. Barbani, *Polymer International* (**1998**), *47*, 491.
11. F.M. Veronese, M. Morpurgo, *Il Farmaco* **1999**, *54*, 497-516.
12. K. Y. Leea, S. H. Yukb, *Prog. Polym. Sci.* **2007** Article in press (available online at www.scincedirect.com).

74

13. C. H. Bamford, *Makromol. Chem.* **1980**, *1*, 52.
14. C. H. Bamford, *Template Polymerisation*, in: Dev. Polym., vol. 2, R. N. Haward, Ed., Applied Science Publishers, London, **1979**, p. 125.
15. G. Challa, Y. Y. Tan, *Pure Appl. Chem.* **1981**, *53*, 627.
16. Y. Y. Tan, G. Challa, *Makromol. Chem., Macromol. Symp.* **1987**, *10/11*, 215.
17. G. Challa, E. J. Vorekamp, Y. Y. Tan, *Polym. Prep.* (Am. Chem. Soc., Div. Polym. Chem.) **1982**, *23*, 31.
18. E. Tsuchida, Y. Osada, *J. Polym. Sci. Polym. Chem. Ed.* **1975**, *13*, 559.
19. Y.Y. Tan, G. Challa, *Encyclopedia of Polymer Science and Engineering*, Wiley, New York **1986**, p 554.
20. V.A. Kabanov, *Makromol. Chem. Suppl.* **1979**, *3*, 41.
21. G. V. Schulz, R. V. Figini, G. Lohr, *Makromol. Chem.* **1966**, *96*, 283.
22. I. Rainaldi, C. Cristallini, G. Ciardelli, P. Giusti, *Polym. Int.* **2000**, *49*, 63.
23. C. Cristallini, A. Villani, L. Lazzeri, G. Ciardelli, A. Sesto Rubino, *Polym. Int.* **1999**, *48*, 63.
24. M.G. Cascone, L. Lazzeri, N. Barbani, C. Cristallini and G. Polacco, *Polymer International*, **1996**, *41*, 17.
25. G. Polacco, M.G. Cascone, L. Petarca, G. Maltinti, C. Cristallini, N. Barbani, and L. Lazzeri: *Polymer International*, **1996**, *41*, 443.
26. C. Cristallini, G. Ciardelli, G. Polacco, A. Villani, L. Lazzeri and P. Giusti, *Polymer International*, **1999**, *48*, 1251.
27. I. Rainaldi, C. Cristallini, G. Ciardelli and P. Giusti, *Macromolecular Chemistry and Physics*, **2000**, *13*, 201.
28. G. Ciardelli, N. Barbani, I. Rainaldi, P. Giusti, and C. Cristallini, *Polymer International*, **2001**, *50*, 588.
29. C. Cristallini, N. Barbani, P. Giusti, L. Lazzeri, M. G. Cascone, and G. Ciardelli, *Macromolecular Chemistry and Physics,* **2001**, *202*, 2104.
30. G. Ciardelli, D. Silvestri, C. Cristallini, N. Barbani, P. Giusti, *Journal of Biomaterials Science. Polymer Edition*, **2005**, *16* (2), 219.
31. G. Ciardelli, C. Cristallini, N. Barbani, G. Benedetti, A. Crociani, L. Travison, P. Giusti, *Macromol. Chem. Phys.* **2002**, *203*, 1666.
32. S. Vijayasekaran, *J. Biomat. Sci., Polym. Ed.* **1996**, *8*, 685.
33. I. Rainaldi, C. Cristallini, G. Ciardelli, P. Giusti, *Macromol. Chem. Phys.* **2000**, *201*, 2424.
34. M. Chaplin and C. Brucke, *Enzyme Technology*. Cambridge University Press, Cambridge, **1990**.
35. L.B. Wingard, *Enzyme Engineering*. Plenum Press, New York, **1980**.
36. M. Chaplin, *Medical Application of Enzymes*. South Bank University, London, **2002**.
37. H. Kumar, A. Kumar, P. Kumari, S. Jyotirmai and N.B. Tulsani, *Biotechnol Appl Biochem* **1999**, *30*, 231.

38. M. Wolf, M. Wirth, F. Pittner and F. Gabor, *Int J Pharm* **2003**, *256*, 141.
39. L. Giorno and E. Drioli, *TIBTECH 18*, August **2000**, *339* (2000).
40. S. Miura, N. Kubota, H. Kawakita, K. Saito, K. Sugita, K. Watanabe and T. Sugo, *Radiat Phys Chem* **2002**, *63*, 143.
41. J. W. Brennan JD, *Analytica Chimica Acta* **2002**, *461*, 1.
42. D. Silvestri, N. Barbani, G. Ciardelli, F. Bertoni, M.L. Coluccio, C. Cristallini and P. Giusti *Polymer International*, **2005**, *54* (10), 1357.
43. Y. Dumoulin, L. H. Cartilier, M. A. Mateescu, *J. Controlled Release*, **1999**, *60*, 161.
44. J. Heller, R. W. Baker, "Controlled Release Bioactive Materials", R.W. Baker, Ed., Academic, Boston, MA1980, p. 1.
45. J. Heller, "Pulse and Self-Regulated Drug Delivery", J. Kost, Ed., CRC Press, Boca Raton, FL 1990, 93.
46. J. Kost, S. Shefer, *Biomaterials* **1990**, *11*, 695.
47. P. Ispas-Szabo, F. Ravenelle, I. Hassan, M. Preda, M. A. Mateescu, *Carbohydr. Res.* **2000**, *323*, 163.
48. M. L. Coluccio, G. Ciardelli, F. Bertoni, D. Silvestri, C. Cristallini, P. Giusti, N. Barbani, *Macromol. Biosci.* **2006**, *6*, 403.
49. M. Rahmouni, F. Chouinard, F. Nekka, Y. Lenaerts, J.C. Leroux, Eur. *J. Pharm. Biopharm.* **2001**, *51*, 191.
50. L. Tuovinen, S. Peltonen, M. Liikola, M. Hotakainen, M. Lahtela-Kakkonen, A. Poso, K. Jarvinen, *Biomaterials* **2004**, *25*, 4355.
51. M.L. Coluccio, N. Barbani, A. Bianchini, D. Silvestri and R. Mauri, *Biomacromolecules*, **2005**, *6*, 1389.
52. F. Bertoni N. Barbani P. Giusti G. Ciardelli, *Biotechnol. Lett.*, **2006**, *28*, 697.

Chapter 5

Enzymatic Products from Modified Soybean Oil Containing Hydrazinoester

Atanu Biswas[1,2], R. L. Shogren[1], J. L. Willett[1], Sevim Z. Erhan[2], and H. N. Cheng[3,*]

[1]Plant Polymers Research Unit and [2]Food and Industrial Oil Research Unit, Agricultural Research Service, U.S. Department of Agriculture, 1815 North University Street, Peoria, IL 61604
[3]Hercules Incorporated Research Center, 500 Hercules Road, Wilmington, DE 19808–1599

Efforts have been made in this work to use soybean oil to produce new, non-petroleum based products. The starting material is the ene reaction product of soybean oil and diethyl azodicarboxylate (DEAD), which can then be hydrolyzed chemically and enzymatically. Chemical hydrolysis gives hydrazino-fatty acids, whereas enzymatic hydrolysis gives the fatty acids modified with the diethyl-azadicarboxylate functionality. Moreover, we have examined enzymatic trans-esterification reactions in order to explore the chemistry and the utility of these materials. Enzymatic transesterification with methanol gives modified methyl soyate, while enzymatic reactions with glycerol and poly(ethylene glycol) give modified fatty acid glycerides and fatty acid ethoxylates. These new materials can potentially be used as ingredients in coatings, cosmetics, biodiesel fuel, and bio-surfactants.

Introduction

One of the current active research areas is green chemistry and sustainability. As part of this thrust, there has been a fair amount of research interest in soybean oil and its derivatives. Soybean oil (SBO) (*1*) is cheap, easily available, renewable, and environmentally friendly. Some examples of SBO derivatives are epoxidized oil (*2*), SBO methyl ester (methyl soyate) (*3*), maleated products (*4*), SBO polymers (*5*), and others (*6*). This area of research has been reviewed (*5a, 7*).

In contrast to previous work, our approach is to incorporate nitrogen into the triglyceride structure in different ways to produce various types of SBO derivatives. For example, Biswas, et al (*8*) have grafted diethylamine onto epoxidized oil, using $ZnCl_2$ as a catalyst (Figure 1).

Another useful reaction was amination with diethyl azodicarboxylate (DEAD) (*9*). This turned out to be a facile reaction that can be conducted in the absence of any catalyst and solvent. It is also versatile and can proceed via conventional heat at 120°C for 2-4 hours or in a microwave oven for about 10 minutes. The main reaction product is SBO-aza-dicarboxylate ester (structure **2**) (Figure 2).

In another publication (*10*), we indicated that compound **2** can undergo hydrolysis and possibly other derivatization reactions. In this paper, we have followed up on that work and carried out several reactions, particularly with enzymes, in order to fully explore the chemistry and the utility of these new bio-based materials.

Results and Discussion

In this work, we shall describe three reactions: hydrolysis, methyl ester formation, and transesterification to form modified SBO materials.

Hydrolysis

In an earlier work (*10*), it has been reported that the SBO-aza-dicarboxylate ester (compound 2), derived from the thermal reaction between SBO and DEAD, can be readily hydrolyzed. The reaction products are different depending on the nature of hydrolysis, as shown in Figure 3.

Through alkaline hydrolysis (NaOH or KOH at 115°C overnight), hydrazino-fatty acid salt (structure **4** in Figure 3) can be obtained at 90% yield. In the enzymatic pathway, a lipase can hydrolyze the triglyceride ester to release glycerol; however, the aza-dicarboxylate ester is not affected. Thus, the

78

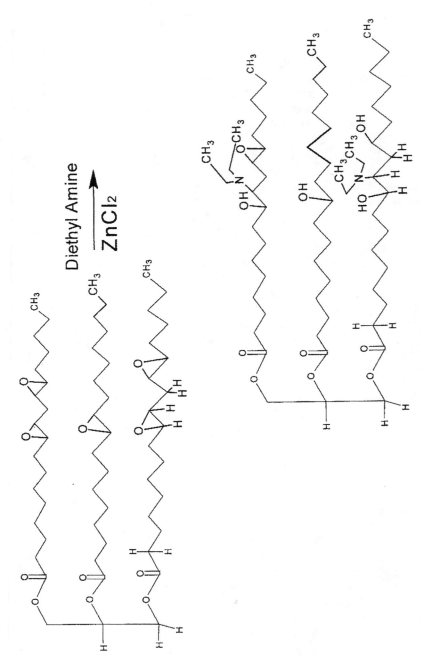

Figure 1. Reaction of epoxidized oil and diethylamine

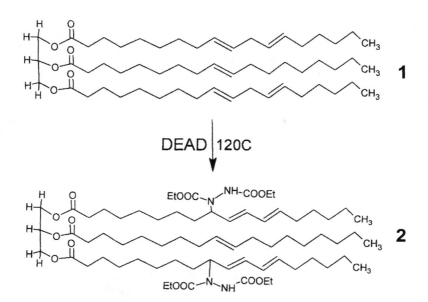

Figure 2. Reaction of soybean oil with DEAD. In Structure **2** we have
arbitrarily placed an unreacted oleic acid moiety on position 2 of glycerol.
In reality, we have a mixture of different fatty acid moieties in soybean oil.

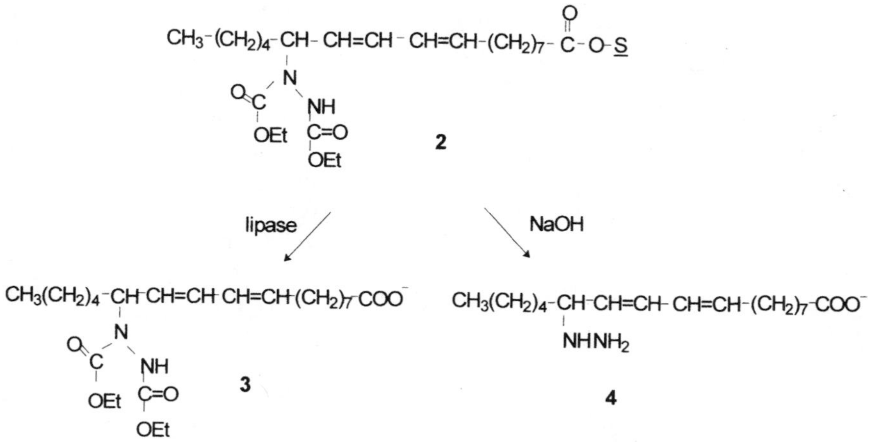

Figure 3. Hydrolysis via chemical and enzymatic pathways.
$\underline{S}$ refers to the rest of the molecule in SBO

enzymatic reaction is a good method to preserve the aza-dicarboxylate ester functionality (structure **3**). For this reaction, at least three commercially available lipases have been found to be satisfactory: Amano 30, Amano 12, and Novozym® 435 lipase.

These reactions are good ways to incorporate the hydrazine functionality into a fatty acid or triglyceride. It may be noted that the hydrazine chemistry is well known (*11*). Hydrazine itself is a chemical intermediate that is used to produce agricultural chemicals, spandex fibers and antioxidants. Hydrazine compounds are sometimes used as oxygen scavengers and corrosion inhibitors. Thus, the new materials shown here perhaps can be used in several applications, e.g., as lubricants and ingredients in coatings, cosmetics, and other oil-based or oil-containing chemical products.

Methyl Esters

For fuel applications, a popular SBO derivative is methyl soyate (SBO methyl ester, also known as Soy Gold) (*3*). This "biodiesel" has several advantages over diesel fuel from petroleum sources because it is biodegradable, has a high flash combustion temperature, contains negligible amount of sulfur, is neutral with respect to carbon dioxide emission, and can potentially reduce many harmful exhaust emissions. Methyl soyate is also an increasingly useful industrial solvent for grease removal.

Currently methyl soyate is made commercially via an alkaline transesterification reaction involving soybean oil. methanol, sodium methoxide at 60-66°C, sometimes under pressure. In this work, we have used an enzyme to carry out a similar reaction with the SBO-DEAD adduct (Figure 4).

In a mixture of SBO, methanol, and water, a lipase can carry out the transesterification reaction at high yields. The end product is similar to methyl soyate, except for the added aza-carboxylate ester. Since methanol is in excess, dimethyl esters are formed. The reaction can be conducted at a low temperature (35°C). Several lipases have been attempted, and the reaction is relatively straightforward. The use of an immobilized lipase (e.g., Novozym® 435) is particularly beneficial because it allows enzyme recycling and cost reduction.

Polyol Esters

Another class of materials can be made via enzyme-catalyzed transesterification with polyols. A number of polyols can be used. Two examples with glycerol and poly(ethylene glycol) (PEG) are shown in Figure 5 below.

The first reaction involves glycerol, which is currently cheap and readily available. Changes in the glycerol level can produce different ratios of mono-

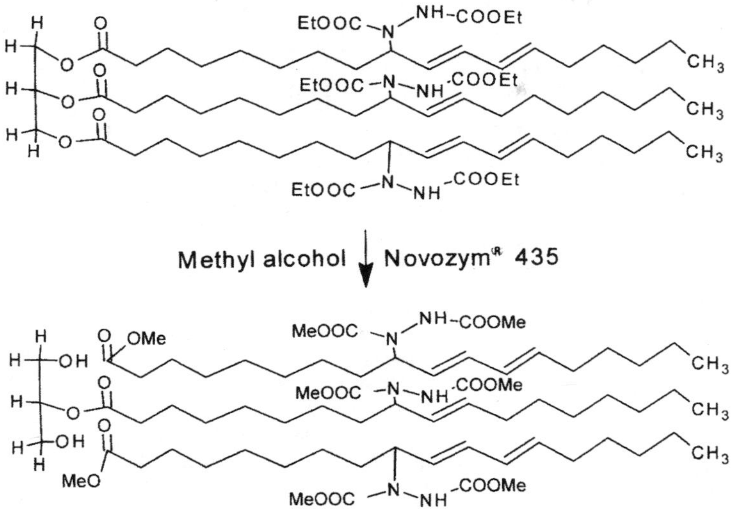

Figure 4. Enzymatic reaction of structure 2 to form methyl soyate containing dimethyl azadicarboxylate (5)

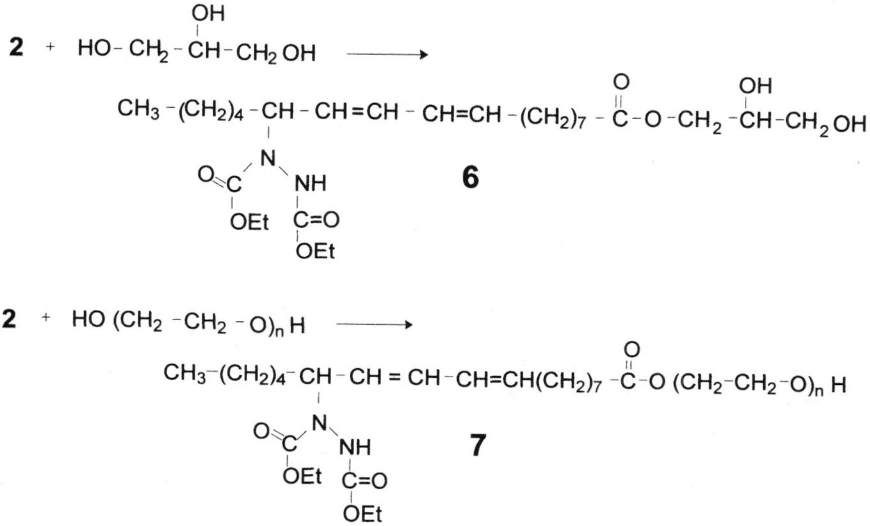

Figure 5. Enzymatic reactions of SBO-DEAD with glycerol and PEG.

and diglycerides (structure **6**). In the second reaction, poly(ethylene glycol) can be grafted onto SBO to give fatty acid ethoxylates (structure **7**). In our experiments, we used glycerol and ethylene glycol trimer, but the reaction for PEG is similar. As before, the same lipase enzymes can be used, but an immobilized enzyme (e.g., Novozym® 435) is preferred because of the advantage of enzyme recycling.

It may be noted that fatty acid glycerides and fatty acid ethoxylates are well-known surfactants (*12*). Thus, the DEAD derivatives may possibly be useful as surfactants as well, with modified properties.

Experimental

Materials

Alkali refined soybean oil was obtained from ADM Packaged Oils, Decatur, IL and was used as received without further purification. The Novozym® 435 enzyme was obtained from Novozymes, Inc. The Amano enzymes were obtained as gifts from Amano Enzyme USA Co., IL. All other materials were acquired from Aldrich Chemical Company.

Reaction of Soybean Oil with DEAD

The reactions have been reported before (*9*). The same procedures have been used, involving either conventional heat or microwave.

Alkaline Hydrolysis

In a round bottom flask 16 ml of 1N NaOH was added to 1 g of soybean/DEAD and heated using the thermowatch with heating mantel and flat surface thermocouple to 115°C, utilizing the reflux condenser. The reactions were heated while stirring for 24 h. Enough 1N hydrochloric acid was added to the sample to acidify the solution to pH 6. There was also a visual indicator that acidity was reached because the reaction became cloudy and oil was formed noticeably. The sample was then transferred to a separatory funnel and 20 ml of chloroform was added to extract the product from solution. The chloroform layer was collected and washed two additional times with water and once with saturated sodium chloride solution. The chloroform layer was then dried with magnesium sulfate, filtered into a round bottom flask and evaporated in a rotatory evaporator. The product was then placed in the vacuum oven overnight at 50°C.

Enzymatic Hydrolysis

In a capped 25-ml flask 200 mg of soybean oil/DEAD adduct, 200 mg of Novozym® 435 lipase and 10 ml of water were placed. About 10 glass beads were added to the flask, which was placed a shaker at 50°C at 200 rpm for 24 hours. The sample was then transferred to a separatory funnel and 30 ml of chloroform was added to extract the product from the aqueous media. The chloroform layer was collected and washed two additional times with 0.1N hydrochloric acid followed by one wash with saturated sodium chloride solution. The chloroform layer was then dried with magnesium sulfate, filtered into a round bottom flask and evaporated in a rotatory evaporator. All the samples were then placed in the vacuum oven overnight at 50°C.

Enzyme-Catalyzed Methyl or Polyol Ester Formation

In a capped 25-ml flask 1 g of soybean oil/DEAD adduct, 400 mg of Novozym® 435 lipase and 500 mg of methanol, 400 mg of glycerol, or 600 mg of triethylene glycol were placed. About 10 glass beads were added to the flask, which was placed a shaker at 50°C at 200 rpm for 48 hours. The sample was then transferred to a separatory funnel and 30 ml of chloroform was added to extract the product from the aqueous media. The chloroform layer was collected and washed two additional times with 0.1N hydrochloric acid followed by one wash with saturated sodium chloride solution. The chloroform layer was then dried with magnesium sulfate, filtered into a round bottom flask and evaporated in a rotatory evaporator. All the samples were then placed in the vacuum oven overnight at 50°C.

NMR Spectroscopy

All ^{13}C NMR spectra were recorded quantitatively with a Bruker ARX-500 spectrometer (Bruker, Rheinstetten, Germany) at a frequency of 125 MHz and a 5-mm dual probe. The sample solutions were prepared in deuterochloroform (CDCl$_3$, 99.8% D, Cambridge Isotope Laboratories, Inc., Andover, MA). Standard operating conditions were used with 30° pulse angle, 3 seconds between pulses, and ^{1}H decoupling.

Conclusions

In this work, we have produced additional bio-based materials from the SBO. The parent compound is the ene reaction product of SBO and DEAD (compound **2**). This product can be hydrolyzed enzymatically or chemically to give interesting new SBO derivatives. Moreover, enzymatic transesterification reaction with methanol gives DEAD-methyl soyate. Enzymatic reaction with glycerol and PEG gives fatty acid-DEAD glycerides and polyol esters of fatty acid-DEAD. A summary of the reactions is given in Table 1. These new materials can be used as synthons for further reactions. Perhaps some of them will find application as ingredients in coatings, cosmetics, and oil-based or oil-containing chemical products. In particular, the DEAD-methyl soyate may possibly be used in biodiesel fuel, and the DEAD-fatty acid glycerides and polyol esters as bio-surfactants.

Table 1. Summary of Modified Soybean Oil Products

Reactant(s)	Reaction/Conditions	Product (Structure #)
SBO + DEAD	Heat	SBO-DEAD adduct (**2**)
2	Lipase hydrolysis	FA-DEAD adduct (**3**)
2	NaOH hydrolysis	FA-hydrazine (**4**)
2 + methanol	Lipase transesterification	Modified methyl soyate (**5**)
2 + glycerol	Lipase transesterification	Modified FA glyceride (**6**)
2 + PEG	Lipase transesterification	Modified FA ethoxylate (**7**)

Acknowledgement

The authors wish to thank Janet Berfield and Kelly Utt for expert technical assistance.

References

1. For example, a) O'Brien, R. D., *Fats and Oils: Formulating and Processing for Applications*, 2nd ed., CRC Press: Boca Raton, 2003. b) Lawson, H. W., *Food Oils and Fats: Technology, Utilization and Nutrition*, Chapman and Hall: New York, 1995.
2. For reviews, see a) Petrovic, Z.S.; Zlatanic, A.; Lava, C. C.; Sinadinovic-Fiser, S. *Eur. J. Lipid Sci. Technol.* **2002**, *104*, 293. b) Kuo, M.-C.; Chou, T.-C. *Ind. Eng. Chem. Res.* **1987**, *26*, 277.
3. For example, a) Knothe, G. *Fuel Processing Technol.* **2005**, *86*, 1059. b) Srivastava, A.; Prasad, R. *Renewable & Sustainable Energy Rev.* **2000**, *4*, 111.

4. For example, a) Warth, H.; Mulhaupt, R.; Hoffmann, B.; Lawson, S. *Angew. Makromol. Chem.* **2003**, *249*, 79. b) Tran, P.; Seybold, K.; Graiver, D.; Narayan, R. *J. Am. Oil Chem. Soc.* **2005**, *82*, 189.
5. For example, a) Sharma, V.; Kundu, P. P. *Prog. Polym Sci.* **2006**, *31*, 983. b) Li, F.; Hanson, M. V.; Larock, R. C. *Polymer* **2001**, *42*, 1567. c) Khot, S. N., et al., *J. Appl. Polym. Sci.* **2001**, *82*, 703.
6. For example, a) Baber, T. M.; Graiver, D.; Lira, C. T.; Narayan, R. *Biomacromolecules* **2005**, *6*, 1334. b) Benecke, H. P., Vijayendran, B. R., Elhard, J.D. *U.S. Patent* 6,797,753, September 28, 2004.
7. Biermann, U, et al *Angew. Chem. Int. Ed.* **2000**, *39*, 2206.
8. Biswas, A. ; Adhvaryu, A.; Gordon, S. H.; Erhan, S.Z.; Willett, J. L. *J. Ag. Food Chem.*, **2005**, *53*, 9485.
9. Biswas, A.; Sharma, B. K..; Willett, J. L.; Vermilion, K.; Erhan, S. Z., Cheng, H. N. *Green Chemistry* **2006**, *9*, 85.
10. Biswas, A.; Shogren, R. L.; Willett, J. L.; Erhan, S. Z., Cheng, H. N. *ACS Polymer Prepr.* **2006**, *47*(2), 259.
11. Clark, C. C., *Hydrazine*, Mathieson Chemical Corporation: Baltimore, MD, 1953.
12. For example, a) Flick, E. W., *Industrial Surfactants* (2nd Ed.), William Andrew Publishing/Noyes, Park Ridge, NJ, 1993. b) Ash, M; Ash, I., *Handbook of Industrial Surfactants* (4th Ed.), Synapse Information Resources, Inc., Endicott , NY, 2005.

Chapter 6

Reinforcement Effect of Soy Protein and Carbohydrates in Polymer Composites

L. Jong

National Center for Agricultural Utilization Research, Agricultural Research Service, U.S. Department of Agriculture, 1815 North University Street, Peoria, IL 61604

The modulus of soft polymer material can be increased by filler reinforcement. A review of using soy protein and carbohydrates as alternative renewable reinforcement material is presented here. Dry soy protein and carbohydrates are rigid and can form strong filler networks through hydrogen bonding and ionic interactions. They are also capable of interacting with polymers that possess ionic and hydrogen bonding groups. Through filler-filler and filler-rubber interactions, the rubber modulus is significantly increased. The variables in the reinforcement effect include the compositions of both soy protein and carbohydrate, polymer composition, particle sizes of dispersions, and co-fillers.

Introduction

For practical applications, rubber material is usually reinforced with fillers. Carbon black is the most common filler. Carbon black is mainly derived from aromatic oil in petroleum or from natural gas. Substitution of carbon black (CB) with renewable filler has been investigated in recent years. Recent studies

---

NOTE: Names are necessary to factually report on available data; however, the USDA neither guarantees nor warrants the standard of the product, and the use of the name by USDA implies no approval of the product to the exclusion of others that may also be suitable.

reported the modulus enhancement of rubbers by natural materials, for example, oil palm wood (*1*), crab shell chitin (*2*), and bamboo fiber (*3*). For the reasons of using renewable resources and protecting our environments, renewable materials such as soy protein and other soybean products have been investigated as a component in plastic and adhesive applications (*4-8*), but have rarely been investigated as a reinforcement component in elastomers. Attempts to use protein in rubber latex can be traced back to the 1930s. A few patents (*9-11*) had claimed the use of protein in rubber composites. For example, Lehmann and coworkers had demonstrated the use of casein (milk protein) in natural rubber latex to achieve approximately a four-fold increase in the modulus (*11*). Protein as an additive in rubber materials has also been claimed to improve the anti-skid properties of winter tread tires (*12-14*).

More recently, different fractions from soybean have been investigated as reinforcements in elastomers (*15-19*). These fractions are defatted soy flour (DSF), soy protein concentrate (SPC), soy protein isolate (SPI), and soy spent flakes (SSF). These fractions are classified in terms of a centrifugal separation method instead of their pure chemical components. Generally, soybean can be processed into soybean oil and defatted soy flour. Defatted soy flour contains soy protein isolate, soluble soy carbohydrates and insoluble soy carbohydrates (*20*). In a conventional industrial process, soy protein is extracted at alkali pH and the insoluble soy carbohydrates are separated by centrifugation. The resulting soy protein dispersion is then precipitated at acidic pH and the soluble soy carbohydrate is separated centrifugally. The products from this separation process are soy protein isolate and soy spent flakes. Soy protein concentrate is obtained by precipitating soy protein and insoluble soy carbohydrates in acidic pH and centrifugally removing the soluble soy carbohydrates. Although pure components are preferred for scientific studies, economically processed fractions are industrially useful. These fractions can be classified in terms of their soy protein concentration. Soy protein isolate, soy protein concentrate, defatted soy flour, and soy spent flakes contain approximately 90%, 70%, 55%, and 15% protein, respectively. Physically, soy protein is a globule protein that forms aggregates while soy carbohydrates have a film-like structure (Figure 1). DSF and SPC are a mixture of globular proteins and carbohydrates (Figure 1). The remaining components in these soy fractions are mainly soy carbohydrates and a small amount of fat, fiber, and ash.

In an aqueous dispersion, these soy products swell to different degrees and show a broad size distribution except for soy protein aggregates, which have a number average size of ~2.2 μm in the wet state (Figure 2). DSF, SPC, and SSF have a similar swollen size and size distribution due to swollen soy carbohydrates. Under ultrasonic dispersion, SPI aggregates tend to breakup while DSF, SPC, and SSF retain a similar size and distribution.

In the dry state these soy products show different rigidities (Figure 3), and varying filler-filler and filler-polymer interactions due to their compositional differences. As a result, they exhibit different reinforcement effect in the same

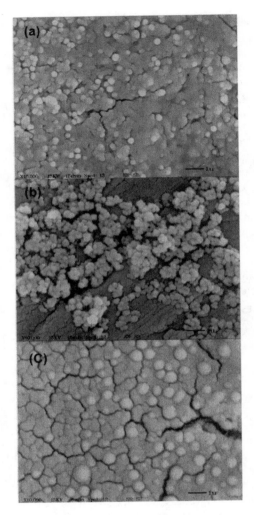

Figure 1. (a) The DSF film on an aluminum substrate showing the soy protein aggregates are embedded in the soy carbohydrate film. (b) SPI aggregates on a gray mottled aluminum substrate. (c) SPC film. The scale of black bar located at the right bottom corner is 1 μm. (Reproduced from reference 16.)

polymer matrix (Figure 4). In rubber reinforcement, factors such as aggregate structure, effective filler volume fraction, filler-rubber interaction and elastic modulus of the filler clusters have an important impact on the modulus of rubber composites (*21*). Mechanically, the elastic modulus of base rubber is not significant when compared with the modulus of the filler network in highly filled elastomeric composites (*22*). One aspect of most natural materials that is

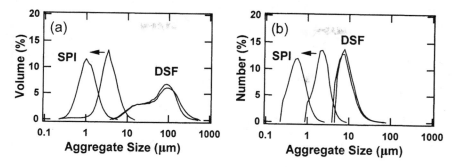

Figure 2. The measurements of volume and number weighted aggregate size in water. Both DSF and SPI aggregates were subjected to one hour of ultrasonic dispersion. Aggregate size of SPI is significantly reduced and the size distribution curve is shifted to the left, whereas the DSF curve only changes slightly. (Reproduced from reference 16.)

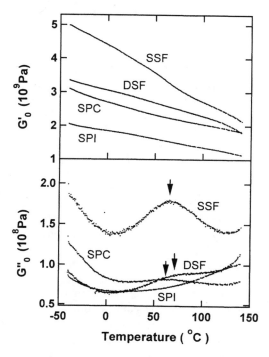

Figure 3. Elastic and loss moduli of four soy products. SSF = Soy Spent Flakes, DSF = Defatted Soy Flour, SPC = Soy Protein Concentrate, SPI = Soy Protein Isolate. Arrows indicate the regions of glass transition. Measured at 0.05% strain and 0.16 Hz. (Reproduced from reference 17.)

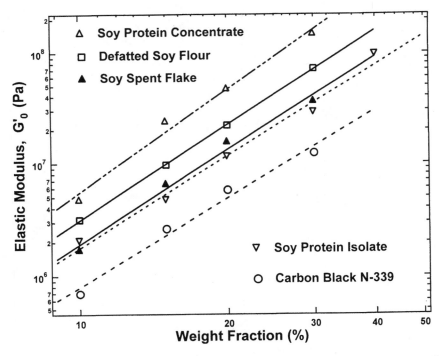

Figure 4. Shear elastic modulus of composites filled with varying amounts of SPC, DSF, SSF, SPI, or CB. The samples were prepared by a casting method. Measured at 0.05% strain, 0.16 Hz, and 140 °C.

different from petroleum based materials is their moisture sensitivity in some applications, but it can be improved through product formulation, processing methods, or selective applications. For example, natural materials may be used as ingredients in multilayered structures, in coated objects, in higher temperature applications or in a rubber part used in greasy/oily environments where the moisture effect is minimal.

The rubber matrix used in these studies is a styrene-butadiene (SB) rubber with a small amount of carboxylic acid-containing monomer units. The carboxylated SB forms a crosslinked rubber by the aggregation of ionic functional groups without the complication of covalent reactions. Carboxylated SB rubber is classified as an ion-containing polymer. Its viscoelastic properties are affected by molecular weight, degree of crosslinking, glass transition temperature (T_g), copolymer composition, the number of ionic functional groups, the size of ionic aggregation, the degree of neutralization, and the size of the neutralizing ions (23,24). Previous studies also have shown honeycomb-like structures in the film of carboxylated latexes due to a higher concentration of carboxylic acid groups on the particle surface (25). For reinforcement effects, previous studies have indicated the importance of the interaction between filler and matrix (22). In this aspect, soy protein contains a significant amount of carboxylic acid and substituted amine functional groups (26) and soy carbohydrates contain mostly hydroxyl functional groups, which can interact with carboxylic functional groups in the SB matrix through hydrogen bonding and ionic interactions. Although ionic interactions can occur between these soy products and the carboxylated SB, the condensation reactions do not occur under alkali conditions between the carboxyl groups of SB and the major functional groups such as hydroxyl, carboxyl, thiol, amine, and amide groups in these soy products.

Small Strain Shear Modulus

The reinforcement effect is most easily measured from the increase of small strain shear modulus (Figure 4). All the measurements shown in Figure 4 were conducted on dry samples prepared by a casting process (27) because biomaterials tend to retain a certain amount of moisture that behaves as plasticizer in protein and carbohydrate. The effect of moisture is shown in Figure 5.

The dehydration of soy protein reinforced carboxylated styrene-butadiene composite causes a significant increase of the shear modulus. The effect comes from the increase in rigidity of the filler network and the increase of filler-rubber

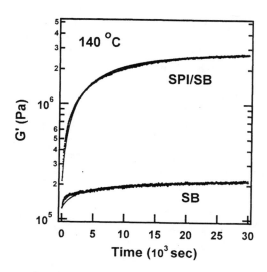

Figure 5. Shear elastic modulus of 20/80 SPI/SB composite with an initial moisture content of 3.5%. SB has an initial moisture content of 1%. Measured at 0.05% strain and 0.16 Hz. (Reproduced from reference 27.)

interactions because these interactions are ionic and hydrogen bonding in nature, which are enhanced when water is absent. A typical temperature and concentration dependent G' of SPI is shown in Figure 6. Carbon black reinforced composites are also shown as a comparison. SPI composites show a percolation threshold between 6-8% volume fraction (Figure 6(b)). Above the percolation threshold, a logarithmic plot of modulus vs. concentration is linear (*19*). The significant increase of the composite modulus above the percolation threshold is mainly due to the formation of a filler network. The strength of the filler network may be estimated from the bulk modulus of the compressed SPI sample (Figure 3). The moduli of these soy composites are influenced by their filler network structures, which in turn are influenced by the composite preparation method. For example, the polymer composites prepared by casting or freeze-drying methods exhibit different mechanical properties due to their different filler network structures. The comparison of these preparation methods is therefore instructive to understand the reinforcement mechanisms of these rubber composites.

Figure 7 shows the comparison of SSF and CB composites, and the comparison of these composites prepared by casting and freeze-drying methods (*28*). Generally, the composites prepared by the casting method have a greater G' than those prepared by the freeze-drying method. The greater G' in composites prepared by casting method can be explained by the strength of the filler network. The casting method is a slow process which allows the fillers to associate with each other as water evaporates. The freeze-drying method, on the other hand, produces a homogeneous mixture of filler and rubber particles, where filler aggregates are surrounded by the rubber particles because the filler has a smaller volume fraction in the mixture. Upon compression molding the freeze-dried crumbs, the filler network structure is more likely to have a polymer layer sandwiched between filler aggregates compared to the filler network structure produced by the casting method. With the same filler, filler aggregate size, and filler volume fraction, a filler network with increasing number of polymer mediated filler structures is softer than a filler network without polymer mediation. Therefore, the polymer mediation model (*29*) is adequate in explaining these differences. Another variable is that the composites prepared by freeze-drying methods may have a thicker polymer layer between filler aggregates and thus have a softer filler network structure because the rigid, immobilized polymer layer is likely to extend only a few nanometers (*30*) outward from the filler surface.

The information of filler-rubber interactions can be obtained from the shifting of loss maximums of these composites in Figure 8, which shows the G" of SSF and CB composites in the glass transition region (*28*). The loss maximums represent the temperature at which the energy loss as heat from the composite structure reaches its maximum and corresponds to a glass transition temperature. There is no significant shifting in the loss maximum of the SSF composites as SSF fraction in the composites is increased. However, the shifting

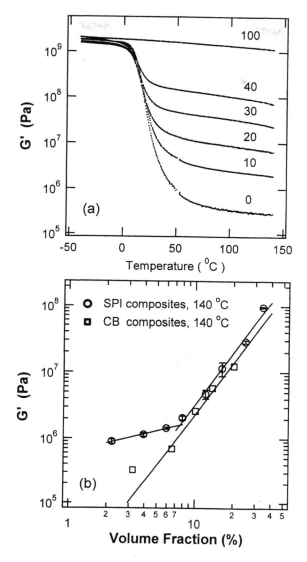

Figure 6. (a) Shear elastic moduli of rubber composites filled with 0 to 40% of SPI. (b) G' vs. filler volume fraction at 140 °C. Measured at 0.05% strain and 0.16 Hz. (Reproduced from reference 19.)

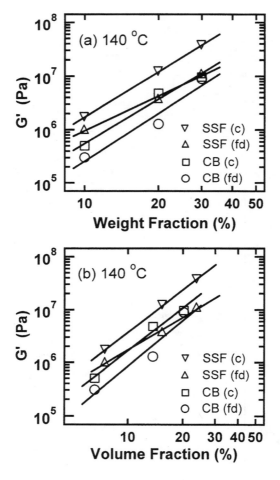

Figure 7. Elastic moduli of SSF and CB composites at small strain region plotted against weight and volume fractions. c = casting and fd = freeze-drying. The measurements were conducted at 0.16 Hz, 0.05% strain, and 140 °C. (Reproduced from reference 28.)

of the loss maximum to a higher temperature in the CB composites is observable as the CB concentration is increased. This is an indication that there is a greater extent of filler-rubber interaction in the CB composites because the fraction of filler immobilized polymer has increased and resulted in a few degrees increase in the average glass transition temperature of SB rubber matrix. Whether the effect is due to a stronger filler-rubber interaction is not known, but the greater

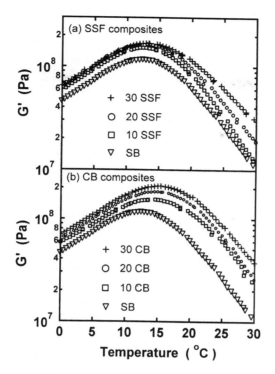

Figure 8. Loss moduli of SSF and CB composites prepared by freeze-drying method. Measured at 0.05% strain and 0.16 Hz. (Reproduced from reference 28.)

filler surface area from the smaller CB aggregates certainly contributes to such an effect. A similar trend was also observed on the same composites prepared by the casting method (*31*).

In some applications, it may be advantageous to combine different fillers in order to achieve a desirable property that can not be achieved by any single filler. For example, Figure 9 shows G' of co-filler composites with three different co-filler ratios prepared by the freeze-drying method (*28*). In these composites, SSF increases the stiffness while having a lower raw material cost. Thus, an adjustment of composite properties may be achieved through the use of co-filler. The combination of fillers may sometimes yield unexpected results such as additional transitions in the temperature dependent tan δ plots (Figure 10). For example, in SSF and carbon black co-filler composites prepared by the freeze-drying method, two tan δ maximums were clearly observed for almost all compositions (*28*). By comparing with single filler composites, it became obvious that the higher temperature tan δ transitions could not be assigned to either the effect of CB or SSF. The higher temperature transitions in the glass

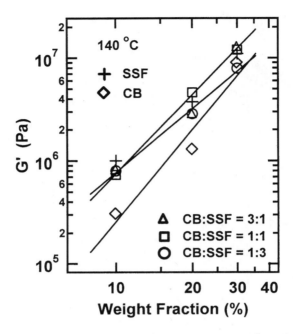

Figure 9. Elastic moduli of co-filler composites at 140 °C. The SSF and CB composites are also included for comparison. Measured at 0.05% strain and 0.16 Hz. (Reproduced from reference 28.)

transition region were attributed to the co-filler effect because neither the SSF or CB composites show such double transitions. The additional higher temperature transitions may be due to the presence of regions of highly immobilized polymer chains between filler aggregates. These regions of highly immobilized polymer chains could be caused by a higher degree of confinement from the combination of SSF and CB filler aggregates.

Stress-Strain properties

Figure 11 shows the stress-strain behavior of DSF, SPI, SPC, and CB composites under ambient conditions.

Compared with CB composites (Figure 11(d)) at the same weight fraction, DSF, SPC, and SPI composites showed a higher modulus in the small strain region, but the elongation at break was also reduced. The elongation of 10% or 20% SPI composites is similar to that of SB rubber but 30% SPI composite has a reduced elongation (Figure 11(b)). For DSF and SPC composites (Figure 11(a)

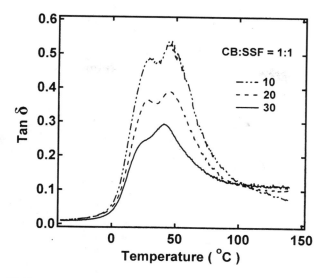

Figure 10. Loss tangent of co-filler composites with 1 to 1 ratio of CB to SSF. Three filler fractions, 10-30 wt %, are indicated. Measured at 0.05% strain and 0.16 Hz. (Reproduced from reference 28.)

and (c)), only composites with 10% filler have a similar elongation as that of SB rubber. This indicates soy carbohydrates in DSF and SPC has an effect of reducing the elongation and increasing the small strain modulus. On the other hand, CB composites show a similar elongation for all compositions. Comparing the tensile stress at break for these composites, the 20% CB composites exhibited both high elongation and high tensile strength at break compared to the 20% SPI composites. The toughness of different composites can be estimated from the area under the stress-strain curve. Within the concentration range measured, it was observed that the toughness of CB composites increases and the toughness of soy composites decreases as the filler concentration increases. It is also noted that 10% SPC composite has a similar toughness as that of 20% CB composite. This indicates that soy carbohydrates and protein significantly change the rigidity of the composites, and they gradually exhibit plastic, rather than rubber like behavior as the DSF, SPC, or SPI concentration increases. These behaviors were also reported for DSF/polyurethane composites (*15*). The low strain behaviors of these composites are consistent with those measured by dynamic mechanical methods. In general, the stress-strain behaviors indicate the filler loading in these soy composites is an important parameter to adjust in order to obtain different material characteristics for different applications.

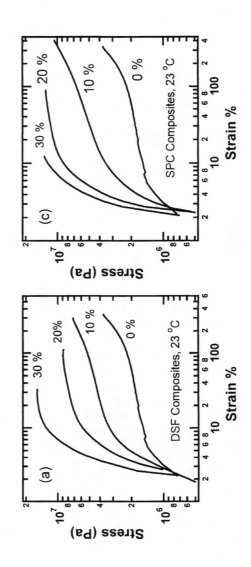

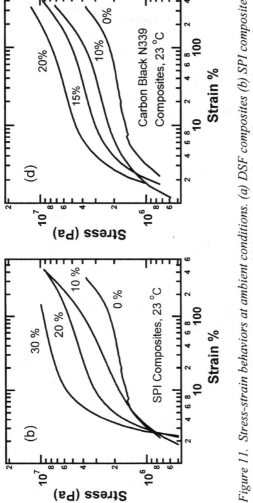

Figure 11. Stress-strain behaviors at ambient conditions. (a) DSF composites (b) SPI composites (c) SPC composites (d) CB composites. Wt% of filler is indicated on each curve. The samples were prepared by the casting method. (Reproduced from reference 16.)

Stress Softening Effect

Nonlinear viscoelastic behaviors of soy composites were investigated by using a series of strain cycles. Figure 12 shows the effect of oscillatory strain cycles on the modulus of the composites. The effect is called stress softening, which occurs in most filled elastomers. It is defined as the stress needed to deform the filled rubber, at a given elongation, is reduced during the second cycle of deformation. The effect is also called the Mullin effect; named for his extensive studies (32,33) on this phenomenon. The stress softening effect is generally considered to be caused by filler related structures and therefore can yield some insight into the filler structures (22).

The effect shown in Figure 12 is a typical example. Other soy composites also show a similar trend with variations in their viscoelastic properties. In Figure 12, the strain sweep curves for both 30/70 SPI/SB and CB/SB composites become reproducible after three cycles of dynamic strain. 100% styrene-butadiene rubber also shows the stress softening effect, but its contribution to the stress softening effect of the composites is not significant. This is evident by comparing the difference between the shear elastic modulus of the first and the eighth strain cycle in Figure 12(a) and 12(b). The contribution of stress softening effect from the rubber is less than 0.5% of the stress softening effect of 30/70 SPI/SB or 30/70 CB/SB composite. The stress softening effect in SPI/SB composites is mostly from the contribution of the protein related structures such as protein network and protein-rubber interactions. The increasing magnitude of strain (deformation) in the first three strain cycles obviously causes the protein network to break down and possibly the polymer chains to detach from the protein aggregates. In this aspect, the current SPI/SB composites are not very different from carbon black filled rubber composites. After three strain cycles, protein related network structures can be weakened and rebuilt and this is an indication of reaching an equilibrium condition. For the loss modulus, the energy dissipation process becomes less pronounced and the maxima are shifted to the lower strain amplitudes. The structure responsible for the energy dissipation process is obviously reduced after the first three cycles. Comparing the SPI composite with the carbon black filled composite, the magnitude of the loss maximum in the SPI composite is not as pronounced as that in the carbon black composite.

The magnitudes of shifting in the position of loss maxima in Figure 12(b) and 11(c) are different. The SPI/SB composite in Figure 11(b) exhibits a loss maximum at 1.68 % strain in the first cycle, while CB/SB composite in Figure 12(c) has a loss maximum at 0.98 % strain in the first cycle. This may indicate the protein network and related structures are stronger and can only be weakened at a larger strain. In the eighth cycle, the loss maximum of protein composite occurs at 0.31 % strain and that of the carbon black composite is at 0.59 % strain. A greater shifting of loss maximum towards the lower strain at the eighth cycle in the protein composite may indicate the protein related

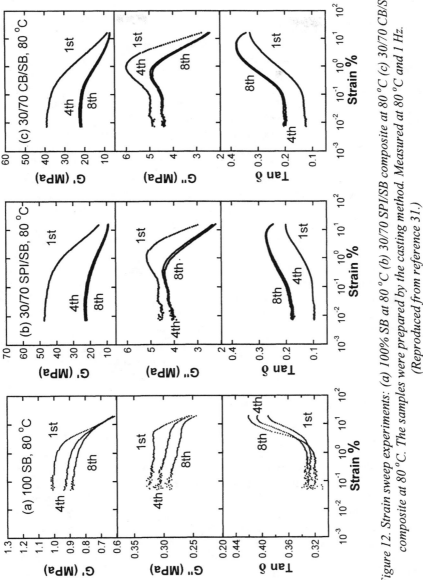

Figure 12. Strain sweep experiments: (a) 100% SB at 80 °C (b) 30/70 SPI/SB composite at 80 °C (c) 30/70 CB/SB composite at 80 °C. The samples were prepared by the casting method. Measured at 80 °C and 1 Hz. (Reproduced from reference 31.)

network structure does not have better recovery behavior than the CB composite within the same period. This is not an effect of filler volume fraction because a 20/80 SPI/SB composite that has a smaller filler volume fraction than a 30/70 CB/SB composite also exhibits the same tendency.

Other information of the composite structure can also be obtained by studying the recovery behavior from the stress softening effect. This is especially interesting in the composites made by the freeze-drying method (28), where the recovery curves of the CB composite lie above the G' of the first strain cycle (Figure 13(b)).

The result indicates the composite structure was changed by the eight consecutive deformation cycles and rearranged to form a stronger structure upon reconditioning at 140 °C for 24 hrs. One possible explanation of this behavior is that the composite structure prepared by the freeze drying method was frozen in a homogeneous state, which was not an equilibrium state compared to one formed by the casting method, which allows a much longer time for a filler network to form. At 20% filler level, the composite is still soft enough for the structure to rearrange itself after the perturbation of the deformation cycles and to reach a more equilibrated state. On the other hand, Figure 13(a) shows that the recovery curve of the SSF composite did not exceed the G' of the first cycle. This may suggest that CB aggregates are more mobile than SSF aggregates due to their smaller size and can diffuse in a softer rubber matrix to form a more connected filler related network.

For the loss tangent properties, the magnitudes of tan δ for both CB and SSF composites are similar. The magnitude of tan δ has practical importance in rubber applications such as tires. A rubber composite that has a smaller tan δ value tends to have a reduced rolling resistance and save energy, while a larger tan δ tends to have an improved skid resistance and wet grip. The ability of SSF to absorb some moisture in a wet state tends to reduce G', increase tan δ, and lead to better wet traction.

Reversible Strain Dependence of Shear Elastic Modulus

The reduction of shear elastic modulus with increasing strain is a familiar phenomenon reported by Payne (34-36) on carbon black filled rubbers in the early 1960s. Later Kraus (37) proposed a phenomenological model based on Payne's postulation of filler networking. The model is based on the aggregation and de-aggregation of carbon black agglomerates. In this model, the carbon black contacts are continuously broken and reformed under a periodic sinusoidal strain. Based on this kinetic aggregate forming and breaking mechanism at equilibrium, elastic modulus was expressed as follows:

$$\frac{G'(\gamma) - G'_{\infty}}{G'_0 - G'_{\infty}} = \frac{1}{1 + (\gamma/\gamma_c)^{2m}} \tag{1}$$

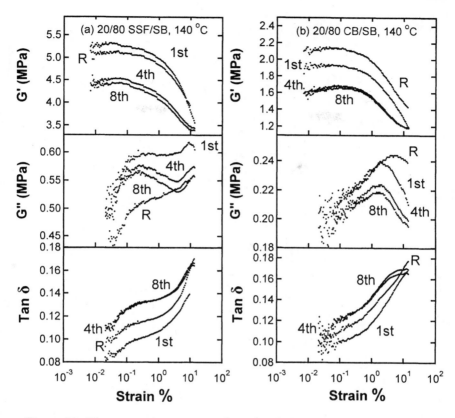

Figure 13. The composites were reinforced with 20 wt % filler and prepared by freeze-drying method: (a) SSF composite; (b) CB composite. Only 1st, 4th and 8th strain cycles are shown. R indicates the recovery curve after the samples were conditioned at 140 °C for 24 hours. Measured at 1 Hz and 140 °C. (Reproduced from reference 28.)

where G'_∞ is equal to $G'(\gamma)$ at very large strain, G'_0 is equal to $G'(\gamma)$ at very small strain, γ_c is a characteristic strain where $G'_0 - G'_\infty$ is reduced to half of its zero-strain value, and m is a fitting parameter related to the filler aggregate structures. Eq. (1) has been shown to describe the behavior of $G'(\gamma)$ in carbon black filled rubber reasonably well (*21*). The loss modulus and loss tangent, however, do not have a good agreement with experiments (*38*), mainly because of the uncertainty in the formulation of the loss mechanism. In this study, an empirical fit is useful to show the difference in strain behaviors between different composites. An example of DSF and SPI composites (*16*) is shown in Figure 14.

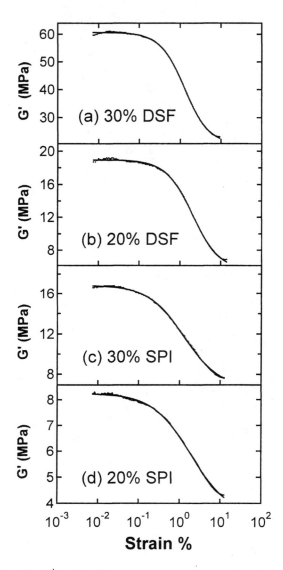

Figure 14. The 8th cycle of strain sweep experiments at 140 °C and 1 Hz. (a) 30/70 DSF/SB composite (b) 20/80 DSF/SB composite (c) 30/70 SPI/SB composite (d) 20/80 SPI/SB composite. The samples were prepared by the casting method. Solid lines are the fit from the Kraus model. (Reproduced from reference 16.)

In general, a smaller fitting parameter m indicates a more rapid and continuous decrease of G' with increasing strain and suggests a more rapid and continuous break-up of the filler network structure as the strain is increased. On the other hand, a larger m indicates a tougher structure, which breaks down at a lower rate and has a plateau-like modulus in the small strain region. Recently, Huber et al. also modeled the Payne effect and gave a similar expression as the Kraus model, but with a physical interpretation of the fitting parameter m in the Kraus model. Based on the cluster-cluster aggregation (CCA) model, Huber et al. (*39*) obtained $m = 1 / (C - d_f + 2)$, where C is a connectivity exponent related to the minimum path along the cluster structure and d_f is the fractal dimension of clusters. Therefore, the fitting parameter m has a physical meaning related to filler structures or filler immobilized rubber structures (a reflection of filler structure).

The fractal dimension of fractal-like protein clusters can be estimated from the slope in Figure 6(b) by using the strong link model (*40*) because the protein aggregates form a network above the percolation threshold (Figure 6). The strong link model indicates log $G'_0 \sim (3 + C)/(3 - d_f)$ log ϕ, where ϕ is the volume fraction of particles and C and d_f have been defined previously. C takes a value from 1 to d_f (*40*). For soy protein aggregates, d_f is estimated to be 1.35 - 1.48 from the slope in Figure 6(b) by substituting C with 1 or d_f in the equation of the strong link model. The parameter m can then be estimated from the fractal dimension of soy protein aggregates and is approximately in the range 0.47 – 0.65. The agreement between the estimation and the best fit of experimental data in Figure 14 and Table 1 is reasonable and is similar to that of carbon black composites, $m = 0.5 - 0.6$ (*21,41*). For DSF composites, the estimated fractal dimension is 1.39 - 1.53 and the parameter m is estimated to be 0.49-0.74. The estimated range for parameter m is somewhat large due to the uncertainty in parameter C. Furthermore, the DSF structure consisting of soy carbohydrates embedded with protein aggregates may no longer fit the definition of particle aggregate. Therefore, it may not be appropriate to use fractal-like descriptions for the DSF composites.

Conclusions

Economical soy products including SPI, DSF, SPC, and SSF can be mixed and coagulated with polymer latex in the aqueous phase to form dry composites with significantly enhanced modulus. These dry soy products have a shear elastic modulus of 1-5 GPa within the temperature range of -40 to 140 C. The carboxylated styrene-butadiene composites filled with these soy products show a significant increase of shear modulus compared to that of the polymer matrix alone. The different compositions of these soy products generate a different reinforcement effect and approximately follow the order: SPC > DSF > SSF~SPI. The dehydration of these soy reinforcement fractions causes the

Table 1. Fit Parameters of Shear Elastic Modulus[a]

Composition	Best fit[b]	
	m	
DSF/SB	4th cycle	8th cycle
20/80	0.73 ± 0.02	0.66 ± 0.02
30/70	0.70 ± 0.02	0.72 ± 0.02
SPI/SB		
20/80	0.51 ± 0.02	0.46 ± 0.02
30/70	0.51 ± 0.02	0.46 ± 0.02

Composition	γ_c (%)	G'_0 (MPa)	G'_∞ (MPa)
DSF/SB	8th cycle	8th cycle	8th cycle
20/80	2.05 ± 0.03	18.9 ± 0.5	5.59 ± 0.23
30/70	1.27 ± 0.03	60.6 ± 0.1	20.5 ± 0.4
SPI/SB			
20/80	1.88 ± 0.13	8.25 ± 0.03	3.53 ± 0.13
30/70	1.39 ± 0.07	16.9 ± 0.06	6.2 ± 0.2

[a] Samples prepared by casting method and measured at 140 °C
[b] Best fit of shear elastic modulus vs. strain with the Kraus Model
SOURCE: Reproduced from reference 16.

increase of composite modulus, indicating the effect of a more rigid filler network and perhaps also the enhanced filler-polymer interactions. SPI composites show a percolation threshold at 6-8% volume fraction. Above the percolation threshold, a logarithmic plot of modulus vs. concentration is linear. The composite modulus is also affected by the method of preparation, where the composites prepared by the casting method have a higher modulus than those prepared by the freeze drying method due to their different filler related network structures produced by these methods. Compared with carbon black, soy fillers have larger aggregate sizes and exhibit no shifting in loss maximums. The composites reinforced by the co-filler of SSF and CB exhibit two tan δ maximums, in which a higher temperature transition is attributed to the co-filler effect. Stress-strain properties indicate soy fillers increase the small strain modulus while reducing the elongation at break. Toughness of soy composites also decreases as the filler concentration increases in opposite to that of CB composites, except for the 10% SPC composite which has a toughness similar to the 20% CB composite. Stress softening effects indicated that the SPI filler network is stronger than CB but CB composite has a better instant recovery behavior. Carbon black filled composites prepared by freeze-drying also showed a recovery modulus greater than the initial modulus after eight strain sweep cycles, indicating a reorganization of filler network structures. Fitting of reversible strain-dependent moduli with the Kraus model indicates SPI filler network structures are more similar to CB networks than other soy products. The model fitting also indicates DSF filler related network structures are tougher than SPI or CB filler network structures.

Acknowledgements

The author thanks A. J. Thomas for Instron measurements; Dr. A. R. Thompson for scanning electron microscopy; and Dr. S. C. Peterson for proofreading the manuscript.

References

1. Ismail, H.; Jaffri, R. M.; Rozman, H. D. *J. Elastomers Plast.* **2003**, *35(2)*, 181-192.
2. Nair, K. G.; Dufresne, A. *Biomacromolecules* **2003**, *4*, 666-674.·
3. Ismail, H.; Shuhelmy, S.; Edyham, M. R. *Eur. Polym. J.* **2001**, *38(1)*, 39-47.
4. Wang, C.; Carriere, C. J. *U.S. Patent* **2001**, 6 310 136 B1.
5. Wang, S.; Sue, H. J.; Jane, J. *J M S-Pure Appl. Chem.* **1996**, *A33(5)*, 557-569.
6. Paetau, L.; Chen, C.; Jane, J. *Ind. Eng. Chem. Res.* **1994**, *33*, 1821-1827.
7. Mo, X.; Sun, X. S.; Wang, Y. *J. Appl. Polym. Sci.* **1999**, *73*, 2595-2602.
8. Wu, Q.; Zhand, L. *Ind. Eng. Chem. Res.* **2001**, *40*, 1879-1883.
9. Coughlin, E. T. A. *U.S. Patent* **1936**, 2 056 958.
10. Isaacs, M. R. *U.S. Patent* **1938**, 2 127 298.
11. Lehmann, R. L.; Petusseau, B. J.; Pinazzi, C. P. *U.S. Patent* **1960**, 2 931 845.
12. Fuetterer, C. T. *U.S. Patent* **1963**, 3 113 605,
13. Beckmann, O.; Teves, R.; Loreth, W. *Ger Offen* **1996**, DE 19622169 A1 19961212.
14. Recker, C. *Eur. Pat. Appl.* **2002**, EP 1234852 A1 20020828.
15. Chen, Y.; Zhang, L.; Du, L. *Ind. Eng. Chem. Res.* **2003**, *42*, 6786-6794.
16. Jong, L. *J. Appl. Polym. Sci.* **2005**, *98(1)*, 353-361.
17. Jong, L. *Polym. Int.* **2005**, *54(11)*, 1572-1580.
18. Jong, L. *Composites: Part A.* **2006**, *37*, 438-446.
19. Jong, L. *J. Polym. Sci., Part B: Polym. Phys.* **2005**, *43(24)*, 3503-3518.
20. Carter, C. M.; Cravens, W. W.; Horan, F. E.; Lewis, C. J.; Mattil, K. F.; Williams, L. D. In *Protein Resources and technology;* Milnre, M.; Scrimshaw, N. S.; Wang, D. I. C.; Eds.; AVI Publishing: Westport, CT, 1978; pp 282-284.

21. Heinrich, G.; Kluppel, M. *Adv. Polym. Sci.* **2002**, *160*, 1-44.
22. Wang, M. J. *Rubber Chem. Technol.* **1998**, *71*, 520-589.
23. Richard, J. *Polymer* **1992**, *33(3)*, 562-571.
24. Zosel, A.; Ley, G. *Macromolecules* **1993**, *26*, 2222-2227.
25. Kan, C. S.; Blackson, J. H. *Macromolecules* **1996**, *29*, 6853-6864.
26. Garcia, M. C.; Torre, M.; Marina, M. L.; Laborda, F. *Crit. Rev. Food Sci. Nutr.* **1997**, *37(4)*, 361-391.
27. Jong, L. *Composites: Part A.* **2005**, *36*, 675-682.
28. Jong, L. *Composites: Part A.* **2007**, *38*, 252-264.
29. Yurekli, K.; Krishnamoorti, R.; Tse, M. F.; Mcelrath, K. O.; Tsou, A. H.; Wang, H. C. *J. Polym. Sci.: Part B: Polym. Phys.* **2001**, *39*, 256-275.
30. Vieweg, S.; Unger, R.; Hempel, E.; Donth, E. *J. Non-Crystalline Solids* **1998**, *235/237*, 470-475.
31. Jong, L. *J. Polym. Environ.* **2005**, *13(4)*, 329-338.
32. Mullins, L. *J. Rubber Res.* **1947**, *16*, 275-89.
33. Mullins, L. *Phys. Colloid Chem.* **1950**, *54*, 239-251.
34. Payne, A. R. *J. Appl. Polym. Sci.* **1962**, *6(19)*, 57-63.
35. Payne, A. R. *J. Appl. Polym. Sci.* **1962**, *6(21)*, 368-372.
36. Payne, A. R. *J. Appl. Polym. Sci.* **1963**, *7*, 873-885.
37. Kraus, G. *J. Appl. Polym. Sci., Appl. Polym. Symp.* **1984**, *39*, 75-92.
38. Ulmer, J. D. *Rubber Chem. Technol.* **1995**, *69*, 15-47.
39. Huber, G.; Vilgis, T. A. *Kautsch Gummi Kunstst* **1999**, *52*, 102-107.
40. Shih, W.; Shih, W. Y.; Kim, S.; Liu, J.; Aksay, I. A. *Phys. Rev. A* **1990**, *42*, 4772-4779.
41. Heinrich, G.; Vilgis, T. A. *Macromol. Chem. Phys., Macromol. Symp.* **1995**, *93*, 253-260.

Enzyme Immobilization and Assembly

Chapter 7

Green Oxidation of Steroids in Nanoreactors Assembled from Laccase and Linear-Dendritic Copolymers

Ivan Gitsov[1,2,*], Arsen Simonyan[2], Albert Krastanov[2,3], and Stuart Tanenbaum[2]

[1]The Michael M. Szware Polymer Research Institute, State University of New York, Syracuse, NY 13210
[2]Department of Chemistry, College of Environmental Sciences and Forestry, State University of New York, Syracuse, NY 13210
[3]Faculty of Biotechnology, University of Food Technologies, Plovdiv, Bulgaria

This chapter describes our recent advances on the utilization of polymer-modified laccase complexes in aqueous systems towards the oxidation/polymerization of naturally hydro-phobic steroidal compounds, Equilin (EQ) and 17-β-estradiol (β-EST). We elucidate the kinetic and synthetic aspects of the process with the model compound 5,6,7,8-tetrahydro-2-naphthol (THN). The nano-reactor system is composed of linear poly(ethylene oxide)-dendritic poly(benzyl ether) diblock copolymer (G3-PEO13k) and laccase isolated from *Trametes versicolor*. Other advantages of the complex in comparison to the native enzyme are its recyclability, enhanced stability, activity, and overall simplicity in product harvesting and isolation. A principle of action of the complex is suggested based on these findings and is further supported by the biphasic solid-liquid nature of the reaction medium, which exhibits continuous influx of starting material and steady solid product expulsion. Comparative experiments with linear-linear poly(styrene)-*block*-poly(ethylene oxide) copoly-mer under identical conditions do not evince formation of a complex that could assist the reactions tested. An assessment is provided of the potential scale-up applicability of the complex towards "green" pharmaceutically and industrially interesting transformations.

Introduction

The biotransformation of steroids to pharmacologically useful intermediates and end products began on a practical scale over fifty years ago (*1*). At the outset, various bacterial and fungal species, exhibiting oxidase, hydrolase, dehydrogenase, aromatase or diverse side-chain clastic activities were grown and exposed to presumptive precursors. Limitations in these procedures were caused by insolubility or cellular toxicity of the starting materials, or by attendant use of organic solvents or dispersal agents. More recently, cell-free enzymes have been used to effect similar conversions (*2*), but suffer also from biocatalyst denaturations under reaction conditions which employ organic phase or detergent emulsifications.

Several dendritic structures, which exhibit intrinsic quasi-enzymatic activities (*3*), are functionalized with antibodies (*4*), or are covalently linked to enzyme-antibody complexes for biochemical diagnostic purposes (*5*) have been described. However, dendrimer-enzyme assemblies heretofore have only been described by Gitsov et al. (*6*), who demonstrated that dendritic-linear-dendritic poly(benzyl ether)-*block*-poly(oxyethylene)-*block*-poly(benzyl ether)s can selectively bind to glycoprotein enzymes, Figure 1. Following prior work on laccase-catalyzed oxidation of steroid hormones with phenolic A rings in organic solvents (*7*) we have reinvestigated the biotransformation of 17-β-estradiol, equilin and the model compound 5,6,7,8-tetrahydro-2-naphthol (Figure 2) by laccase/linear-dendritic copolymer. This complex leads to their rapid oxidation in aqueous buffers with high product yields. It is further detailed here that such complexes can be manipulated by the reaction conditions to achieve specific product distributions and can be recycled for sequential modes of operation.

Experimental

Materials

The linear-dendritic copolymer [G-3]-PEO13k has a linear poly-(oxyethylene) block with molecular weight of 13,000 Da (PEO13k) and was formed via "living" anionic polymerization of ethylene oxide initiated by a

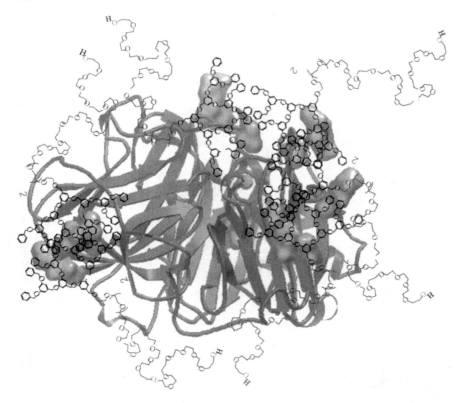

*Figure 1. Schematic representation of G3-PEO13k-laccase complex – the grey regions are the glycan residues that serve to anchor the dendritic blocks**
*- Protein image obtained from AstexViewer™@MSD-EBI Database (8)

third-generation poly(benzyl ether) monodendron, [G-3] (9). The linear-linear block copolymer PSt1.8k-PEO2.1k having poly(styrene) block with molecular weight 1,800 Da (PSt1.8k) and poly(oxyethylene) block of 2,100 Da (PEO2.1k) was purchased from Polymer Source, Inc. (Montreal, Canada). Equilin, EQ (p.a.), β-Estradiol, β-EST (97%) and 5,6,7,8-tetrahydro-2-naphthol, THN (98%) were purchased from Sigma-Aldrich and used without further purification. 2,2'-Azino-bis(3-ethylbenzothiazoline-6-sulfonic acid), ABTS, was obtained from Sigma (St. Louis, USA) in 98 % purity and used as received. Laccase was produced in a three stage procedure using the basidiomycete white rot fungus *Trametes versicolor*. Isolation and purification details are provided elsewhere (6,10). Deionized (DI) water, (18.2MΩ) was produced in a Barnstead Nanopure system. Tetrahydrofuran (THF) was purchased from Aldrich and distilled under nitrogen.

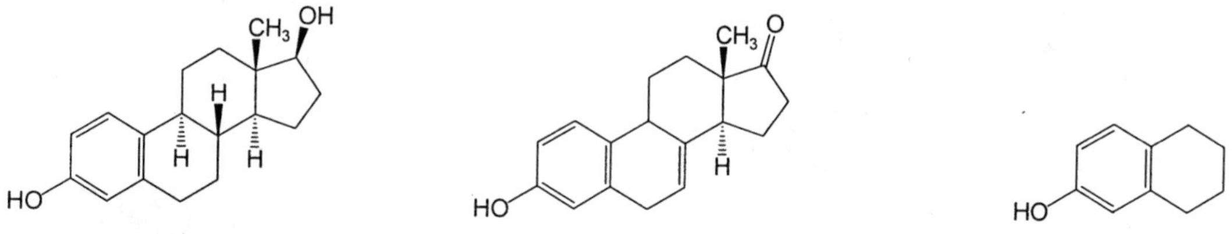

17-ß-Estradiol Equilin 5,6,7,8-Tetrahydro-2-naphthol

Figure 2. Structures of the substrates oxidized by G3-PEO13k-laccase.

Instrumentation

Size-exclusion chromatography (SEC) in THF was performed on a line consisting of Waters 510 pump, Waters U6K injector, an Applied Biosystems 785A programmable UV-Vis detector and a Viscotek Model 250 instrument with dual refractive index and viscometry detection. Three PLgel 5μ columns (50 Å, 500 Å and a mixed C) from Polymer Laboratories were used for the separations at 40°C. Calibrations were made with poly(styrene) standards. Data was collected and manipulated with an OmniSEC 3.1 software package from Viscotek. ^{1}H-NMR spectra were taken on a Bruker Avance instrument at 600 MHz in $CDCl_3$. GC-Mass spectra of the dimeric products were obtained on a Thermo Polaris Q instrument in chloroform on a DB-1 column, temperature gradient from 60°C to 300°C at 10°C/min, hold time 30 min and at 70 eV. UV-Vis spectra were obtained in water on a Beckman 680B spectrophotometer.

Laccase Modification

All polymer-enzyme complexes were prepared by the following protocol: solid diblock copolymer of the type G3-PEO13k (LD) was introduced into a solution of hundred-fold diluted purified laccase (specific activity = 6.7 μkat/mg) to a final concentration of 1.4 ± 0.1 mg/mL. This solution was stirred at room temperature and pH of 6.2-6.7 for 3 hours prior to its usage.

Oxidation Reactions

The solid substrates were introduced into the laccase-LD complex solutions placed in a 50 mL round bottom flasks and the mixtures were magnetically stirred at room temperature. After the end of the process the reaction products were separated and purified by the following procedure: initially the mixtures were centrifuged at 2800 G force for 60 min, the clear aqueous solution of polymer-enzyme complex was filtered through 0.45 μm Whatman cellulose filter and kept at 4°C for further oxidations. The yellow to brown precipitate was collected, washed twice with DI water and dried at room temperature under vacuum. It was analyzed by SEC in THF. The separation of the oxidation product(s) was achieved by preparative fractionation on the same SEC system. The THF solvent in each fraction was evaporated and the dry contents were analyzed spectroscopically. The general sequence of procedures is depicted on the flow chart in Scheme 1.

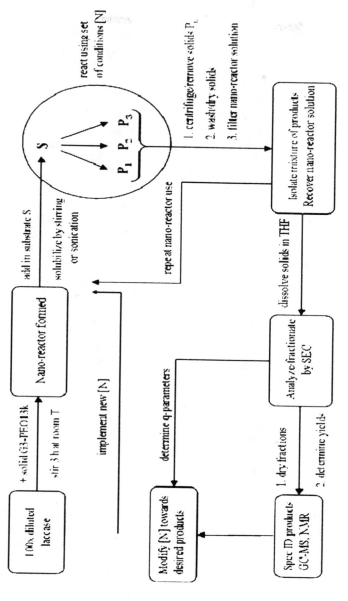

Scheme 1. Flow Chart of Procedures

Results and Discussion

The mechanism of action of the polymer-enzyme complex is presumably based on the selective binding of substrates with very low solubility in water by the hydrophobic dendritic "pockets" in close proximity to the active site of the complex as shown in our earlier studies (6,10). Their encapsulation is driven by the favorable lowering of the free energy of the system due to the enhanced hydrophobic-hydrophobic interactions (11). The fact that we can subject otherwise water insoluble organic compounds to selective oxidation by a water-soluble enzyme system is one of the significant advantages of our system. Moreover, experiments show that the solubilized solid substrates turn into solid reaction products that can be easily removed by simple filtration and analyzed. The optimal concentration of LD copolymer in the complex (1.4 g/L or ~1.10^{-4} M) is determined against ABTS (ε_m = 36,000 at λ = 420 nm), Figure 3. The attempted complexation with the linear-linear analogue of our G3-PEO13k, PSt1.8k-PEO2.1k, leads to a significant decrease of laccase activity. There are two plausible explanations for the observed behavior of the linear-linear analogue: the formation of a denser packing of the polystyrene chains in the proximity of the active center and supply channels of the enzyme leading to their blockage or the lack of sorption of this copolymer on the glycan pockets causing the formation of separate micelles, which deprive the laccase of ABTS by selectively binding the substrate. The adverse effect of formation of polymeric micelles has also been observed with the linear-dendritic copolymer at concentrations larger than 1.4 g/L and evidenced by aqueous SEC.

On the other hand, lower LD copolymer concentrations decrease the binding capability of the complex and thus impair the overall oxidation of the substrates studied. When no LD copolymer was used only traces of oxidation products were observed by UV-Vis and SEC in THF due to the limited surface contact between the dissolved enzyme and the solid β-EST, EQ and THN. Figure 4 contains comparative SEC profiles for all three substrates after 48 hours of reaction time. Representative SEC chromatograms of model THN reaction mixtures at different times are shown in Figure 5.

SEC with poly(styrene) calibration shows that the apparent molecular weight of the reaction products is in the range of 600-2500 Da, indicating that the oxidation yields oligomers with low degree of polymerization (DP=2-7). It should be noted that these reaction products are isolated after centrifugation and simple filtration, which implies that, after the biotransformation at the active center inside the complex, the derivatives formed are expelled in the aqueous phase. This may be due to their increased polarity or to their increased size or both.

Another remarkable aspect of this phenomenon is the repetitive nature of action of the complex, which suggests its recyclability potential. Using one and

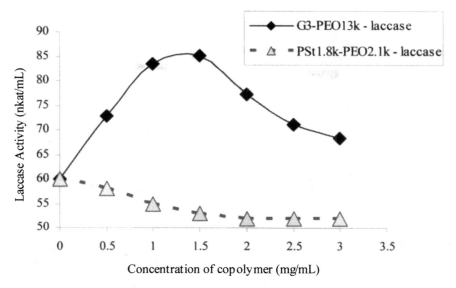

Figure 3. Polymer-enzyme complex activities with ABTS oxidation

the same complex solution we performed 4 consecutive reactions with THN, and the elution profiles of the reaction mixtures are shown on Figure 6.

It is clear that the complex retains nearly 100% of its activity at least through the fourth cycle. In addition, the ratio between dimers, trimers and higher oligomers seems to change in favor of the dimers, suggesting that the complex is initially highly reactive and promotes the formation of higher-molecular-weight products. If dimeric products are sought its activity can be reduced by simple dilution, addition of an inhibitor or by lower concentration of the LD-laccase complex. After a 72 h reaction of adjusted-reactivity complex with THN we obtain a product mixture, which does not contain starting material, i.e. conversion is 100%.

The distribution of oligomers as calculated in mole % from the peak areas is shown in Table 1 (only peaks for dimers and trimers elute separately, tetramers and probably higher oligomers appear as a shoulder). Peak molecular weights, polydispersity indices and corresponding number of THN units, as calculated from a polystyrene calibration curve, are also given.

Based on the fact that 1.4 g/L LD copolymer concentration produces the highest activity towards ABTS, a trifactorial experiment of a THN oxidation is developed in order to steer the reaction towards either dimers or higher

118

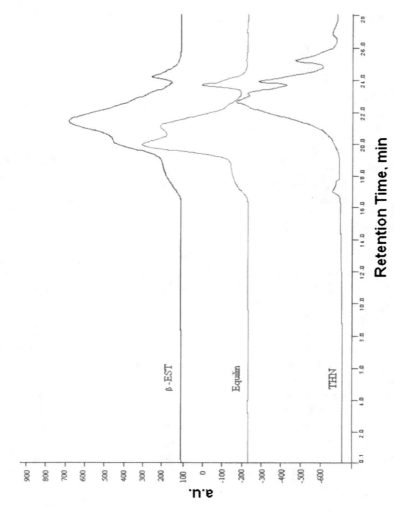

Figure 4. Chromatograms of the precipitate isolated after 48 hours of β-EST, EQ and THN oxidation assisted by LD-laccase.

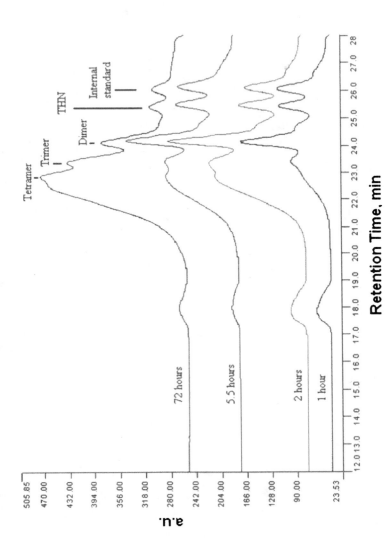

Figure 5. Oxidation of THN by G3-PEO13k-laccase.

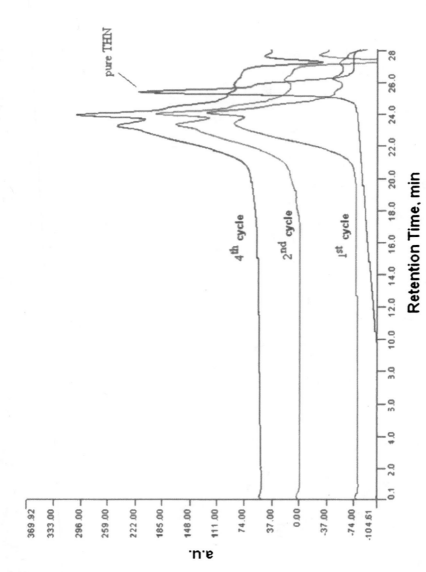

Figure 6. SEC traces of reaction mixtures from THN oxidations with reused
G3-PEO13k-laccase complex, 72 h, 22°C

oligomers. The initial concentrations for both substrate and complex are varied thrice, while the reaction time is changed twice: the concentration of LD copolymer-laccase complex increases from 0.35 g/L through 0.7 g/L up to 1.4 g/L copolymer equivalent, the substrate concentration increases from 1.5 g/L through 3 g/L up to 6 g/L, and reaction times of 1 hour and 2 hours are tested.

Table 1. Distribution of the THN Oxidation Products after 72 h at 22°C

Retention Volume mL	Mole %	$M_p{}^a$ g/mol	PDI^b	# of Repeating units
22.96	36.8	560	1.09	≥4
23.32	25.6	451	1.01	3
24.02	34.2	294	1.03	2

[a]Molecular weight at the top of the peak in the SEC trace; [b]Polydispersity Index

This scheme produces 18 separate combinations, whose results can be summarized and are best understood in terms of an efficiency q-parameter introduced hereto. When efficiency towards dimers is sought it can be easily defined as:

$$q2 = \chi \,(di\text{-}THN)^2 / \chi THN) * \chi \,(olig\text{-}THN),$$

where χ signifies the mole fraction of dimers (di-THN), monomer (THN) and oligomers (olig-THN), respectively, as obtained from peak area calculations. It can be seen that the higher the mole fraction of dimers and the lower the mole fraction of unreacted THN and higher oligomers, the higher the q-value for the given experiment. All dimer q-values are summarized in Table 2. When the shorter reaction time is employed the best synthetic conditions for obtaining dimers are 1.5 g/L THN and 0.7 g/L LD, while for the longer reaction time these can be doubly increased to 3 g/L THN and 1.4 g/L LD.

Table 2. Values of the experimental dimer efficiency q-parameter at 1 h and 2 h

1 hour	THN, g/L			2 hours	THN, g/L		
LD,g/L	1.5	3	6	LD,g/L	1.5	3	6
0.35	0.96	0.73	0.29	0.35	0.68	0.44	0.45
0.7	1.53	0.58	0.52	0.7	1.94	0.55	0.48
1.4	1.33	0.62	0.60	1.4	1.97	2.36	0.58

A clear and expected trend is shown – high THN concentrations at longer times produce more oligomer products that drive q down. The effect of LD-laccase concentration is not that clear, but in the longer run it can be seen to increase the q value within each set of THN concentrations. The optimized conditions enable the selective formation of THN dimer at significantly higher yields (28 % from the overall oxidation products) than the previously reported data obtained both by laccase- or chemically induced oxidations (7-14 %) (*7c*).

A desirable consequence of the cyclic nature of the material transport in and out of the complex is the achievement of high conversions. In the case of THN oxidations we observe yields from ~ 76% up to 100% depending on the selected conditions. β-EST and EQ conversions are also within the same 70-100% yield range. Sonication at low power, instead of magnetic stirring, produces higher conversion rates for the same reaction times. Higher power sonication has an unfavorable effect on the complex stability and thus hampers the oxidation. The oxidation of β-EST and EQ by laccase could afford a complex mixture of reaction products both through C-O and C-C alkylation pathways (*7b,c*). However, after the separation and isolation of the products from the model THN oxidation, NMR data show the presence of three isomers with C-C connectivities only (Figure 7), in contrast to the previously published data, where just two C-C isomers were reported (*7c*). Surprisingly the GC-Mass analyses reveal the presence of 6 species all of them having identical molecular ions of 294.3 Da, Figure 8A,B. It is assumed that these 6 peaks are caused by the existence of 3 pairs of rotational isomers with prohibitively high rotation barriers (*12*) – 3 of them have the hydroxyl groups at the 1-1', 1-3' and 3-3' bonds oppositely oriented in a *trans* configuration (Figure 7) and 3 of them have these OH moieties facing each other (*cis* configuration). Due to the possibility of hydrogen bond formation the second group would have different polarities, dipole moments and presumably different boiling temperatures that will affect their retention times on the GC column, Table 3.

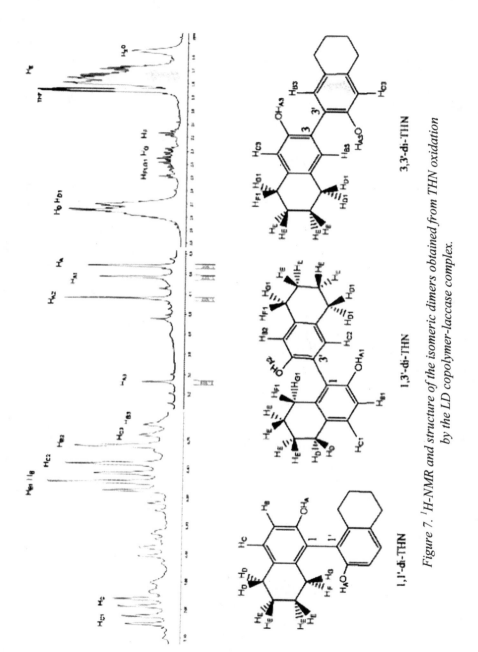

Figure 7. ^{1}H-NMR and structure of the isomeric dimers obtained from THN oxidation by the LD copolymer-laccase complex.

124

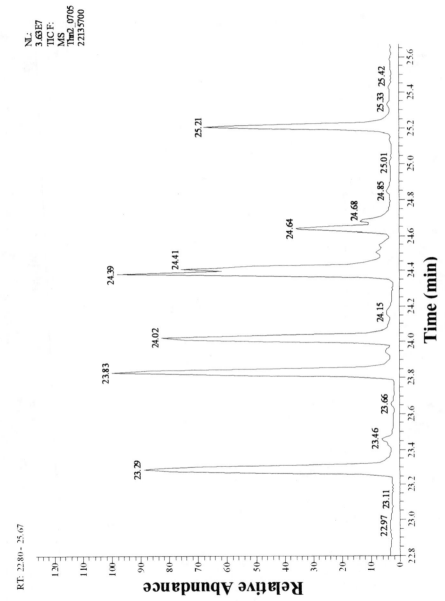

Figure 8A. Model oxidation of THN by G3-PEO13k-laccase. GC trace.

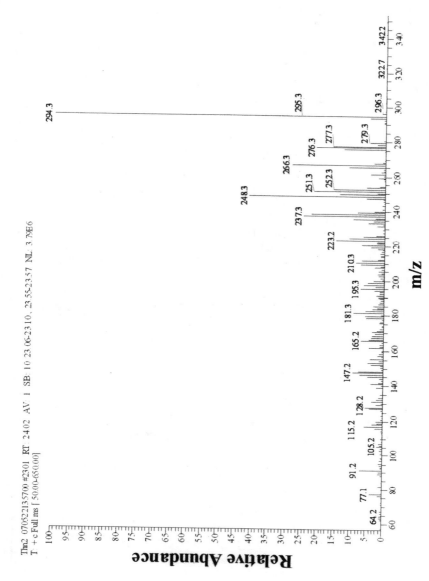

Tm2 07052213570) #2301 RT 24.02 AV 1 SB 10 23.06-23.10 , 23.55-23.57 NL: 3.79E6
T + c Full ms [50.00-650.00]

Figure 8B. Model oxidation of THN by G3-PEO13k-laccase. Mass Spectrum of the fraction at 24.02 min.

**Table 3. GC Retention Times, Dipole Moments and Assigned
Structures of Rotation Isomers (*12*)**

Time, min	μ, D	Isomer
23.29	1.614	trans-1,3'
23.83	1.778	trans-1,1'
24.02	1.878	cis-1,1'
24.39	2.376	cis-1,3'
24.63	2.639	trans-3,3'
25.21	2.734	cis-3,3'

β-EST behaves in a similar fashion and yields only C-C alkylation products. The ^{1}H-NMR spectrum reveals the typical chemical shifts (*7b*) for the protons in 1-1' and 1-3' dimers - CDCl$_3$, ppm, δ = 7.33 (d, Ph-*H meta*, 1-1'cis), 7.31 (d, Ph-*H meta*, 1-1'trans), 6.78 (d - Ph-*H ortho*, 1-1'), 4.96 (br s, Ph-O*H*, 1-1'), 3.73 (s, HO-C-*H*, 1-1'), 0.78 (s, C*H*$_3$ 1-1'); 7.40, (d, Ph-*H meta*, A ring, 1-3'), 7.00-6.91 (s, Ph-*H* 1 *meta* - 1 *para*, A' ring, 1-3'), 6.72 (d, Ph-*H ortho*, A ring, 1-3'), 4.96 (br s, Ph-O*H*, 1-3'), 3.74 (HO-C-*H*, 1-3'), 0.77 (C*H*$_3$, 1-3'). The content of the 3-3' dimer is probably below the detection limit. The ^{1}H-NMR spectra recorded with EQ oxidation products show similar structures, as well.

Conclusion

This research is a logical continuation of our efforts to build nano-sized reactors able to mediate organic transformations of hydrophobic substrates under "green" conditions, i.e. in water and at room temperature. In contrast to the previously concluded lack of practical benefits from laccase immobilization (*13*) the linear-dendritic/laccase nanoreactor discussed here has enhanced enzyme reactivity and improved stability due to the regioselective surface envelopment and increased accessibility of the active site for hydrophobic substrates. Modification of the reaction conditions (starting concentrations, reaction time, low-power sonication or magnetic stirring) allowed us to optimize the system to yield predominantly dimeric products and minimize the formation of higher oligomers while the overall conversion still reached 100%. This might be of particular importance for the construction and production of novel oligomeric ligands under environmentally benign conditions. In addition our system fully retained its enzyme catalytic activity over four consecutive reaction cycles, thus potentially improving the maximum transformation of substrate fourfold. Overall, the presented system and its enhanced qualities infer promising application in remote functionalization and/or polymerization of steroids or smaller molecules which have glycoenzyme-susceptible moieties.

Acknowledgments

This research was supported in part by the National Science Foundation (MCB-0315663). A.K. acknowledges the Council for International Exchange of Scholars for the Senior Fulbright Fellowship that enabled his residency at SUNY ESF.

References and Notes

1. Crueger, W.; Crueger, A. "Biotechnology: A Textbook of Industrial Microbiology", Second Edition, Sinaur: Sunderland, 1990, pp. 292-297
2. Alexander, D.L.; Fisher, J.F. *Steroids* **1995**, *60*, 290
3. Kofoed, J.; Reymond, J.L. *Curr. Opin. Chem. Biol.* **2005**, *9*, 656
4. Cordova, A.; Janda, K.D. *J. Am. Chem. Soc.* **2001**, *123*, 8248
5. Singh, P. In *Dendrimers and Other Dendritic Polymers;* Fréchet, J.M.J. and D.A. Tomalia, Eds., John Wiley & Sons, Ltd: Chichester, 2001, pp. 463-484
6. Gitsov, I.; Lambrych, K.; Lu, P.; Nakas, J.; Ryan, J.; Tanenbaum, S. In *Polymer Biocatalysis and Biomaterials;* Gross, R.A. and H.N. Cheng, Eds. *ACS Symp. Ser. 900,* American Chemical Society: Washington, DC **2005**, pp. 80-94
7. a) Lugaro, G.; Carrea, G.; Cremonesi, P.; Casellato, M.M.; Antonini, E. *Arch. Biochem. Biophys.* **1973**, *159*, 1; b) Nicotra, S.; Intra, A.; Ottolina, G.; Riva, S.; Danieli, B. *Tetrahedron (Asymmetry)* **2004**, *15*, 2927; c) Intra, A.; Nicotra, S.; Riva, S.; Danieli, B. *Adv. Synth. Catal.* **2005**, *347*, 973
8. Hartshorn, M. J., *J. Comp. Aided Mol. Des.* **2002**, *16*, 871
9. Gitsov, I.; Simonyan, A.; Vladimirov, N.G. *J. Polym. Sci.: Part A: Polym. Chem.* **2007**, accepted
10. Gitsov, I.; Hamzik, J.; Ryan, J.; Simonyan, A.; Nakas, J.P.; Krastanov, A.; Omori, Sh.; Cohen, T.; Tanenbaum, S., *Biomacromolecules* **2007,** submitted
11. Gitsov, I.; Lambrych, K.R.; Remnant, V.A.; Pracitto, R. *J. Polym. Sci., Part A: Polym. Chem.* **2000**, *38*, 2711
12. Calculations of the potential barriers for rotation were performed with HyperChem, ver.7 using a semi empirical method AM1. The dipole

moments of all 6 isomers were computed by the same software package. Initial structures were geometrically optimized by a Polak-Ribiere conjugate gradient algorithm at the same level.

13. Brandi, P.; D'Annibale, A.; Galli, C.; Gentili, P.; Pontes, A.S.N. *J. Mol. Catal. B: Enzym.* **2006**, *41*, 61

Chapter 8

Lipase Immobilization on Ultrafine Poly(acrylic acid)–Poly(vinyl alcohol) Hydrogel Fibers

Lei Li and You-Lo Hsieh[*]

Fiber and Polymer Science, University of California, Davis, CA 95616
[*]Corresponding author: ylhsieh@ucdavis.edu; 530–752–0843

Ultra-fine fibrous membranes were prepared by electro-spinning of aqueous mixtures of poly(acrylic acid) (PAA) and poly(vinyl alcohol) (PVA) at 0.17, 0.58 and 0.83 PAA weight fractions or 0.14, 0.97 and 3.5 COOH/OH molar ratios followed by heat-induced crosslinking (140°C for 5 min). The averaged diameters of these hydrogel fiber ranged from 270 to 450 nm and increased with increasing PAA contents. Swelling of these hydrogel fibrous menbranes also increased with increasing PAA contents as well as pH from 2 to 7, reaching 31 times for the membrane with a 3.5 COOH:OH ratio and at pH 7. Lipase was immobilized on these ultra-fine hydrogel fibers by physical adsorption and covalent bonding mechanisms. The optimal lipase immobilized was 0.65mg/mg dry fibers (3.5 COOH/OH and pH 4). The hydrogel bound lipase enzyme exhibited the same activity as the free lipase. The reuse efficiency of the immobilized lipase was improved by covalent bonding mechanism, however, at the expense of lowered activity.

Introduction

Enzymes have been used in a variety of biotechnological, biomedical and pharmaceutical applications due to their high specificity, non-toxicity, and water solubility, which are major advantages over inorganic catalysts (1). Because separation and purification of enzymes often requires complicated processing and retrieving them can be costly, immobilization of enzyme has attracted a lot of research (1-4).

Fibrous materials are among the most suitable solid supports for enzyme immobilization due to their intrinsically high specific surface and porous structures, which offer the binding capacity for the enzymes and accessibility to their active sites to achieve high reaction rates and conversions. Enzymes have been immobilized on fibers, either on the surfaces (3, 5) or within the bulk (4). Lipase has been immobilized on coarse nylon fibers (around 0.4 mm diameter) using glutaraldehyde and adipic acid dihydrazide as the spacers (5). Although the specific activity of immobilized lipase was found to be 5 times higher than that of crude lipase, the improvement appeared to be over estimated due to the fact that the assay did not allow for the full activity of the crude enzyme to be expressed. Ultra-fine cellulose fibers with diameter around 500 nm have also been used as the solid support for lipase immobilization (3). The cellulose fiber surfaces were reacted with polyethylene glycol (PEG) diacylchloride to simultaneously add amphiphilic spacers and reactive end groups for coupling with the lipase enzymes. Whereas the free lipase retained little of its activity following exposure to organic solvents, the fiber-bound lipase possessed much superior retention of catalytic activity after exposure to cyclohexane (81%) and toluene (62%) and hexane (34%). The fiber-bound lipase also exhibited significantly higher catalytic activity at elevated temperatures than the free form, i.e., 10 times at 70°C. Lipase has also been entrapped in poly(vinyl alcohol) (PVA) or poly(ethylene oxide) (PEO) ultra-fine fibers at a 20 to 30% loading level (4). The lipase in PVA/lipase ultra-fine fibers was 6 times more active than that in the cast film from the same solution.

Hydrogels possess some of the very attractive features for enzyme immobilization. The high water content inside the polymer networks provides biocompatible environment for immobilized enzymes, and the highly porous structure helps with mobility and diffusion of substrates and reaction products (6). Lipase is one of the most extensively investigated enzymes because of its ability in fat splitting, esterification, trans-esterification, and other reactions of industrial importance (2). Lipase has been successfully entrapped in 2.5-mm diameter chitosan gel beads by dropping the acid mixture of the enzyme with citosan into a sodium tripolyphosphate coagulation solution (1). The entrapped lipase retained about 42% of its original enzymatic activity. We have generated polyelectrolyte grafts and hydrogels onto ultra-fine cellulose fiber surfaces to bind lipase enzymes (7). The lipase bound to the hydrogel on the fiber surfaces exhibited greatly improved organic solvent stability over the free lipase.

The intrinsic properties of hydrogels in fibrous forms, including swelling behavior and high specific surface, should offer ideal substrate characteristics for enzyme immobilization. We have prepared a series of ultra-fine hydrogel fibers by electrospinning of binary polymer systems followed by crosslinking (8, 9). These fibers have much higher specific surface areas than conventional fibers and their swelling behaviors can be tuned by polymer compositions. The fibrous membranes distinguished themselves as being far stronger and re-absorbing significantly faster than the cast-films of the same chemical compositions (10). This paper reports the immobilization of lipase from *Candida Rugosa* (Type VII) on ultra-fine PAA/PVA hydrogel fibers by both physical adsorption and covalent bonding mechanisms. The swelling of hydrogel fibers on the loading of lipase was studied at different pH. The activity and stability of the immobilized lipase were investigated. The reuse efficiency of the immobilized lipase was improved by coupling the lipase with the hydrogel fibers using a water-soluble carbodiimide.

Materials and Methods

Chemicals and Reagents

Poly(acrylic acid) (PAA) (average M_v ~450 kDa) and polyvinyl alcohol (PVA) (average M_w 124-186 kDa, 87-89% hydrolyzed) were obtained from Aldrich Chemical Company. Lipase, type VII, from *Candida rugosa*, was purchased from Sigma, and used as received. The regents needed for lipase assay including olive oil (substrate), gum Arabic (emulsion regent), sodium deoxycholate (emulsion regent), triethanolamine hydrochloride (buffer), sodium diethyldithiocarbanate (color indicator), stearic acid (standard) and 1-Ethyl-3-(3-Dimethylaminopropyl) carbodiimide Hydrochloride (EDC), all from Acros, were used without further treatment. pH buffers were purchased from EM Science. All PAA and PVA solutions were prepared with purified water (Millipore Milli-Q plus water purification system).

Preparation of ultra-fine PAA/PVA hydrogel fibrous membrane and cast film

Ultra-fine PAA/PVA fibrous membranes were prepared by electrospinning of aqueous PAA/PVA mixtures by a previously described procedure (9). The molar ratios between PAA carboxylic acid and PVA hydroxyl (assuming 88% hydrolysis) were 0.14, 0.97 and 3.5, corresponding to 0.17, 0.58 and 0.83 PAA weight fractions, respectively.

PAA/PVA electrospun fibrous membranes were heated at 140°C for 5 minutes to induce crosslinking. Films were cast from the mixture solution on a

Teflon surface by evaporating water at 60°C under vacuum for 24 hrs, and then thermally crosslinked under the same condition as electrospun fibrous membranes.

Measurement and Characterization

The solution viscosities were measured according to ASTM D445 using a Cannon-Fenske viscometer (Cannon instrument company, USA). Conductivities of the solutions were determined using a digital conductivity meter (VWR International, Inc.). The fiber and fibrous membrane morphology was observed with a scanning electron microscope (SEM) (XL30-SFEG, FEI/Philips) at 2kV accelerating voltage.

The swelling degree (q) of the crosslinked fibrous membrane was measured by immersing the membrane in pH buffers (pH 2 to 7) without or with lipase (5mg/ml) at 30°C for 24 hrs and calculated by the following equation.

$$q = (W_s - W_o)/W_o$$

where W_o and W_s are the masses of the samples before and after fully swelling, respectively.

Immobilization of lipase on the ultra-fine PAA/PVA hydrogel fibrous membrane

Both physical absorption and covalent bonding mechanisms were investigated for immobilizing the lipase onto the PAA/PVA hydrogel fibrous membrane. For the covalent bonding mechanism, EDC was used as the coupling reagent. All measurements were performed in triple.

Physical absorption

About 10.0mg of thermally crosslinked PAA/PVA electrospun fibrous membrane was immersed in 2.0ml of the selected pH buffer solution with the lipase concentration at 5.0mg/ml. After the mixture was incubated in the water bath at 30°C for 24 hrs, the fibrous membrane was taken out and rinsed with the same pH buffer. The washing pH buffer was added to the residue solution left from the adsorption to make up the total solution volume to be exactly 10.0ml in a volumetric flask. The concentration of lipase in the above solution was then determined by Bio-Rad (Hercules, CA) protein assay (7, 11). The amount of lipase physically adsorbed by the electrospun membrane was calculated by deducting the lipase amount in the buffer solution from the original total amount. The average from measurements of three samples was reported.

Covalent bonding

Method I. 10 mg of PAA/PVA electrospun membrane was swollen at 30°C in 2.0 ml of pH4 buffer solution containing both EDC (EDC: COOH of PAA = 2:1 in molar ratio) and lipase (5mg/ml). After 24 hrs, the membrane was washed thoroughly with pH4 buffer and deionized water, and dried in vacuum at room temperature.

Method II. PAA/PVA ultra-fine fibrous membrane was first activated by EDC in pH4 buffer solution for one hour. After the residual EDC was removed by thoroughly rinsing with deionized water, the treated membrane was immersed in another pH4 buffer solution containing lipase for 24 hrs. The quantities of the materials used were the same as in Method I.

Method III. A reversed two-step procedure as described in Method II was used. The lipase adsorption was performed in a pH4 buffer solution for 24 hrs first, and then the PAA/PVA ultra-fine fibrous membrane with lipase was immersed in another pH4 buffer solution with EDC for one hour. Again, the quantities of the materials used were the same as in Method I.

Assay of Lipase Activity

The catalytic activity of the lipase-containing electrospun PAA/PVA hydrogel fibers on hydrolyzing olive oil was assayed using a standard photometric method (*12*). The molar ratio between the PAA carboxylic acid groups and the PVA hydroxyl groups in the hydrogel fibers was 3.5. Briefly, a stabilized olive oil emulsion substrate and lipase bound PAA/PVA ultrafine fibers (~1.0mg) were added to a buffer (pH=8.5) and incubated at 30°C under constant shaking (60 rpm). At a designed assaying time (10min for crude lipase, and 1hr for adsorbed lipase), the incubation solution was heated to 100°C to denature the lipase. Copper (II) nitrate aqueous solution and chloroform were added to extract the liberated fatty acids from hydrolysis of olive oil to chloroform layer in the form of the copper salts. The same procedure was applied to the olive oil emulsion alone to prepare a blank solution. The amount of copper (II) ion, or COO⁻ of the fatty acid product, was determined spectrophotometrically at 436 nm (Hitachi U-2000 spectrophotometer) with sodium diethyldithiocarbamate as a color indicator. The absorbance was converted to concentration using a calibration curve from stearic acid standard solutions. Lipase activity was expressed as unit per milligram lipase (U/mg lipase), where U represented the amount (μmol) of fatty acid released from the catalytic hydrolysis of olive oil per hour.

Results and Discussion

Formation of PAA/PVA hydrogel fibers

Three mixtures of PAA and PVA with 17 wt%, 58 wt%, and 83 wt% PAA or at 0.14, 0.97, and 3.5 COOH/OH molar ratios were prepared at a fixed 6 wt% total polymer concentration. At 6 wt%, individual PVA and PAA solutions had viscosities of 94.7 and 74.6 cp, respectively. The mixtures showed significantly higher viscosities especially at close to 50% PAA composition, which is probably due to the strong and abundant hydrogen bonding between PAA carboxylic and PVA hydroxyl groups (Table 1). The average diameter was about 450 nm for the bicomponent fibers electrospun from the mixtures with more than 50% PAA and about 270 nm when PAA was only 17% in the mixture (Figure 1). When PVA was the majority, the bicomponent fibers was much thinner. This was attributed to the much lower molecular weight of PVA when compared with PAA (9).

The ultra-fine PAA/PVA fibrous membranes that were heated at 120°C or 140°C for 5 min became insoluble in water, indicating inter-molecular crosslinking under these heating conditions. However, only those crosslinked at 140 °C could retain the fibrous structure after extended (24hr) water immersion (Figure 2a). The water immersion caused those membranes heated at 120°C for 5 min to partially lose their fibrous integrity and became film-like at random locations of the membrane (Figure 2b). PAA/PVA ultra-fine fibrous membranes and cast films discussed from here on were all thermally crosslinked at 140 °C for 5 min.

The crosslinked fibrous membranes exhibited pH-responsive swelling behavior (Figure 3a). At pH 2 and 4, the membranes took up to 6 and 11 times of liquid, respectively. Significantly more liquid was absorbed at pH 5, up to 29 times, followed by a slight increase to 31 times at pH 7. This is consistent with the pK_a of PAA being about 4.7. Higher absorption was observed with fibrous membranes that contained more PAA at all pH.

Physical adsorption of lipase in PAA/PVA hydrogel fibers

The polymer compositions of PAA/PVA hydrogel fibrous membranes showed obvious effect on the lipase adsorption efficiency. The amount of lipase adsorbed in the fibrous membranes increased from 0.64 to 0.75 mg/mg fibers when COOH:OH the molar ratio in the fibers decreased from 3.5 to 0.14 (Table 2). The fibrous membranes with higher PAA contents consisted of larger fibers (Table 1) and swelled to higher degrees in the buffer solutions (Table 2), both of which were expected to lower the specific surface of the hydrogel fibers. Although the hydrogel fibers containing more PAA swelled to higher degrees which facilitated the diffusion of lipase molecules into the hydrogel fibers, the reduced specific surfaces lowered the lipase adsorbed on the fiber surfaces. Consequently, the overall lipase adsorption efficiency was reduced.

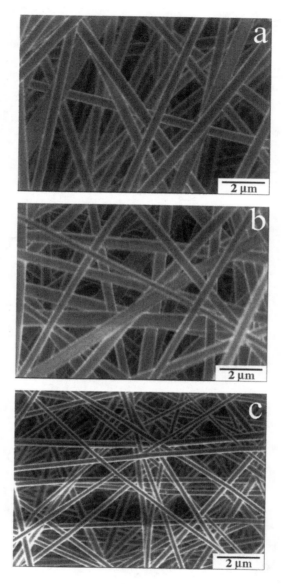

*Figure 1. SEM of PAA/PVA bicomponent fibers (bar = 2 μm) electrospun from
6 wt% total polymer concentration at varying COOH/OH molar ratio: (a) 3.5,
(b) 0.97, and (c) 0.14.*

Table 1. Properties of PAA/PVA mixture solutions and the average diameters of the bicomponent fibers.

Composition (PAA wt fraction)	Viscosity (cp)	Conductivity (mS/cm)	Average diameter (nm)
0.83	219.8	2.02	450
0.58	1310.7	1.79	450
0.14	1131.4	0.10	270

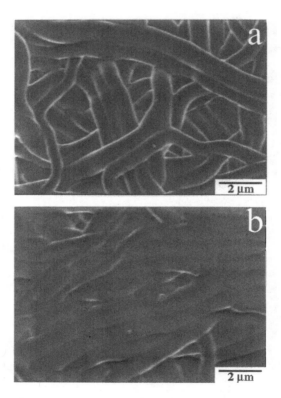

Figure 2. SEM (bar = 2 μm) of PAA/PVA bicomponent fibers (COOH/OH molar ratio = 0.97) cross-linked at different temperatures for 5 min, immersed in water for 24 hrs and dried in vacuum at ambient temperature: (a) 140°C and (b) 120°C.

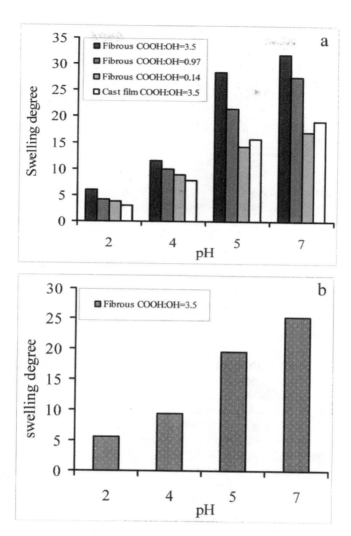

Figure 3. pH-responsive swelling behavior of hydrogel fibers and cast film with varying COOH/OH molar ratio in pH buffers: (a) without or (b) with lipase (5.0 mg/ml).

The lipase adsorption efficiency was also affected by the pH of the adsorption medium. The swelling degrees of the crosslinked PAA/PVA fibrous membrane (3.5 COOH/OH molar ratio) in lipase solution (5.0 mg/ml) increased from 6 to 25 with increasing pH from 2 to 7 (Figure 3b), which were 10 to 20% lower than those in pH buffer alone (Figure 3a). This may be because the lipase

Table 2. Effect of fiber composition on the amount of adsorbed lipase.

Fiber composition (in COOH:OH molar ratio)	Swelling degrees in pH4 buffer	Adsorption of lipase (mg/mg fiber)
3.5	12.0	0.64
0.97	10.0	0.73
0.14	8.9	0.75

molecules competed with water molecules for adsorption. The amount of lipase adsorbed in the hydrogel fibers increased from 0.33 to 0.64 mg/mg fibers when pH increased from 2 to 4, and then decreased to 0.47 and 0.24 mg/mg fibers as pH increased further to 5 and 7, respectively (Figure 4). When pH was 2, the very low swelling of the hydrogel fibers limited the diffusion of the lipase molecules into the hydrogel structure than in buffers with higher pH. On the other hand, if the solution pH increased to be significantly higher than the pKa (~4.7) of the PAA carboxylic acid, for example in the pH 7 buffer, the swelling of the hydrogel fibers increased significantly. The larger swelling degree resulted in the largest fiber sizes or the lowest specific surfaces, which caused an obvious decrease in the amount of lipase adsorbed in the fibers. Meanwhile, the stronger ionic attraction between the carboxylate groups of PAA in the hydrogel fibers and the ammonium cations of lipase molecules, compared with the case in pH4 buffer, may also hinder the diffusion of lipase into the hydrogel fibers (*13*). In summary, the amount of lipase adsorbed in the ultra-fine fibrous hydrogel membranes is determined by the extent of swelling as well as the specific surface of the hydrogels, both as a result of the pH of the absorption medium.

As expected, the amount of lipase adsorbed in hydrogel fibers was much higher than that in the cast films with the same polymer composition because of the much larger specific surface of the fibers. For example, when the adsorption was performed in a pH 7 buffer, the amount of lipase immobilized in the cast film (3.5 COOH/OH molar ratio) was only about 58% of that in the hydrogel fibers of the same polymer composition.

Activity of immobilized lipase

The activities of crude lipase and physically adsorbed lipases in the PAA/PVA fibrous hydrogel membrane (3.5 COOH/OH molar ratio) in pH 4 buffer solution were monitored over time (Figure 5). It took about 5min for the crude lipase (Figure 5a), and about one hour for the adsorbed lipases (Figure 5b) to reach their maximum activities, respectively. The highest activity of adsorbed lipases was 5.5 U/mg lipase, which was about 75% of that of crude lipase.

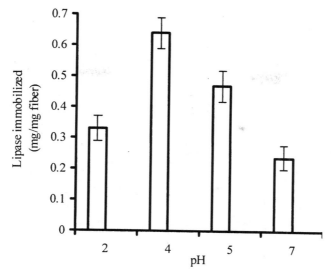

Figure 4. Amount of lipase physically adsorbed in PAA/PVA (COOH/OH molar ratio = 3.5) hydrogel fibers in buffers with different pH values.

    The reuse efficiency of the adsorbed lipase was evaluated by consecutive cycles of assay without rinsing in between (Figure 6). The residual activity of the adsorbed lipase in the second cycle decreased to only about 30% of that in the first cycle, and further decreased to 10% and 5% in the third and fourth cycles. The sharply decreased activity after the first cycle was due to the loss of lipase during the previous assay, which was verified by a separate measurement on lipase release from the hydrogel fibers. In order to imitate the lipase assay process, about 1.0 mg hydrogel fibers (3.5 COOH/OH molar ratio) with lipase (adsorbed previously in pH4 buffer) was immersed in 1.0ml pH 7 buffer solution and incubated in water bath at 30°C for 1hr. Afterwards, the hydrogel fibers were transferred to 1.0ml of fresh pH7 buffer for another release cycle. The amount of lipase released from the hydrogel fibers were measured by Bio-Rad protein assay (Figure 7). The reason that the pH value for the lipase release measurement (pH=7) was different from that for lipase activity assay measurement (pH=8.5) is because the Bio-Rad protein assay is not accurate under basic conditions. As shown in Figure 7, about 65% and 23% of the adsorbed lipase were lost in the first and the second release cycle, respectively. After the second release cycle, the concentration of the lipase released was too low to be measured. The activities of the adsorbed lipase in the first and the second reuse cycles (Figure 6), normalized to the amount of lipase released in the corresponding cycles of the release measurement (Figure 7), were 8.3 and 7.4 U/mg lipase, respectively. These were very close to actitivty of the crude lipase (7.3 U/mg lipase). It indicated that the hydrolysis of olive oil was catalyzed by

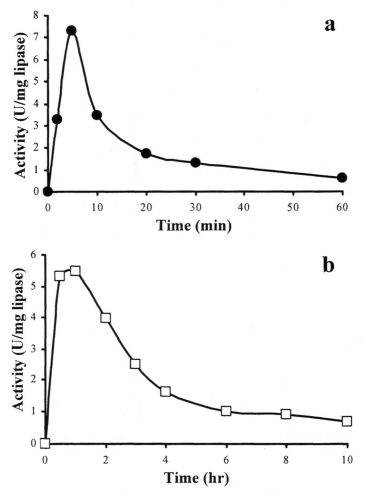

Figure 5. Kinetics of catalytic hydrolysis (30°C, pH 8.5) of olive oil by (a) crude lipase; (b) lipases adsorbed by PAA/PVA hydrogel fibers (COOH/OH molar ratio = 3.; the lipase adsorption was conducted in pH 4 buffer solution).

the lipase released to the assay solution, and the released lipase still kept its original activity.

In order to prevent or slow the release of lipase enzymes from the hydrogel fibers, a coupling reagent, EDC, was used to couple the functional groups of PAA (COOH) and PVA (OH) in the hydrogel fibers with those having active hydrogen from lipase molecules, such as COOH, OH, and NH_2. The activities of lipase immobilized on PAA/PVA hydrogel fibers (COOH/OH molar ratio = 3.5)

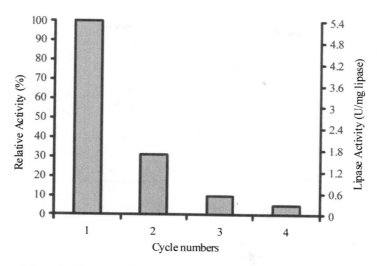

Figure 6. Reuse efficiency of lipase physically adsorbed in PAA/PVA hydrogel fibers (COOH/OH molar ratio = 3.5). Relative activity is defined as the ratio of activity over the first-use activity.

by such covalent bonding mechanism were very different depending on the immobilization method. The activity of lipase immobilized by method I was only about 0.5 U/mg lipase, and no activity could be detected in subsequent cycle. In this method, the hydrogel fibers, lipase and the coupling reagent, EDC, were mixed together in the pH buffer during the immobilization. The crosslinking reaction was expected to happen mainly between lipase molecules since both lipase and EDC were soluble in the buffer solution but the hydrogel fibers existed as a solid. The low activity of the immobilized lipase was thought to be caused by the severe crosslinking, which deactivated the lipase molecules. The lipase immobilized by method II, however, showed the very similar activities and release behavior as those of the physically adsorbed, and thus, the results were not shown here. This suggested that the lipase molecules immobilized in method II were just physically adsorbed on the hydrogel fibers rather than covalently coupled with the fibers. Since both PAA and PVA have the active hydrogen, EDC could be completely consumed during the first activation step in the method II, and could not function as a coupling reagent any more in the second immobilization step.

The lipase immobilized on the hydrogel fibers by method III showed an activity of 2.4 U/mg lipase in the first cycle, which was about 33% of that of the free lipase. The activities then lowered gradually to 1.5, 1.0, and 0.3 U/mg lipase in the subsequent cyclic measurements (Figure 8). The activity of the second cycle was nearly doubled with method III than that of the physically

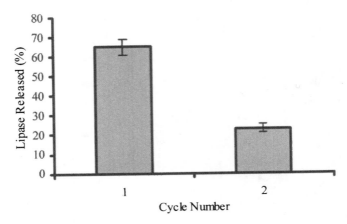

Figure 7. The amount of lipase released from PAA/PVA hydrogel fibers (COOH/OH molar ratio = 3.5) in pH7 buffer.

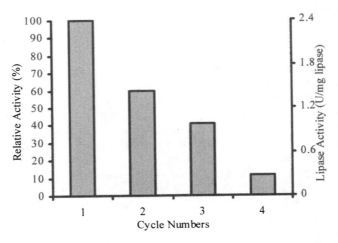

Figure 8. Reuse efficiency of lipase covalently immobilized (Method III) on PAA/PVA hydrogel fibers (COOH/OH molar ratio = 3.5). Relative activity is defined as the ratio of activity over the first-use activity.

adsorbed lipase (Figure 6). This lessened cyclic reduction was attributed to the conformational change to the lipase molecules caused by crosslinking. It is possible that the surface layer of hydrogel fibers was also crosslinked by EDC, which slowed the release of the adsorbed lipase molecules, lessening the reduction in the lipase activity.

Conclusions

A series of ultra-fine hydrogel fibers have been prepared by electrospinning of the aqueous mixtures of PAA and PVA followed by thermal crosslinking to achieve varying degrees of crosslinking and pH-responsive swelling behaviors. Lipase enzyme was bound to these ultra-fine hydrogel fibers by physical adsorption and covalent bonding mechanisms. Physical adsorption of lipase in the hydrogel fibers was significantly influenced by solution pH or fiber swelling during the adsorption process. The low swelling of the hydrogel fibers at pH 2 hindered the diffusion of lipase molecules into the hydrogel fibers and resulted in a low level of bound enzymes. When the swelling was too much, such as at pH 7 buffer, low lipase adsorption was also observed due to the significantly enlarged fiber sizes or lowered specific surface of the hydrogel fibers. As a result, the highest adsorption of lipase (0.65 mg/mg dry fibers) in the hydrogel fibers (3.5 COOH/ OH molar ratio) was achieved at pH 4. Physically adsorbed lipase could be released from the hydrogel fibers during the assay reaction, and exhibited the same activity as the free lipase. The reuse efficiency of the immobilized lipase was improved by covalent bonding using a coupling agent. However, the lipase activity decreased about 67% due to the possible conformation change of lipase molecules from crosslinking.

References

1. Betigeri, S. S.; Neau, S. H. *Biomaterials* **2002**, *23*, 3627-3636.
2. Chauhan, G. S.; Mahajan, S.; Sddiqui, K. M.; Gupta, R. *J. Appl. Polym. Sci.* **2004**, *92*, 3135-3143.
3. Wang, Y.; Hsieh, Y.-L. *J. Polym. Sci. Part A Polym. Chem.* **2004**, *42*, 4289-4299.
4. Xie, J.; Hsieh, Y.-L. *J. Mater. Sci.* **2003**, *38*, 2125-2133.
5. Braun, B.; Klein, E.; Lopez, J. L. *Biotechnol. Bioeng.* **1996**, *51*, 327-341.
6. Ivekovic, D.; Milardovic, S.; Grabaric, B. S. *Biosens. Bioelectron.* **2004**, *20*, 872-878.
7. Chen, H.; Hsieh, Y. L. *Biotechnol. Bioeng.* **2005**, *90*, 405-413.
8. Li, L.; Hsieh, Y.-L. *Polymer* **2005**, *46*, 5133-5139.
9. Li, L.; Hsieh, Y.-L. *Nanotechnology* **2005**, *16*, 2852-2860.
10. Jin, X.; Hsieh, Y. L. *Polymer* **2005**, *46*, 5149-5160.
11. Bradford, M. M. *Anal. Biochem.* **1976**, *72*, 248-54.
12. Schmidt, F. H.; Stork, H.; Von Dahl, K. In *Methods of Enzymatic Analysis* 2nd ed.; Bergmeyer, H. U.; Ed.; Verlag Chemie: Weinheim, 1974; Vol. 2, pp 819.
13. Cleary, J.; Bromberg, L. E.; Magner, E. *Langmuir* **2003**, *19*, 9162-9172.

Chapter 9

Immobilized Porcine Pancreas Lipase for Polymer Synthesis

Feng He

Key Laboratory of Biomedical Polymers of the Ministry of Education,
College of Chemistry and Molecular Sciences, Wuhan University,
Wuhan 430072, People's Republic of China

Porcine Pancreas Lipase (PPL), a common lipase with low cost, could be immobilized on silica particles with good stability and recyclablity. Then the immoblized lipase (IPPL) was employed as the catalyst for polymer synthesis, such as polyesters, polycarbonates, polyphosphates, and their copolymers. Here we present a mini-review of some works in our lab within this area.

Introduction

Biodegradable polymers are receiving more and more attentions for their wide application in biomedical uses, such as drug carriers, matrices in tissue engineering, surgical sutures, etc (1).

Sn(II) 2-ethylhexanoate, which has been approved for surgical and pharmacological applications by the FDA, is generally employed as the catalyst for the synthesis of biomedical polymers. However, it has been reported that Sn(II) 2-ethylhexanoate cannot be removed by a purification process such as the dissolution/precipitation method, thus the residual Sn may be concentrated within matrix remnants after hydrolytic degradation (2). To avoid the potential harmful effects of metallic residues in biomedical polymer materials, enzymatic polymerization is one of the powerful candidates for polymer synthesis (3). Enzymes, natural kinds of protein without toxicity, have remarkable properties such as high catalytic and high selectivity under mild reaction conditions. Up to now, various kinds of biodegradable polymers have been synthesized by enzymatic polymerization, such as polyesters (4,5), polycarbonates (6,7), polyphosphates (8) and their copolymers. Among them, enzymatic ring-opening polymerization has received much attention as a new methodology of biodegradable polymer synthesis for lactones, cyclic carbonates and other cyclic monomers (9-12).

However, to our knowledge, most previous studies of enzyme-catalyzed polymerizations have avoided temperatures > 90 oC, which is likely due to thermal deactivation of enzyme catalyst (13-15). It has been found that enzyme immobilization can improve the stability and recyclablity of native enzyme (16). Silica particles, activated by methanesulfonic acid, are effective and economic inorganic carriers for enzyme immobilization (17). Herein, we present a mini-review of our works about immobilized porcine pancareas lipase on silica particles (IPPL) for polymer synthesis, such as polycarbonates, polyesters, polyphosphates and their copolymers.

IPPL For Polycarbonates Synthesis

Aliphatic polycarbonates are a class of surface erosion biodegradable polymers attracting great interests due to their good biocompatibility, favorable mechanical properties and low toxicity (18,19). The polymerization of aliphatic cyclic carbonates such as trimethylene carbonate (TMC) have been extensively studied (20,21).

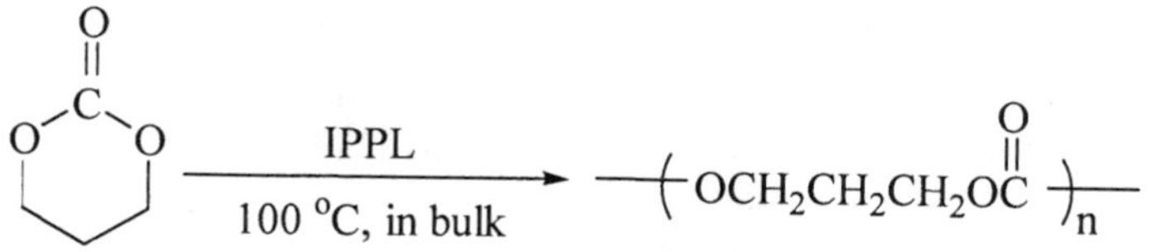

Figure 1. Ring-opening polymerization of TMC catalyzed by IPPL.

In our study (6), porcine pancreas lipase (PPL) immobilized on silica particles (narrow distributed micron particles) was employed for ring-opening polymerization of TMC. No evidence of decarboxylation occurring during the polymerization. The results showed that silica microparticles improved immobilization efficiency much more. The most preferable polymerization temperature of TMC was 100 °C during 24h polymerization. The M_n of the resulting polymers was significantly increased compared with that catalyzed by

PPL while the yield had no marked change. Furthermore, the recovered IPPL could be repeatedly used for many times. It is very interesting that the catalytic activity of recovered IPPL would increase and tend to keep constant after repeated uses. The highest M_n of PTMC 87400 was obtained at around 0.1 wt% of the seventh recovered IPPL. The excellent recyclablity of IPPL is very helpful to its further industry applications.

Table 1. TMC polymerization catalyzed by recycled IPPL at 100 °C for 24h.

Recycled Time	IPPL Conc.(wt%)	M_n	M_w/M_n	Yield(%)
1[a]	1.0	14700	1.92	70
2[b]	1.0	25400	1.77	75
3	1.0	30800	2.11	81
4	1.0	45200	2.01	78
5	1.0	61700	1.94	83
6	1.0	63000	2.53	81
7	1.0	64000	2.20	72
7[c]	0.1	87400	2.06	69

[a] IPPL used for the first time.
[b] Recycled IPPL used for the second time. The rest is the same as this method.
[c] Carried out by 0.1wt% recycled IPPL which was the seventh used.

IPPL For Polyesters Synthesis

In recent years, the enzymatic synthesis of biodegradable polyesters was focused on the polycondensation method (22,23). Among the very few successful example of enzymatic ring-opening polymerization for polyesters synthesis, Novozyme-435 (immobilized lipase B from *Candida antartica*) has been proved an effective catalyst for polycaprolactone (PCL) synthesis in toluene (24). Considering the low cost and high recyclablity of IPPL, we also

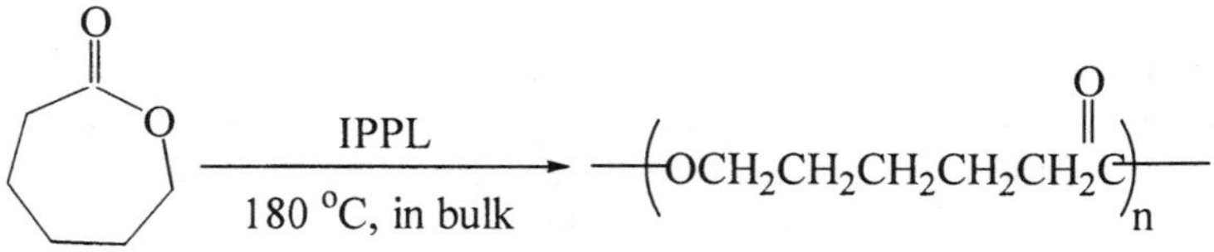

Figure 2. Ring-opening polymerization of ε-CL catalyzed by IPPL.

employed the IPPL for ring-opening polymerization of ε-caprolactone (CL) with/without solvent.

However, compared with Novozym-435, IPPL presented very low catalytic activity for PCL synthesis in toluene, and almost no PCL could be obtained. Thus, the IPPL-catalyzed polymerization of CL was carried out without solvent (in bulk). The results showed that M_n of resulting PCL was significantly increased compared with that catalyzed by native PPL. Higher temperature and longer reaction time both contributed to gain PCL with higher molecular weight, while the yield had almost no change (25). In addition, for evaluating the recyclablity of IPPL for the polymerization of CL, the most severe reaction conditions (180 °C, 240 h) were adopted in the recycling experiments. It was found that the recovered IPPL could be used again with compatible high catalytic activity. The highest M_n of 21300 of PCL could be obtained at 5.18 wt% of the reused IPPL at 180 °C for 240 h.

IPPL For Polyphosphates Synthesis

Polyphosphates are one of the most promising biodegradable materials for their biocompatibility, low toxicity and biodegradability in biomedical use. They have the similar structure to nucleic acid and teichoic acids which are major components of cell walls, particularly in some bacteria and responsible for a number of biological functions. Polyphosphates and their copolymers have been used for gene therapy, tissue engineering and controlled drug delivery (26-30). In our lab, IPPL was also employed successfully for ring-opening polymerization of cyclic phosphate (ethylene isobutyl phosphate, EIBP) (8). A good coupling yield of IPPL was achieved to 41.88 mg of native lipase/g of silica particles. After incubation in water at 90 °C for 1h, the activity of IPPL can retain above 70%. The optimum polymerization temperature was 70 °C. The enzymatic ring-opening polymerization was achieved in bulk with M_n values ranging from 1600 to 5800 g/mol. It was found recovered IPPL worked more actively for the polymerization of EIBP and M_n was significantly increased.

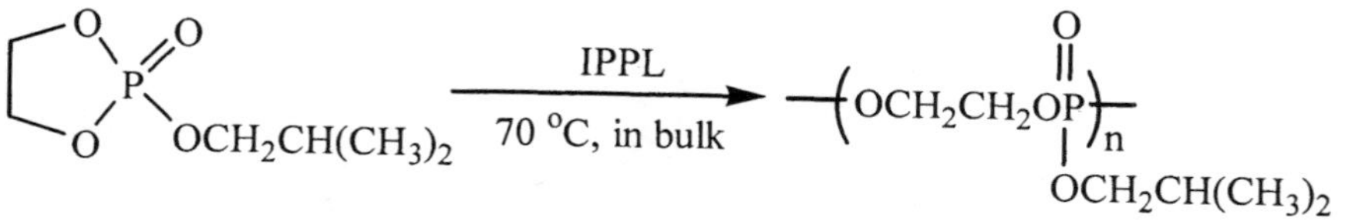

Figure 3. Ring-opening polymerization of EIBP catalyzed by IPPL.

IPPL For Copolymers Synthesis

Although various kinds of biodegradable polymers have been studied widely and then used in biomedical field, there are also many properties of these materials would be improved to fit various applications. One useful strategy for modifying the properties of biodegradable polymers is copolymerization (*31*). Another effective method is to introduce pendant functional groups to polymer materials (*32*). Though few studies on enzymatic copolymerization have been reported, it would be likely to attract more and more attentions in the future, due to its combined advantages of copolymerization and enzymatic polymerization. Herein we introduced some results on the enzymatic ring-opening copolymerization in our lab.

IPPL-catalyzed ring-opening copolymerization of BTMC with DTC

IPPL with different size were employed for ring-opening copolymerization of 5-benzyloxy-trimethylene carbonate (BTMC) with 5,5-dimethyl-trimethylene carbonate (DTC) in bulk (*33*).

Table 2. Synthesis of poly(BTMC-*co*-DTC) catalyzed by IPPL with different size[a]

IPPL[b]	Conc.(wt%)	Yield%	M_n^c	M_w/M_n^c
IE-1	0.1	82	13600	2.05
IE-1	0.2	80	11400	2.88
IE-1	0.5	82	6500	2.36
IE-1	1.0	86	5900	2.75
IE-2	0.1	83	26400	1.74
IE-2	0.2	77	16500	1.84
IE-2	0.5	82	10000	2.22
IE-2	1.0	83	6900	2.66
IE-3	0.1	94	19600	1.80
IE-3	0.2	92	15300	3.16
IE-3	0.5	73	5500	2.75
IE-3	2.0	39	4700	2.28

[a] Carried out in bulk at 150 °C for 24h with equal feed molar ratio of BTMC/DTC.
[b] Carrier sizes of IPPL: 150-250 μm (IE-1), 75-150μm (IE-2) and 1μm (IE-3).
[c] Determined using GPC.
SOURCE: Reproduced with permission from reference 33. Copyright 2003 Elsevier.

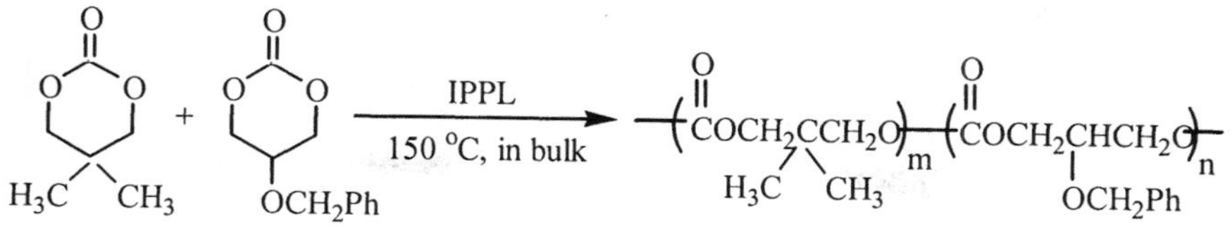

Figure 4. Ring-opening copolymerization of BTMC with DTC catalyzed by IPPL with different size

Three kinds of silica particles with different sizes (150-250 μm, 75-150 μm and 1 μm) were selected as carriers for PPL immobilization to study the relationships between the carrier size of immobilized enzyme, the catalytic activity for ring-opening copolymerization and the polymer yield. The highest M_n of 26400 of poly(BTMC-co-DTC) was obtained at around 0.1 wt% of IPPL with size of 75-150 μm. Moreover, the M_n of poly(BTMC-co-DTC) decreased rapidly with the increasing of IPPL concentration.

It is important that the results of enzymatic activity assay could not reflect the real catalytic capability in the copolymerization reaction. The coupling yields were equivalent to 26.55 mg of native PPL/1g of 150-250 μm silica particles (IE-1), 37.34 mg of native PPL/1g of 75-150 μm silica particles (IE-2), and 53.20 mg of native PPL/1g of 1 μm silica particles (IE-3). However, in the range of selected experimental conditions, the IPPL with medium size (75-150μm, IE-2) exhibited the highest catalytic activity for the copolymerization of BTMC with DTC. The reasons maybe include three aspects: (I) the different substrate used in the enzymatic activity assay (olive oil) and in the enzymatic ring-opening copolymerization (beginning with BTMC/DTC and subsequently with the chain end of copolymers); (II) the different reaction media used in the enzymatic activity assay (phosphate buffer solution) and in the enzymatic ring-opening copolymerization (in bulk); (III) two different monomers used in the copolymerization reaction. It is well know that enzyme has many characteristic features, such as regio-, stereo-, and chemoselectivities. Enzyme should exhibit different catalytic activity to different monomer, such as BTMC and TMC, which would lead to unusual results in the copolymerization reaction.

IPPL-catalyzed ring-opening copolymerization of BTMC with DON

Poly(1,4-dioxan-2-one) (PDON), a well-known aliphatic polyester with outstanding biocompatibility, bioabsorbability and good flexibility (*34*), has received the approval of the FDA to be used as suture material in gynecology (*35*). However, Haruo Nishida (*36*) reported that the M_n of PDON was 1800 g/mol with PPL as the catalyst. In our studies (*37*), only some powders could be obtained (about 6.5 %) with IPPL as the catalyst, which can't be dissolved in common solvents. On the other hand, because of the poor solubility of PDON with high molecular weight obtained by other catalyst, the copolymers containing DON contents have been synthesized to meet different needs (*38,39*).

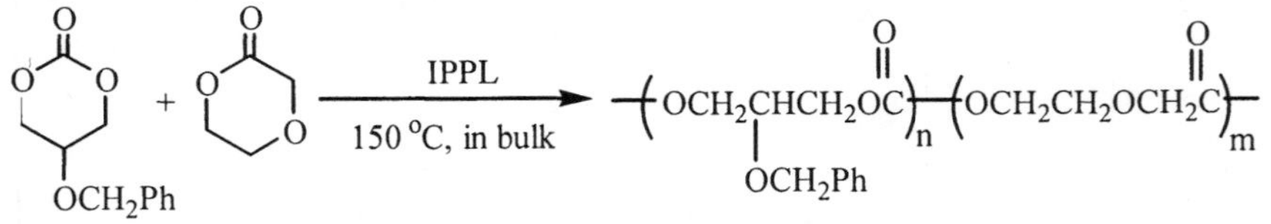

Figure 5. Ring-opening Copolymerization of BTMC with DON catalyzed by IPPL

In our studies, a series of poly(BTMC-*co*-DON) copolymers with different compositions were successfully synthesized in bulk at 150 °C for 24h by ring-opening copolymerization using 0.45 wt% IPPL as the catalyst. DON content was introduced to adjust the flexibility and degradation rate of aliphatic polycarbonates, while BTMC content was employed to improve the solubility of PDON and to introduce functional side group for further modifications.

Table 3. Synthesis of poly(BTMC-*co*-DON) with different compositions

Entry[a]	BTMC:DON (feed ratio)[b]	BTMC:DON (composition)[c]	Yield (%)	M_n	M_w/M_n[d]
1	100: 0	100: 0	84	83100	1.51
2	79:21	86:14	80	63400	1.85
3	64:36	78:22	64	22300	1.84
4	50:50	54:46	80	20200	1.82
5	33:67	40:60	51	9400	1.71
6	20:80	29:71	42	5600	1.52

[a] Carried out in bulk at 150 °C for 24 h using 0.45 wt% IPPL as the catalyst.
[b] Monomer feed ratio in mol/mol.
[c] Copolymer molar composition measured by ^{1}H NMR.
[d] Determined by GPC.

The resulting copolymers were characterized by ^{1}H NMR, ^{13}C NMR and GPC. For all the copolymers, GPC chromatograms showed symmetric and narrow molecular weight distributions. There was no peak in the zone of low molecular weights, thus indicating the absence of residual BTMC or DON monomer.

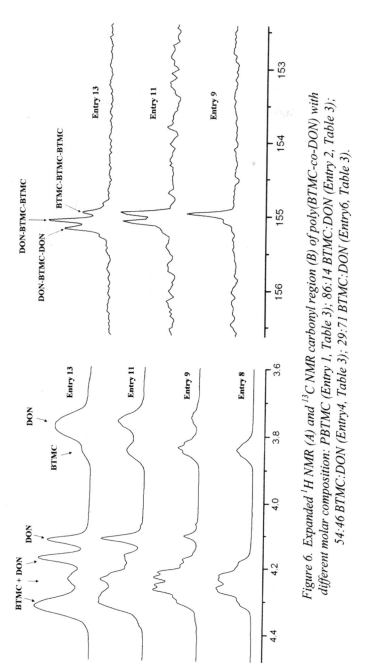

Figure 6. Expanded ^1H NMR (A) and ^{13}C NMR carbonyl region (B) of poly(BTMC-co-DON) with different molar composition: PBTMC (Entry 1, Table 3); 86:14 BTMC:DON (Entry 2, Table 3); 54:46 BTMC:DON (Entry4, Table 3); 29:71 BTMC:DON (Entry6, Table 3).

In the ^{1}H NMR spectra of the copolymers, signals from both the BTMC units and DON units can be clearly observed. The signal intensity at 3.74-3.78, 4.12 and 4.2-4.3 ppm increased along with DON contents in the copolymers (shown in Figure 6 (A)). The expanded BTMC carbonyl group region (152-157 ppm) of the ^{13}C NMR spectrum is shown in Figure 6 (B). With the BTMC:DON compositions of 54:46 and 86:14, the BTMC carbonyl resonance separates into two peaks. However, changing the composition to 29:71 in favor of DON, the BTMC carbonyl resonance separates into three peaks. It could be due to the lower polymerability of DON. Only the copolymers with richer DON contents can lead to the triad sequences. Therefore the signal at 155.19, 155.09 and 154.97 ppm are assigned to the DON-BTMC-DON, DON-BTMC-BTMC and BTMC-BTMC-BTMC respectively. The ^{13}C NMR spectra suggested that the enzymatic copolymerizations of BTMC with DON catalyzed by IPPL resulted in the formation of random copolymers.

On the other hand, the BTMC monomer had higher reactivity in comparison with the DON monomer, which leading to higher BTMC contents in the copolymers than that in feed. With the increase of BTMC molar feed ratio from 20% to 79%, the M_n of the resulting copolymers increased from 5600 to 63400 g/mol. The hydrophilicity of copolymers increased along with the DON contents. *In vitro* degradation rate and drug release rate of poly(BTMC-*co*-DON) (ibuprofen as the model drug) also increased along with the DON contents and can be tailored by adjusting the copolymer compositions. It is expected to obtain a kind of new biodegradable copolymer with improved degradation property and to further apply in drug delivery system.

Conclusions

Immoblized Porcine Pancreas Lipase on silica particles (IPPL) have been successfully employed as the catalyst for biodegradable polymer synthesis. IPPL showed good thermal stability and could be recycled for many times with high catalytic activity. Under the optimum reaction conditions, the targeted polymers could be obtained with high molecular weight and good product yeild. The obvious benefits of IPPL, such as good processbility and low cost, will greatly improve it as a powerful candidate for push enzymatic polymerization methods from the laboratory to industry.

Acknowledgements

We are grateful for the following financial support: National Natural Science Foundation of China (No. 20674056, 20104005), Hubei Province Natural Science Foundation of China (No. 2001B053), Wuhan Chen Guang

Youth Project of China (No. 20015005041), and a grant from National Key Fundamental Research Program of China (No. 2005CB623903).

References

1. Uhrich, K. E.; Cannizzaro, S. M.; Langer, R. S.; Shakesheff, K. M. *Chem. Rev.* **1999**, *99*, 3181.
2. Schwach, G.; Coudance, J.; Engel, R.; Vert, M. *Polym. Bull* **1996**, *37*, 771.
3. Gross, R. A.; Kumar, A.; Kalra, B. *Chem. Rev.* **2001**, *101*, 2097.
4. Varma, I. K.; Albertsson, A. C.; Rajkhowa, R.; Srivastava, R. K. *Prog. Polym. Sci.* **2005**, *30*, 949.
5. Sivalingam, G..; Madras, G. *Biomacromolecules* **2004**, *5*, 603.
6. Feng, J.; He, F.; Zhuo, R. X. *Macromolecules* **2002**, *35*, 7175.
7. Bisht, K. S.; Svirkin, Y. Y.; Henderson, L. A.; Gross, R. A. *Macromolecules* **1997**, *30*, 7735.
8. He, F.; Zhuo, R. X.; Liu, L. J.; Jin, D. B.; Feng, J.; Wang, X. L. *React. Funct. Polym.* **2001**, *47*, 153.
9. Sabino, M.A.; Ronca, G.; Müller, A. J. *J. Mater. Sci.* **2000**, *35*, 5071.
10. Nishida, H; Yamashita, M; Hattori, N; Endo, T. *Polym. Degrad. Stab.* **2000**, *70*, 485.
11. Bhattarai, N.; Bhattarai, S. R.; Yi, H. K.; Lee, J. C.; Khil, M. S.; Hwang, P. H.; Kim, H. Y. *Pharm. Res.* **2003**, *20*, 2021.
12. Raquez, J. M.; Degee, P.; Narayan, R.; Dubois, P. *Macromol. Rapid Commun.* **2000**, *21*, 1063.
13. Mei, Y.; Kumar, A.; Gross, R. A. *Macromolecules* **2002**, *35*, 5444.
14. Kobayashi, S.; Kikuchi, H.; Uyama, H. *Macromol. Rapid. Commun.* **1997**, *18*, 575.
15. Al-Azemi, T. F.; Harmon, J. P.; Bisht, K. S. *Biomacromolecules* **2000**, *1*, 493.
16. Matsumura, S.; Tsukada, K.; Toshima, K. *Macromolecules* **1997**, *30*, 3122.4.
17. Liu, L. J.; Zhuo, R. X.; *Ion Exchange and Adsorption* **1995**, *1*, 541-544.
18. Zhu, K. J.; Hendren, R. W.; Jensen, K.; Pitt, C. G. *Macromolecules* **1991**, *24*, 1736.
19. Rokicki, G. *Prog. Polym. Sci.* **2000**, *25*, 259.
20. Kricheldorf, H. R.; Jenssen, J. *J. Macromol. Sci., Pure Appl. Chem.* **1989**, *A26(4)*, 631.
21. Keul, H.; Bacher, R.; Hocker, H. *Makromol.Chem.* **1990**, *191*, 2579.
22. Kobayashi, S.; Ohmae, M. *Adv. Polym. Sci.* **2006**, *194*, 159-210.
23. Uyama, H.; Kobayashi, S. *J. Mol. Catal. B: Enzym.* **2002**, *19*, 117-127.
24. Kumar, A.; Gross, R. A. *Biomacromolecules* **2000**, *1*, 133.

154

25. Wang, Y. X.; He, F.; Li, F.; Feng, J.; Zhuo, R. X. *Chem. J Chinese Universities* **2006**, *27*, 982-984.
26. Wang, S.; Mao, H. Q.; Leong, K. W. *Biomaterials* **2001**, *22*, 1147-1156.
27. Wang, J.; Mao, H. Q.; Leong, K. W. *J. Am. Chem. Soc.* **2001**, *123*, 9480-9481.
28. Xu, X. Y; Yu, H.; Gao, S. J.; et al. *Biomaterials* **2002**, *23*: 3765-3772.
29. Zhao, Z.; Wang, J.; Mao, H.Q.; Leong, K. W. *Adv. Drug Deliv. Rev* **2003**, *55*, 483-499.
30. Troev, K.; Tsatcheva, I.; Koseva, N.; Georgieva, R.; Gitsov I. *J. Polym.Sci. Part A: Polym. Chem.* **2007**, *45*, 1349.
31. Zhou, Y.; Zhuo, R. X.; Liu, Z. L.; Xu, D. *Polymer* **2005**, *46*, 5752.
32. Wang, X. L.; Zhuo, R. X.; Liu, L. J.; He, F., Liu, G. *J. Polym. Sci. Part A: Polym. Chem.* **2002**, *40*, 70.
33. He, F.; Wang, Y. X.; Feng, J.; Zhuo, R. X.; Wang, X. L. *Polymer* **2003**, *44*, 3215.
34. Yang, K. K.; Wang, X. L.; Wang, Y. Z. *J Macromol Sci Polym Rev* **2002**, *42*, 373.
35. Sabino, M. A.; Ronca, G.; Müller, A. J. *J. Mater. Sci.* **2000**, *35*, 5071.
36. Nishida, H.; Yamashita, M.; Nagashima, M.; Endo, T.; Tokiwa, Y. *J. Polym. Sci. Part A: Polym. Chem.* **2000**, *38*, 1560.
37. He, F.; Jia, H. L.; Liu, G.; Wang, X. L.; Feng, J.; Zhuo, R. X. *Biomacromolecules* **2006**, *7*, 2269-2273.
38. Bhattarai, N.; Bhattarai, S. R.; Yi, H. K.; Lee, J. C.; Khil, M. S.; Hwang, P. H.; Kim, H. Y. *Pharm. Res.* **2003**, *20*, 2021.
39. Raquez, J. M.; Degee, P.; Narayan, R.; Dubois, P. *Macromol. Rapid Commun.* **2000**, *21*, 1063.

Chapter 10

Effects of Porous Polyacrylic Resin Parameters on *Candida antarctica* Lipase B Adsorption, Distribution, and Polyester Synthesis Activity

Bo Chen[1], Elizabeth M. Miller[2], Lisa Miller[3], John J. Maikner[2], and Richard A. Gross[1,*]

[1]NSF I/UCRC for Biocatalysis and Bioprocessing of Macromolecules, Polytechnic University, Six Metrotech Center, Brooklyn, NY 11201
[2]Rohm and Haas Company, P.O. Box 904, Spring House, PA 19477
[3]National Synchrotron Light Source, Brookhaven National Laboratory, Upton, NY 11973

Polyacrylic resins were employed to study how immobilization resin particle size influences *Candida antarctica* Lipase B (CALB) loading, fraction of active sites, and catalytic properties for polyester synthesis. CALB adsorbed more rapidly on smaller beads. Saturation occurred in less than 30 seconds and 48 h for beads with diameters 35 and 560-710 μm, respectively. Infrared microspectroscopy showed that CALB forms protein loading fronts for resins with particle sizes 560-710 and 120 μm while CALB appears evenly distributed throughout 35 μm resins. The fraction of active CALB molecules adsorbed onto resins not influenced by particle size was less than 50 %. At about 5% w/w CALB loading, decrease in the immobilization support diameter from 560-710 to 120, 75 and 35 μm increased conversion of ε-CL to polyester (20 to 36, 42 and 61%, respectively, at 80 min). Similar trends were observed for condensation poly-merizations of 1,8-octanediol and adipic acid.

Introduction

Adsorption is a simple and straightforward route for biocatalyst immobilization which offers unique advantages over soluble enzymes, such as enhanced activity, increased selectivity, improved stability and reusability. For example, Assemblase, the commercial name of immobilized pencillin-G acylase from *E. coli*, has been used by industry for manufacture of the semi-synthetic β-lactam antibiotic cephalexin.[1,2]

Candida antarctica Lipase B (CALB) is attracting increasing attention as a biocatalyst for the synthesis of low molar mass and polymeric molecules.[3-6] Almost all publications on immobilized CALB use the commercially available catalyst Novozym® 435, which consists of CALB physically adsorbed onto a macroporous acrylic polymer resin (Lewatit VP OC 1600, Bayer). Primarily, commercial uses of CALB are limited to production of high-priced specialty chemicals[7-9] because of the high cost of commercially available CALB preparations: Novozym 435 (Novozymes A/S) and Chirazyme (Roche Molecular Biochemicals). Studies to better correlate enzyme activity to support parameters will lead to improved catalysts that have acceptable price-performance characteristics for an expanded range of industrial processes.

Most research on immobilization has focused on choice of matrix materials and optimization of immobilization conditions,[10-21] such as hydrophobicity of the support surface[17-19] and pH[20,21] of the enzyme solution Physical properties of supports also showed significant influence on enzyme loading and catalytic behavior. For example, loading and the specific activity of penicillin-G acylase increased with decreased particle size.[22] Vertegel and Dordick reported that, relative to larger particles, adsorption of lysozyme onto nanoparticles resulted in less loss of lysozyme α-helicity and higher catalytic activity.[23] CALB, adsorbed onto mesoporous silica, functionalized by octyltriethoxysilane, showed high enzyme loading (200 mg protein/g of silica) and catalytic activity for acylation of ethanolamine with lauric acid.[24] CALB, adsorbed on octyl-agarose, was highly enantioselective (E=25) for hydrolysis of the R-isomer of chiral R,S-mandelic acid esters whereas, the non-immobilized enzyme showed much lower enantioselectivity (E<2).[25] Our laboratory have found that higher activity for polyester synthesis was observed for CALB-matrix systems that had: *i*) increased density of CALB molecules within matrix pores, and *ii*) CALB distributed throughout the matrix.[26]

The distribution of enzyme in carriers is also critical to understanding catalyst activity. Jeroen van Roon, et al. quantitatively determined intraparticle enzyme distribution of Assemblase using light microscopy.[27] They reported the enzyme was distributed heterogeneously and its concentration was radius-dependent. Our laboratory reported the use of infrared microspectroscopy for analysis of protein spatial distribution within macroporous polymer supports.[28] With a synchrotron beam light source, spatial resolutions down to 5 by 5 μm were achieved.[26,28]

Relative to small molecules, the effects of mass transfer on conversion and catalytic constants will be greater for macromolecular substrates. Since diffusion of substrates and products to and from the catalyst will be largely determined by the size of macroporous particles, this variable is particularly important when assessing catalytic supports for enzyme-catalyzed polymer synthesis and modification reactions. To our knowledge, no systematic studies have been reported on how size of macroporous resins influences enzyme activity.

The effect of matrix particle size on CALB adsorption, catalytic activity, the fraction of active CALB and enzyme distribution has been studied that will provide a basis to others that seek to design optimal immobilized enzyme catalysts for low molar mass and polymerization reactions.

Results and Discussions

CALB Adsorption on Particles

The shape of an adsorption isotherm provides information on the rate of adsorption as well as the maximum equilibrium loading of the adsorbed substance. Figure 1 shows that CALB adsorption onto MMA beads was strongly dependent on bead size. Adsorption occurred more rapidly for smaller beads. All resins showed typical saturation behavior and the saturation time increased as bead particle size increased. Saturation occurred in less than 30 seconds for sample #4 (Amberchrom™CG-71S, 35 μm) compared to 48 h for sample #1 (Amberchrom™CG-71S, 560-710 μm). The dependence of adsorption rate on particle size is due to the pore size that is limiting protein transport to the inside of particles. In other words, the small size of pores slows protein diffusion into beads so that smaller beads more rapidly were saturated in protein. Adsorption yields for all bead sizes studied were about 55% when the initial enzyme and bead concentrations used for immobilization were 1 mg/mL and 10 mg/mL, respectively. As long as sufficient time is given to reach the maximum equilibrium loading value, total enzyme loading is not controlled by the particle size. Thus, similar numbers of adsorption sites are available within resins which correspond to identical values of surface area through the broad range of bead particle sizes studied. In a previous study by Jeroen L. van Roon et al,[22] loading of penicillin-G acylase (Assemblase) onto MMA beads was inversely related to with bead size. In light of the results herein, lower loading for larger size beads may be explained by insufficient incubation time so that saturation was not reached.

Table 1. Candida antartica Lipase B (CALB) immobilization on MMA resins of differing particle size: enzyme loading, fraction of active lipase, and catalytic activity.

resin #	sample name	pore diameter (Å)	particle size (μm)	surface area (m^2/g)	enzyme loading (% w/w)	enzyme loading (mg/m^2)	f (%) [a]
1	AmberliteTM XAD-7HP	250	560-710	500	5.1	0.102	40.4
2	AmberchromM CG-71C	250	120	500	5.2	0.104	44.2
3	AmberchromM CG-71M	250	75	500	5.4	0.106	45.0
4	AmberchromT MCG-71S	250	35	500	5.7	0.114	43.0

[a] fraction active lipase
[b] kinetic constant

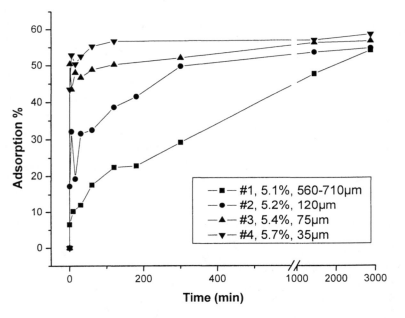

Figure 1. Adsorption isotherms of CALB onto resins that differ in particle size. (Reproduced from Langmuir ***2007***, 23, 1381–1387. Copyright 2007 American Chemical Society.)

CALB Distribution within Immobilized Support

Figure 2 displays infrared images of protein distribution for CALB on MMA resins that differ in particles size. CALB forms a protein loading front for sample #1, Amberlite™ XAD-7HP, and sample #2, Amberlite™ XAD-71C, with particle sizes 560-710 and 120 μm respectively. The thickness of protein loading fronts for these beads with particle sizes 560-710 and 120 μm is 100 μm and 40 μm, respectively. Thus, approximately 50 % and 30 %, respectively, of the internal bead volumes do not contain CALB. CALB penetrates to the inside of sample #3, Amberchrom™CG-71M (75 μm). For sample #4, the infrared image shows that CALB is distributed throughout this polymer resin. Previously, we reported that narrow polydispersity polystyrene with M_n 46 kD readily diffuses throughout Novozyme 435,[28] which consists of CALB physically immobilized within a macroporous resin of PMMA (Lewatit VP OC 1600, average particle and pore size 315-1000 and 150 μm, respectively). Thus, for larger beads where CALB forms a protein loading front, substrates and products diffuse in and out of bead regions that lack the immobilized enzyme which leads to non-optimal catalyst performance.

Active Lipase Fraction

MNPHP is a well-known irreversible inhibitor of lipases that is highly specific for reaction with active-site serine residues. Thus, MNPHP was selected as the inhibitor to determine, by titration, the fraction of catalytic sites that are accessible and active. Since we are concerned with CALB activity in organic media, inhibition was studied in heptane. LC-MS was used to determine the release of p-nitrophenol (pNP) which corresponds with accessible active sites. To ensure that adsorption of pNP by resins was taken into consideration, pNP concentration was corrected as follows. A fixed quantity of enzyme-free resin was incubated overnight in acetonitrile with different concentrations of pNP. Standard curves of pNP adsorption by each resin as a function of pNP concentration were constructed from LC-MS measurements. MNPHP-inhibited immobilized enzymes were used for ε-caprolactone ring-opening polymerizations in toluene (70 °C). No conversion of monomer was observed in 30 minutes. Hence, MNPHP titration resulted in complete inhibition of CALB activity.

By inspecting fraction of active site values in Table 1 we conclude that CALB adsorption onto resins #1 to #4 results in >50% deactivation of immobilized CALB molecules. This may be due to enzyme denaturation and/or inaccessibility of active sites upon CALB adsorption to MMA support surfaces. At enzyme-loadings of 5 to 6%, the fraction of active sites for immobilized CALB ranged from 40 to 45 %. Thus, the fraction of active CALB molecules

160

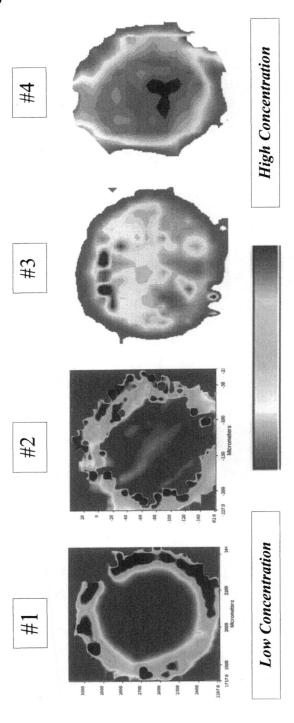

Figure 2. Infrared microspectroscopy images to analyze CALB distribution within a series of MMA resins that differ in particle size (#1 is Amberlite[TM] XAD-7HP, 560–710 µm diameter; #2 is Amberlite[TM] XAD-71C, 120 µm diameter; #3 is Amberchrom[TM]CG-71M, 75 µm diameter; #4 is Amberchrom[TM] CG-71S, 35 µm diameter). (Reproduced from Langmuir 2007, 23, 1381–1387. Copyright 2007 American Chemical Society.) (See page 1 of color insert.)

was independent of resin size for supports that ranged from about 600 to 35 μm. This fraction of active immobilized lipase molecules is very close to results determined by titration with MNPHP of *Candida antarctica* lipase A (CALA) and *Humicola Lanuginosa* lipase (HLL) immobilized on Accurel in heptane.[29]

Immobilized CALB activity

The activity of CALB immobilized on resins #1 to #4 (see Table 1) was assessed by ε-CL ring-opening polymerizations. All immobilized resins have similar CALB loading (5.1 to 5.7 %-by-wt) and water content (1.4 to 1.9%). As shown previously[30], different chemical shifts are observed for methylene protons (-OCH_2-) of ε-CL monomer, PCL internal repeat units and chain terminal -CH_2-OH moities. Hence, by *in situ* NMR monitoring, monomer conversion and polymer number average molecular weight (M_n) were determined.

As Figure 3 illustrates, the polymerization rate strongly depends on the matrix particle size used for CALB immobilization. For example, at ~80 minutes reaction time, as the particle size decreased from 560-710, 120, 75 and 35 μm, ε-CL %-conversion increased from 20 to 36, 42 and 61%, respectively.

Conclusions

Candida antarctica lipase B (CALB) was immobilized by physical adsorption onto cross-linked poly(methyl methacrylate) resins with high yield. Immobilized CALB displayed high activity and similar mechanism in chain propagation towards ring-opening polymerization of ε-CL in toluene. Decrease of the resin diameter did not influence the fraction of active CALB molecules. However, by decreasing resin diameter at nearly constant protein loading, increases in: *i*) CALB adsorption onto resins, *ii*) rate of ε-CL conversion in ring-opening polymerization, and *iii*) growth of poly(octanyl-adipate) chains were observed. These results were explained by decrease in diffusional constraints with smaller diameter resins. A nonuniform distribution with most enzyme present in the outer region of particles was found by IR microspectroscopy with 560-710 and 120 μm diameter resins. In contrast, as the resin particle size was decreased, the protein distribution became increasingly uniform throughout resins.

These results showed the benefits of systematic investigations of immobilization parameters to achieve enhanced enzyme-catalyst activities. However, much work remains to understand on a molecular level how, for example, changes in enzyme loading influences the fraction of active lipase. Details of enzyme conformation and orientation as function of surface chemistry and morphology on model systems taken together with better control of surface

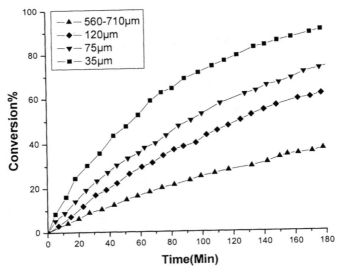

Figure 3. Effect of immobilized CALB particle size (given in µm, see legend box in plot) on the time course of ε-caprolactone ring-opening polymerizations performed at 70 °C in toluene. (Reproduced from Langmuir *2007, 23, 1381–1387. Copyright 2007 American Chemical Society.)*

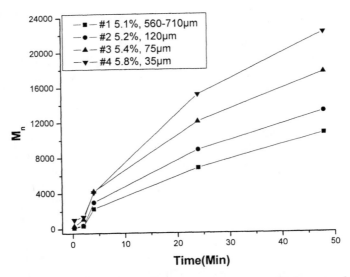

Figure 4. Immobilized CALB catalysis of adipic acid/1,8-octanediol condensation polymerization at 90 °C in bulk. (Reproduced from Langmuir *2007, 23, 1381–1387. Copyright 2007 American Chemical Society.)*

chemistry and morphology when designing macroporous resins will lead to important catalyst improvement.

Acknowledgement

The authors thank the NSF and Industrial members (BASF, Novozymes, Johnson & Johnson, Rohm and Haas, Genencor, Estée Lauder, DNA 2.0, W.R. Grace, Grain Processing Corporation and DeGussa) of the NSF-Industry/University Cooperative Research Center (NSF I/UCRC) for Biocatalysis and Bioprocessing of Macromolecules at Polytechnic University for their financial support, intellectual input and encouragement during the course of this research.

References

1. Schroën, C. G. P. H.; Nierstrasz, V. A.; Moody, H. M.; Hoogschagen, M. J.; Kroon, P. J.; Bosma, R.; Beeftink, H. H.; Janssen, A. E. M.; Tramper, J. *Biotechnol. Bioeng.* **2001**, *73*, 171-178.
2. Schroën, C. G. P. H.; Nierstrasz, V. A.; Kroon, P. J.; Bosma, R.; Janssen, A. E. M.; Beeftink, H. H.; Tramper, J. *Enzyme Microb. Technol.* **1999**, *24*, 498-506.
3. Gross RA and Kalra B. *Science* **2002**, *297*, 803-806.
4. Gross, R. A.; Kumar, A.; Kalra, B. *Chem. Rev.* **2001**, *101*, 2097-2124.
5. Kobayashi S and Uyama H, Kimura S. *Chem. Rev.* **2001**, *101*, 3793.
6. Cheng HN and Gross RA, eds., *Polymer Biocatalysis and Biomaterials.* ACS Symp Ser 900, 2005
7. Kirk O. and Christensen M.W. *Organic Process Research & Development* **2002**, *6*, 446-451.
8. Anderson E.M.; Larsson K.M. and Kirk O *Biocatalysis and Biotrans-formation* **1998**, *16*, 181-204.
9. Alain Houde; Ali Kademi and Danielle, *Applied Biochemistry and Biotechnology* **2004**, *118*, 155-170.
10. Dyal, A.; Loos, K.; Noto, M.; Chang, S. W.; Spagnoli, C.; Shafi, K. V. P. M.; Ulman, A.; Cowman, M.; Gross, R. A. *J. Am. Chem. Soc.* **2003**, *125*, 1684-1685.
11. Dessouki, A. M.; Atia, K. S. *Biomacromolecules* **2002**, *3*, 432-437
12. Maury, S.; Buisson, P.; Pierre, A. C. *Langmuir*, **2001**, *17*, 6443-6446.
13. Soellner, M. B.; Dickson, K. A.; Nilsson, B. L.; Raines, R. T. *J. Am. Chem. Soc.* **2003**, *125*, 11790-11791.
14. Duracher, D.; Elaissari, A.; Mallet, F.; Pichot, C. *Langmuir,* **2000**, *16*, 9002-9008.

15. Lei, C.; Shin, Y.; Liu, J.; Ackerman, E. J. *J. Am. Chem. Soc.* **2002**, *124*, 11242-11243.

16. Gill, I.; Pastor, E.; Ballesteros, A. *J. Am. Chem. Soc,* **1999**, *121*, 9487-9496.

17. Bastida, Agatha; Sabuquillo, Pilar; Armisen, Pilar; Fernandez-Lafuente, Roberto; Huguet, Joan; Guisan, Jose M. *Biotechnology and Bioengineering* 1998, 58, 486-493.

18. George B. Sigal, Milan Mrksich, and George M. *J. Am. Chem. Soc.* **1998**, *120*, 3464-3473.

19. Koutsopoulos, S.; van der Oost, J.; Norde, W. *Langmuir* **2004**, *20*, 6401-6406.

20. Xu, K.; Klibanov, A. M. *J. Am. Chem. Soc.* **1996**, *118*, 9815-9819.

21. Pancera, S. M.; Gliemann, H.,Schimmel, T.; Petri, D. F. S. *J. Phys. Chem. B.* **2006**, *110*, 2674-2680.

22. Jeroen L. van Roon; Michiel Joerink; Marinus P. W. M. Rijkers; Johannes Tramper; Catharina G. P. H. Schroe1n; and Hendrik H. Beeftink *Biotechnol. Prog.* **2003**, *19*, 1510-1518.

23. Vertegel, Alexey A.; Siegel, Richard W.; Dordick, Jonathan S. *Langmuir.* **2004**, *20*, 6800-6807.

24. Blanco, Rosa M.; Terreros, Pilar; Fernandez-Perez, Monica; Otero, Cristina; Diaz-Gonzalez, Guadalupe *Journal of Molecular Catalysis B: Enzymatic* **2004**, *30*, 83-93.

25. Fernandez-Lorente, G.; Fernandez-Lafuente, R.; Palomo, J.M.; Mateo, C.; Bastida, A.; Coca, J.; Haramboure, T.; Hernandez-Justiz, O. et. al. *Journal of Molecular Catalysis B: Enzymatic* **2001**, *11*, 649-656.

26. T. Nakaoki; Y. Mei; L.M. Miller; A. Kumar B. Kalra; M.E. Miller; O. Kirk; M. Christensen and R.A. Gross *Industrial Biotechnology* **2005**, *1*, 126-134.

27. Jeroen van Roon; Rik Beeftink; Karin Schroën and Hans Tramper *Current Opinion in Biotechnology* **2002**, *13*, 398-405.

28. Mei, Y.; Miller, L.; Gao, W.; Gross, R. *Biomacromolecules* **2003**, *4*, 70-74.

29. Rotticci, Didier; Norin, Torbjorn; Hult, Karl; Martinelle, Mats *Biochimica et Biophysica Acta (BBA)/Molecular and Cell Biology of Lipids* **2000**, *1483*, 132-140.

30. Henderson, L. A.; Svirkin, Y. Y.; Gross, R. A.; Kaplan, D. L.; Swift, G. *Macromolecules* **1996**, *29*, 7759-7766.

Chapter 11

Immobilization of *Candida antarctica* Lipase B on Porous Polystyrene Resins: Protein Distribution and Activity

Bo Chen[1], M. Elizabeth Miller[2], and Richard A. Gross[1,*]

[1]NSF I/UCRC for Biocatalysis and Bioprocessing of Macromolecules, Polytechnic University, Six Metrotech Center, Brooklyn, NY 11201
[2]Rohm and Haas Company, P.O. Box 904, Spring House, PA 19477

We studied the effects of polystyrene resin particle size and pore diameter on *Candida antarctica* Lipase B (CALB). CALB adsorbs rapidly on polystyrene (saturation time ≤ 4 min) for particle sizes ≤ 120 μm (pore size 300 Å). Through infrared microspectroscopy, CALB was shown to form protein loading fronts regardless of resin particle size at similar enzyme loadings (~ 8%). From IR images, fractions of total surface area available to enzyme are 21, 33, 35, 37 and 88%, respectively, for particle sizes 350-600, 120, 75, 35 μm (pore size 300 Å) and 35 μm (pore size 1000 Å). Titration with *p*-nitrophenyl n-hexyl-phosphate (MNPHP) showed the fraction of active CALB molecules adsorbed onto resins was about 60 %. The fraction of active CALB molecules was invariable as a function of resin particle and pore size. At about 8% w/w CALB loading, by increasing the immobilization support pore diameter from 300 to 1000 Å, turnover frequency (TOF) of ε-CL to polyester increased from 12.4 to 28.2 S^{-1}. However, the ε-CL conversion rate was not influenced by changes in resin particle size. Similar trends were observed for condensation polymerizations between 1,8-octanediol and adipic acid.

Introduction

Candida antarctica lipase B (CALB) exhibits a high degree of selectivity over a broad range of substrates[1-4] and is attracting increasing attention as a biocatalyst for chemical synthesis. It is well known that enzyme immobilization can improve biocatalyst performance. For example, enzyme immobilization can improve enzyme thermal stability, activity and recyclability.[5] Hydrophobic binding of lipases by adsorption has proven successful due to its affinity for water/oil interfaces. Furthermore, if the lipase has poor solubility in substrate-solvent reaction systems, little leakage (i.e. desorption) of lipases in non-aqueous media may be attained[6]. Many literature reports describe the high utility of immobilized-CALB for chemical transformations of low molar mass compounds.[3,7] More recently, immobilized CALB has been shown to efficiently catalyze lactone ring-opening and condensation polymerizations.[8-13] Many of these publications[1,8-13] describing CALB-catalyzed syntheses of low-molar mass and polymeric molecules use commercially available immobilized CALB marketed by Novozymes as Novozym® 435. This catalyst consists of CALB physically immobilized onto a macroporous acrylic polymer resin (Lewatit VP OC 1600 from Bayer).

Both the nature of protein-surface interactions and inherent properties of a specific enzyme will contribute to the catalytic activity of an immobilized biocatalyst. Adsorption of an enzyme onto a surface can induce conformational changes which affect the rate and specificity of the catalyst[14]. The total amount of enzyme loading, enzyme distribution within the immobilization support, and microenvironment surrounding the supported enzyme can all influence enzyme-catalyst activity, specificity and stability.[9,10,15,16]

Immobilization research has largely focused on matrix selection and optimizing immobilization conditions.[17-28] For example, work has addressed support surface hydrophobicity[24-25] and enzyme solution pH.[27,28] Influence of support physical parameters has also been studied. For example, by decreasing the particle size of a macroporous carrier, both loading and specific activity of penicillin-G acylase increased.[29] Vertegel and Dordick reported that, relative to larger particles, adsorption of lysozyme onto nanoparticles resulted in less loss of lysozyme α-helicity and higher catalytic activity.[30] CALB, adsorbed onto octyltriethoxysilane functionalized mesoporous silica, gave high enzyme loading (200 mg protein/g of silica) and activity for acylation of lauric acid with ethanolamine.[31] Excellent catalytic properties of this system was attributed to the support's high porosity and hydrophobic character. Our laboratory physically adsorbed CALB onto commercially available macroporous supports that differed in surface composition, hydrophobicity, pore diameter and surface area. CALB-matrix systems with relatively higher activity for polyester synthesis had: *i*)

CALB distributed throughout the matrix, and *ii*) increased density of CALB molecules within matrix pores.[32]

Previous studies generally did not consider enzyme distribution within carriers. However, this information can be of great value when seeking to better understand differences in enzyme activity observed by changing immobilization matrix parameters. For example, knowledge that an enzyme is primarily located within outer regions of an immobilization support instead of being uniformly distributed throughout the matrix enables meaningful calculations of enzyme density along matrix surfaces as well as improved predictive models of substrate/product diffusion to and from immobilized protein. Jeroen van Roon et al.[33] quantitatively determined the intraparticle distribution of Assemblase using light microscopy. They sectioned the matrix and penicillin-G acylase was labeled with specific antibodies. The enzyme was heterogeneously distributed and its concentration was radius-dependent. Our laboratory used infrared microspectroscopy to analyze protein spatial distribution within macroporous polymer supports.[34] By using a synchrotron beam light source, spatial resolution to 5 by 5 μm was achieved.[32,34] IR absorbance bands corresponding to CALB and the polymer matrix distinguish these two components and were used to generate semi-quantitative maps of protein distribution throughout the matrix.

Macromolecular substrates provide significant challenges relative to small molecules due to their high solution viscosities that cause diffusion constraints limiting catalytic rates. The particle and pore sizes of macroporous enzyme supports will largely determine diffusion rates of substrates and products to and from catalysts. This variable is particularly important when assessing catalytic supports for enzyme-catalyzed polymer synthesis and modification reactions. To our knowledge, systematic studies have not been reported on how macroporous resin particle and pore size influences enzyme activity.

Recently, our laboratory immobilized CALB on a series of methyl methacrylate resins with identical average pore diameter (250 Å) and surface area (500 m^2/g) but varied particle size (35 to 560-710 μm).[35] CALB adsorbed more rapidly onto smaller beads. Infrared microspectroscopy revealed CALB forms protein loading fronts for resins with particle sizes 560-710 and 120 μm. In contrast, CALB appeared evenly distributed throughout 35 μm resins. Titration with *p*-nitrophenyl n-hexylphosphate (MNPHP) showed the fraction of active CALB molecules adsorbed onto resins was <50 %. By increasing loading of CALB from 0.9 to 5.7 % (w/w) onto 35 μm methyl methacrylate beads, an increase in the fraction of active CALB molecules from 30 to 43% was observed. Furthermore, by decreasing the immobilization support diameter, a regular increase in ε-caprolactone (ε-CL) conversion to polyester resulted. Similar trends were observed for condensation polymerizations between 1,8-octanediol and adipic acid.

A series of close-to-spherical styrene/DVB resins of varying particle size and pore diameter were employed as supports for non-covalent adsorptive attachment of CALB by hydrophobic interaction. The effect of matrix particle and pore size on CALB: i) adsorption isotherms, ii) fraction of active sites, iii) distribution within supports, and iv) catalytic activity for ε–CL ring-opening polymerizations and adipic acid/1,8-octanediol polycondensations is reported. Important differences in the above for CALB immobilized on methyl methacrylate and styrene/DVB resins were found. The lessons learned herein provide a basis to others that seek to design optimal immobilized enzyme catalysts for low molar mass and polymerization reactions.

Results and Discussions

Lipase adsorption

Styrenic resins with systematically varied physical properties were employed as supports for CALB immobilization (Table 1) and rates of CALB adsorption were measured (Figure 1). For polystyrene resin samples 2-4, each with 300 Å pore size, the saturation time for CALB adsorption was ≤ 4 min making it difficult to compare their adsorption rate. Resins 4 and 5, both with 35 μm diameter beads, differ in pore size (300 to 1000 Å) and, therefore, resin surface area (700 to 200 $m^2\ g^{-1}$). Despite these differences, resins 4 and 5 showed similarly rapid CALB adsorption. Only when the polystyrene resin particle size was increased from 120 μm to 350-600 μm (pore size 400 Å) did the absorption rate show a large decrease with saturation at about 24 h. In contrast, with methyl methacrylate resins of similar pore (250 Å) and particle sizes (35 to 120 μm),[35] the rate of CALB adsorption decreased to large extents as the resin particle size was increased. Furthermore, the loading saturation time for the 120 μm methyl methacrylate resin was about 300 min as opposed to ≤ 4 min. The enhanced adsorption rate of polystyrene resins relative to methyl methacrylate resins with similar physical parameters is attributed to stronger hydrophobic interactions between styrenic surfaces and CALB.

Active lipase fraction

Methyl p-Nitrophenyl n-hexylphosphate (MNPHP) was selected as the inhibitor to determine, by titration, the fraction of catalytic sites that are accessible and active. Since we are concerned with CALB activity in organic media, inhibition was studied in heptane. LC-MS was used to determine the release of p-nitrophenol (pNP) which corresponds with accessible active sites. Further details on the method used are described in the Experimental and

Table 1. Matrix parameters and loading of *Candida antartica* Lipase B (CALB) on the styrenic beads.

Resin	Samples	Average Pore size (Å)	Particle diameter (μm)	Surface area (m²/g)	Enzyme Loading (wt%)[a]	Enzyme Loading (mg/m²)	Efficiency (%)[b]
1	Amberlite™ XAD 1180	400	350-600	500	7.9	0.160	87
2	Amberchrom™CG 300C	300	120	700	8.7	0.124	95
3	Amberchrom™CG 300M	300	75	700	8.4	0.120	92
4	Amberchrom™CG 300S	300	35	700	8.4	0.120	92
5	Amberchrom™CG 1000S	1000	35	200	8.2	0.420	91

[a] The enzyme loading was determined as the weight of CAL-B units on the carriers divided by the total weight of immobilized enzyme including CAL-B units and carriers.
[b] Efficiency was determined as the ratio of the percent of CAL-B added on the carriers to the total enzyme used during immobilization.

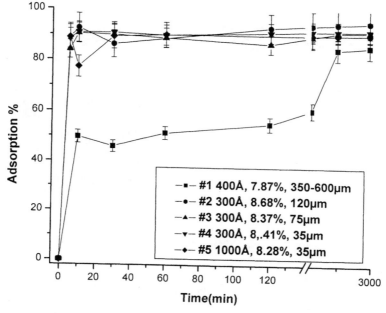

Figure 1. The isotherm of CAL-B adsorption as a function of time on different resins. Sample numbers correspond to entries in Table 1.

Table 2. *Candida antartica* Lipase B (CALB) immobilization on styrene resins of differing particle size and pore size: enzyme loading, fraction of active lipase, and catalytic activity.

Resin #	Protein Loading (wt%)	Water Content (wt%)	Reaction constant (±6%)	Active Site Titration: Fraction of active lipase (%)	TOF (s⁻¹)*
1	7.9	1.7	0.0090	57.7±4.7	10.9
2	8.7	1.7	0.0095	61.3±5.4	14.2
3	8.4	1.2	0.0090	64.2±4.3	12.4
4	8.4	1.0	0.0093	62.5±4.5	12.4
5	8.3	1.2	0.0225	63.2±5.0	28.2

* defined as the number of ε-caprolactone molecules reacting per active site per second during the course of 30 minutes in ring-opening polymerizations performed at 70 °C in toluene.

SOURCE: Reproduced from *Langmuir* **2007**, *23*, 6470. Copyright 2007 American Chemical Society.

Reference 34. MNPHP-inhibited immobilized enzymes were used for ε-caprolactone ring-opening polymerizations in toluene (70 °C). No conversion of monomer was observed in 30 minutes. Hence, MNPHP titration resulted in complete inhibition of CALB activity.

At enzyme-loadings of 7.9 to 8.7%, the fraction of active sites for immobilized CALB ranged from 57.7 to 64.2%. Thus, the fraction of active CALB molecules is independent of resin size for supports ranging from about 600 to 35 μm. Also, increase in the pore size of 35 μm supports from 300 to 1000 Å (resins 4 and 5, respectively) at similar enzyme loadings (~8 %) had no effect on the fraction of active CALB molecules. Compared to the fraction of active CALB molecules (40 to 45%) when CALB was immobilized on polymethyl methacrylate resins with similar particle and pore size values,[35] styrenic resins not only have high affinity for CALB adsorption but also provide CALB with a surface environment that enables CALB to orient and take on conformations that retain a high degree of catalyst active site reactivity.

Enzyme distribution by Infrared (IR) microspectroscopy.

Figure 2 displays IR images of protein distribution for CALB on styrene/DVB resins 1 to 5 at protein loadings given in Table 1. IR images show the formation of protein loading fronts that vary in thickness as a function of resin physical parameters. The formation of a protein loading front was also observed by IR microspectroscopy for CALB physically immobilized on Lewatit

171

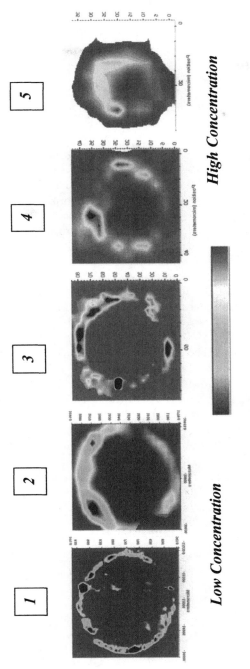

Figure 2. The distribution of CAL-B as function of matrix variables by IR microscopy. Synchrotron light source was only used on the images of 3(75 µm), 4(35 µm) and 5(35 µm). The number of samples is in corresponding to the number of entries in Table 1. (Reproduced from Langmuir 2007, 23, 6467–6474. Copyright 2007 American Chemical Society.) (See page 1 of color insert.)

(i.e. Novozym 435).[32,34] If we assume that CALB homogeneously occupies a spherical zone within resin beads, CALB occupied surface area is proportional to the volume in a spherical zone. The surface area occupied by CALB can be achieved by using eq 1:

$$S= A[1-(d-2b)/d]^3 \qquad (1)$$

S is the surface area occupied by CALB, A is the total surface area of the particles, B is the thickness of particles occupied by CALB, and d is the particle diameter. Consequently, CALB is distributed on surface areas of about 90, 230, 245, 260 and 175 m^2, respectively, This represents, for resins 1 to 5, values of percent total surface area available to enzyme of 21, 33, 35, 37 and 88%, respectively. Thus, at CALB loadings of about 8% for resins 1-4, > 60% of resin surfaces are devoid of catalyst (see Table 1). In contrast, resin 5 is largely occupied by enzyme. Resins 1 to 4 have similar surface areas at which CALB is found (~250 m^2), loading (~8.5%), enzyme density along surfaces, and %-area of beads at which CALB is found. The relatively greater ability of CALB to diffuse further to interior regions of 35 μm polystyrene beads when the pore size was increased from 300 to 1000 Å (see images 4 and 5, Fig. 2) largely increased the %-area of beads at which CALB is found (37 to 88%).

For comparison, the same analysis as above was performed using previously published data by our laboratory for CALB immobilized on a similar series of polymethyl methacrylate (PMMA) resins with average pore size 250 Å.[35] Decrease in PMMA resin particle size from 560-710 to 120, 75 and 35 μm resulted in large changes in protein distribution. As a consequence, percent accessible area of CALB on PMMA resins with particle size 560-710, 120, 75 and 35 is approximately 51, 81, 95 and 100%, respectively. That is, as the resin size decreased, CALB diffused throughout the bead. Even 120 μm PMMA beads had a large extent of CALB within internal bead regions. Since styrenic resins have high affinity for CALB adsorption from the results of adsorption, the protein is immediately attached by styrenic surface once they contact surface, however, they can proceed further in PMMA resins due to their relatively weak interaction. The accumulated CALB molecules can block others, so CALB tends to be distributed on the shell of styrenic resins while diffuses throughout methyl methacrylate resins. CALB also can reach the center of styrenic resin 5 with a pore size of 1000 Å when enzyme loading was increased to 10.6%. These differences in protein distribution, and their relationship to catalyst activity, will be discussed below.

Immobilized CALB activity

The activity of CALB immobilized on resins #1 to #5 (see Table 1) was assessed by ε-CL ring-opening polymerizations. All immobilized resins have similar CALB loading (7.9 to 8.7 %-by-wt) and water content (1.0 to 1.7%). As shown previously[35], different chemical shifts are observed for methylene protons ($-OCH_2-$) of ε-CL monomer, PCL internal repeat units and chain terminal $-CH_2-$OH moieties. Hence, by *in situ* NMR monitoring, monomer conversion and polymer number average molecular weight (M_n) were determined.

Figure 3 illustrates that the polymerization rate is independent of resin diameter. During 30 min reactions, CALB immobilized on resins 1 to 4 gives turnover frequency (TOF) of ε-CL of about 12 s^{-1}. In contrast, our previous work of CALB immobilized on PMMA resins showed a large dependence of ε-CL %-conversion on resin particle diameter. For example, in 30 minutes reaction time, as the particle size decreased from 560-710, 120, 75 and 35 μm, turnover frequency (TOF) of ε-CL increased from 3.8 to 5.3, 7.5 and 11.2, respectively. However, by increasing the resin pore size from 300 (resin 4) to 1000 Å (resin 5) for 35 μm beads, the TOF reached 28.2 s^{-1}. As discussed above, increase in resin pore diameter also corresponds to an increase in %-area of beads at which CALB is found (37 to 88%).

The activity of CALB immobilized on resins #1 to #5 was also assessed by condensation polymerizations of adipic acid and 1, 8-octanediol (Figure 4). As above, all immobilized resins have similar CALB loading (7.9 to 8.7 %-by-wt). As the particle size decreased from 350-600, 120, 75 and 35 μm at constant pore diameter, plots of M_n versus reaction time were nearly identical. In contrast, CALB immobilized on a similar series of polymethyl methacrylate resins showed large increases in M_n as resin particle size was decreased.[35] However, increasing the pore size of 35 μm beads from 300 to 1000 Å resulted in substantial increases in polyester molecular weight. Large differences in polyester molecular weights were observed at short reaction times. For example, at 4 h, poly(octanyl adipate) M_n for CALB immobilized on Resins 4 and 5 were 2100 and 4700, respectively. As above for ε-CL polymerizations, increased chain growth during poly(octanyl adipate) synthesis as a result of increasing the pore size of 35 μm polystyrene resins from 300 to 1000 Å is explained by increased diffusion of substrates and products to and from immobilized enzyme.

From the above results on CALB activity as a function of particle size for polystyrene and PMMA resins, we believe %-surface area occupied by CALB is a critical factor that can be used to improve immobilized CALB activity. Increased %-accessible surface area will increase the probability of collisions between substrates and CALB. As %-accessible surface area for CALB increased for PMMA resins a corresponding increase in polyester synthesis reaction rates was observed (see above). CALB immobilized on styrenic particles of variable size showed little differences in both %-accessible surface area and polyester synthesis catalyst activity. The potential benefit of decreasing bead particle size is to decrease diffusion constraints that lead to productive collisions between enzyme and substrate. However, for polystyrene resins, as particle size decreased, the percent of resin area in which reactions can occur does not change. In contrast, decreasing PMMA particle size dramatically

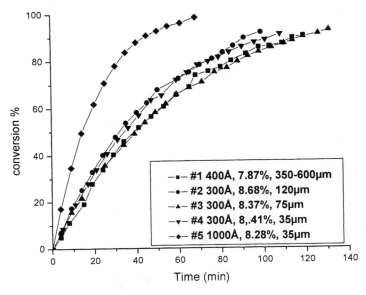

Figure 3. Conversion rate of CALB as a function of matrix variables. The number of samples is in corresponding to the number of entries in Table 1.

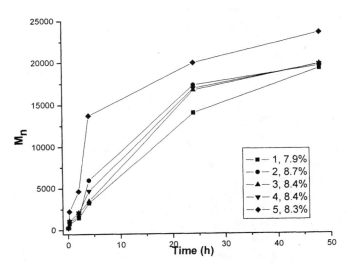

Figure 4. Immobilized CALB catalysis of adipic acid/1,8-octanediol condensation polymerization at 90 °C in bulk. The number of samples is in corresponding to the number of entries in Table 1. (Reproduced from Langmuir ***2007****, 23, 6467–6474. Copyright 2007 American Chemical Society.)*

increased CALB coverage of available resin surfaces. Thus, as PMMA particle size is decreased, substrate and product diffusion in and out of particles increases along with surfaces in which CALB is found. Increased catalyst activity for polystyrene resin 5 relative to 4 was achieved by increasing pore size to 1000 Å. Increasing resin particle surface area caused a large increase in %-resin occupied by CALB. Furthermore, by increase in resin pore size it is expected that limitations caused by substrate-product diffusion into and out-of the resin will be eased. Compared to styrenic resin 4 with with a particle size of 35 μm and a pore size of 300 A, a methyl methacrylate resin with a particle size of 35 μm and 250 A has higher %-accessible surface area but lower activity due to the relatively hydrophilic surface, smaller pore size and lower fraction of active sites of the methyl methacrylate resin. The hydrophilic surface and smaller pore size will limit the diffusion of hydrophobic substrate, ε-CL. The different fraction of active sites can be attributed to the protein orientation influenced by resin surface chemistry, which is currently under investigation.

Conclusions

CALB showed a high affinity for adsorption onto polystyrene resins. Except for the 350-600 μm (pore size 400 Å) resin, the saturation time for CALB adsorption was ≤ 4 min. In contrast, adsorption to methyl methacrylate resins occurred much more slowly.[35] For instance, the loading saturation time for a 120 μm particle size methyl methacrylate resin was about 300 min as opposed to ≤ 4 min for 120 μm polystyrene resin. Enhanced adsorption rate of polystyrene resins relative to methyl methacrylate resins with similar physical parameters is attributed to stronger hydrophobic interactions between styrenic surfaces and CALB. The above results were consistent with those from a Scatchard analyses that gave values for the limiting adsorption of CALB onto polystyrene and polymethyl methacrylate beads.

At enzyme-loadings of 7.9 to 8.7%, the fraction of active sites for immobilized CALB on polystyrene resins ranged from 58 to 64%. The fraction of active CALB molecules is independent of polysytrene resin size for the range of supports studied herein.

In contrast to a previous study with polymethyl methacrylate resins,[35] ε-CL conversion rate was not influenced by changes in resin particle size. We believe that an important contributing factor to this remarkable difference in behavior of immobilized CALB is variation in the distribution of CALB within resins. The fraction of total surface area where CALB is found in resins was determined from IR microspectroscopy generated images. High %-total surface area where CALB is found within resins enables collisions between substrate and CALB leading to product formation. In contrast, resins with large areas devoid of CALB results in substrate diffusion within resin regions where substrate-catalyst

176

collisions cannot occur. For CALB immobilized on polystyrene, catalytic activity was not influenced by particle size since this series of catalysts had similarly small (about 30%) fractions of their surface areas where CALB is found. In contrast, the activity of CALB on PMMA resins increased as particle size decreased which corresponds to large increases in the fraction of resin surface area where CALB was found.

Acknowledgement

The authors thank the NSF and Industrial members (BASF, Novozymes, Johnson & Johnson, Rohm and Haas, Genencor, Estée Lauder, DNA 2.0, W.R. Grace, Grain Processing Corporation and DeGussa) of the NSF-Industry/University Cooperative Research Center (NSF I/UCRC) for Biocatalysis and Bioprocessing of Macromolecules at Polytechnic University for their financial support, intellectual input and encouragement during the course of this research. We are also grateful to Dr. Lisa Miller and colleagues for providing access to the FTIR facilities at the National Synchrotron Light Source (Brookhaven National Laboratory, Upton, New York 11973).

References

1. Gross, R. A.; Kumar, A.; Kalra, B. *Chem. Rev.* **2001**, *101*(7), 2097-2124.
2. Patel, R.N.; Banerjee, A.; Ko, R. Y.; Howell, J. M.; Li, W.S.; Comezoglr, F. T.; Partyka, R. A.; Szarka, F. T. *Biotechnol. Appl. Biochem.* **1994**, *20*, 23-33.
3. Kirk, O.; Christensen, M. W. *Org. Process Res. Dev.* **2002**, *6*(4), 446-451.
4. Kirk, O.; Bjo¨rkling, F.; Godtfredsen, S. E.; Larsen, T. O. *Biocatalysis* **1992**, *6*, 127-134.
5. Kumar, A.; Gross, R. A.; Jendrossek, D. *J. Org. Chem.* **2000**, *65*(23), 7800-7806.
6. Anderson, E. M.; Larsson, K. M.; Kirk, O. *Biocatal. Biotransform.* **1998**, *16*, 181-204.
7. Mahapatro, A.; Kumar, A.; Kalra, B.; Gross, R. A. *Macromolecules* **2004**, *37*(1), 35-40.
8. Sarda, L.; Desnuelle, P. Biochim. *Biophys. Acta* **1958**, *30*, 513-521.
9. Pieterson, W. A.; Vidal, J. C.; Volwerk, J. J.; de Haas, G. H. *Biochemistry* **1974**, *13*, 1455-1460.
10. Svendsen, A.; Clausen, I. G.; Patkar, S. A.; Kim, B.; Thellersen, M. *Methods Enzymol.* **1997**, *284*, 317-340.
11. Hu, J.; Gao, W.; Kulshrestha, A.; Gross, R. A. *Macromolecules* **2006**, *39*, 6789.

12. van der Mee, L.; Helmich, F.; de Bruijn, R.; Vekemans, J. A. J. M.; Palmans, A. R. A.; Meijer, E. W. *Macromolecules* **2006**, *39*, 5021.
13. Peeters, J.; Palmans, A. R. A.; Veld, M.; Scheijen, F.; Heise, A.; Meijer, E. W. *Biomacromolecules* **2004**, *5*, 1862.
14. Roach, P.; Farrar, D.; Perry, C. C. *J. Am. Chem. Soc.* **2005**; *127*, 8168-8173.
15. Wannerberger, K.; Arnebrant, T. *Langmuir* **1997**, *13*(13), 3488-3493
16. Dyal, A.; Loos, K.; Noto, M.; Chang, S. W.; Spagnoli, C.; Shafi, K. V. P. M.; Ulman, A.; Cowman, M.; Gross, R. A. *J. Am. Chem. Soc.* **2003**, *125*, 1684-1685.
17. Dessouki, A. M.; Atia, K. S. *Biomacromolecules* **2002**, *3*, 432-437.
18. Maury, S.; Buisson, P.; Pierre, A. C. *Langmuir*, **2001**, *17*, 6443-6446.
19. Soellner, M. B.; Dickson, K. A.; Nilsson, B. L.; Raines, R. T. *J. Am. Chem. Soc.* **2003**, *125*, 11790-11791.
20. Duracher, D.; Elaissari, A.; Mallet, F.; Pichot, C. *Langmuir,* **2000**, *16*, 9002-9008.
21. Lei, C.; Shin, Y.; Liu, J.; Ackerman, E. J. *J. Am. Chem. Soc.* **2002**, *124*, 11242-11243.
22. Gill, I.; Pastor, E.; Ballesteros, A. *J. Am. Chem. Soc,* **1999**, *121*, 9487-9496.
23. Bastida, A.; Sabuquillo, P.; Armisen, P.; Fernandez-Lafuente, R.; Huguet, J.; Guisan, J. M. *Biotechnology and Bioengineering* **1998**, *58*, 486-493.
24. Sigal, G. B.; Mrksich M; George M.; *J. Am. Chem. Soc.* **1998**, *120*, 3464-3473.
25. Koutsopoulos, S.; van der Oost, J.; Norde, W. *Langmuir* **2004**, *20*, 6401-6406.
26. Xu, K.; Klibanov, A. M. *J. Am. Chem. Soc.* **1996**, *118*, 9815-9819.
27. Pancera, S. M.; Gliemann, H.,Schimmel, T.; Petri, D. F. S. *J. Phys. Chem. B.* **2006**, *110*, 2674-2680.
28. Fernandez-Lorente, G.; Fernandez-Lafuente, R.; Palomo, J.M.; Mateo, C.; Bastida, A.; Coca, J.; Haramboure, T.; Hernandez-Justiz, O. et. al. *Journal of Molecular Catalysis B: Enzymatic* **2001**, *11*, 649-656
29. van Roon, J. L.; Joerink, M.; Rijkers, M. P. W. M.; Tramper, J.; Schroe ln, C. G. P. H.; Beeftink, H. H. *Biotechnol. Prog.* **2003**, *19*, 1510-1518
30. Vertegel, A. A.; Siegel, R. W.; Dordick, J. S. *Langmuir.* **2004**, *20*, 6800-6807
31. Blanco, R. M.; Terreros, P.; Fernandez-Perez, M.; Otero, C.; Diaz-Gonzalez, G. *Journal of Molecular Catalysis B: Enzymatic* **2004**, *30*, 83-93
32. Nakaoki T., Mei, Y.; Miller, L. M.; Kumar A.; Kalra; B.; Miller, M. E.; Kirk, O.; Christensen, M.; Gross, R.A. *Industrial Biotechnology* **2005**, *1*, 126-134,
33. van Roon, J.; Beeftink, R.; Schroën, K.; Tramper, H. *Current Opinion in Biotechnology* **2002**, *13*, 398-405
34. Mei, Y.; Miller, L.; Gao, W.; Gross, R. *Biomacromolecules* **2003**, *4*, 70-74.

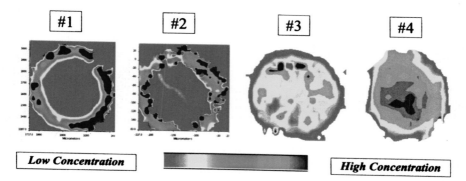

Figure 10.2. Infrared microspectroscopy images to analyze CALB distribution within a series of MMA resins that differ in particle size (#1 is Amberlite^{TM} XAD-7HP, 560-710 μm diameter; #2 is Amberlite^{TM} XAD-71C, 120 μm diameter; #3 is Amberchrom^{TM}CG-71M, 75 μm diameter; #4 is Amberchrom^{TM} CG-71S, 35 μm diameter). (Reproduced from Langmuir **2007***, 23, 1381–1387. Copyright 2007 American Chemical Society.)*

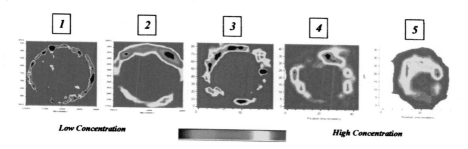

Figure 11.2. The distribution of CAL-B as function of matrix variables by IR microscopy. Synchrotron light source was only used on the images of 3(75 μm), 4(35 μm) and 5(35 μm). The number of samples is in corresponding to the number of entries in Table 1. (Reproduced from Langmuir **2007***, 23, 6467–6474. Copyright 2007 American Chemical Society.)*

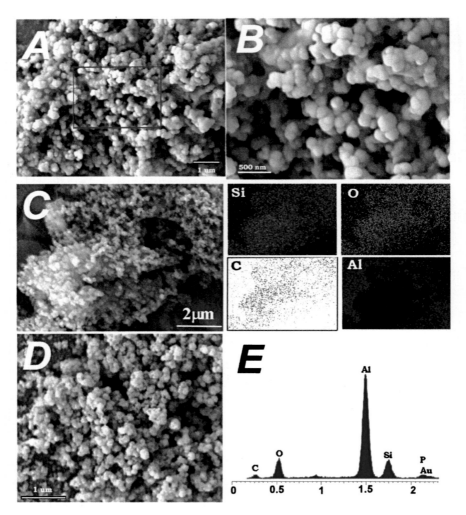

Figure 26.2. (a), (b) SEM micrographs of silica formed using R1. Highlighted area in (a) is presented at higher magnification in (b). (c) Elemental mapping of silica formed by using R1 peptide. (d) SEM micrographs of silica formed using R2 peptide. (e) Typical EDS of silica formed in the presence of R2 peptide. Scale bar = 1 μm (a), (d); 500 nm (b) and 2μm (c).

New Synthetic Approaches

Chapter 12

Biocatalysis on Surfaces: A Microreview

Anil Mahapatro

Center for Biotechnology and Biomedical Sciences and Department of
Chemistry, Norfolk State University, 700 Park Avenue, Norfolk, VA 23504
(email: amahapatro@nsu.edu)

Interfacial reactions on surfaces are becoming increasingly
important due to their widespread applications. Self assembled
monolayers (SAMs) provide excellent platforms to study these
reactions. This review focuses on the various organic reactions
carried out on these SAMs. Challenges and limitations of
organic reactions at surfaces are discussed. Emerging
biocatalytic techniques for carrying out surface reactions are
reviewed.

Introduction

Interfacial reactions are becoming an increasingly important subject for
studies with wide spread applications such as catalysis (*1*), electronics (*2*),
chemical sensing (*3,4*), and many other applications (*5,6*). Understanding the
rules that govern these surface reactions provides important information for
fundamental studies in chemistry and biochemistry (*7,8*). Also the availability of
numerous analytical techniques capable of detecting chemical changes in films
that are few nanometers thick (*9*), have made studies of interfacial reactions a
viable and important area of modern science.

Self-assembled monolayers (SAMs) are perhaps the best model for studying
these interfacial reactions. Figure 1 shows an idealized view and structural model
of a SAM on a metal substrate. SAMs are defined as monomolecular films of a
surfactant formed spontaneously on a substrate upon exposure to a surfactant
solution. Virtually any functional group can be introduced in these monolayers

as a tail group, and this ability to precisely control surface composition makes them an invaluable tool for studying interfacial reactions. Co-adsorption of two different types of SAMs leads to formation of a mixed monolayer, thus enhancing control over surface composition. SAMs can also be prepared on highly curved surfaces, such as colloids, and it is it possible to use conventional analytical techniques for characterization (*10*). Because of the wide use of SAMs in surface science and technologies, this review focuses predominately on results based on SAMs.

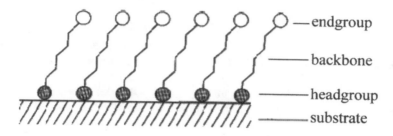

Figure 1. Idealized view and structural model of SAM on metal substrate

Self Assembled Monolayers (SAMs)

Self assembled monolayers (SAMs) have aroused wide spread interest as they provide an opportunity to define the chemical functionality of surfaces with molecular precision, thus creating potential applications related to the control of wettability, biocompatibility and the corrosion resistance of surfaces of a wide range of materials (*11*). Two families of SAMs have received the most attention, SAMs of alkanethiols on gold and alkylsilanes on silicon.

Thiols on gold have been particularly well studied and used as a model system for a variety of applications including biomaterial and biosensor surfaces (*12*). The preparation of SAMs on gold is simple. The clean substrate is immersed in a 1-10 mM solution of the desired alkanethiol at room temperature and after approximately 1 h, the surface is covered with a near perfect monolayer. It is generally believed that the thiol group binds to the gold as a thiolate (*13*), resulting in an extremely densely packed, crystalline monolayer. Detailed information regarding the formation of thiol monolayers on gold are discussed elsewhere (*14*). The choice of the head group thus determines the surface topography, as the underlying substrate becomes completely inaccessible to the molecules in solution. For example, clean gold is naturally hydrophilic, but the formation of SAMs makes it possible to control the contact angle of water on the surface to any value between 0° (-OH and -CO$_2$H groups) and 118° (-CF3 groups) depending on the functional group on the surface (15). The surface energy of a given SAM can also be altered by making a "mixed" SAM with two (or more) components. The problem of bulky head groups is avoided by mixing with less bulky thiols in the feed solution. When the alkanethiols are of equal chain length, the ratio of thiols in the SAM will resemble the ratio in solution (16).

The formation of SAMs of alkylsilanes on silicon or glass is more complex. These monolayers are covalently bound to surface hydroxy groups through Si-O bonds. The molecules used in the formation of such monolayers are either chlorodimethyl long chain alkylsilanes, alkyltrichlorosilanes, or trialkoxy-(alkyl)silanes (17). The alkylchlorosilane derivatives react spontaneously with clean Si/SiO$_2$ or glass, whereas the alkoxysilanes need to be heated, in order to convert the alcohols into leaving groups. The more commonly used alkylchlorosilanes are either deposited from the vapour phase or from solution. These molecules partially hydrolyze in solution, forming oligomers before settling down on the surface into a polymeric network. The hydrolysis of trichlorosilane derivatives can, however, also result in polymer networks "dangling off" the surface (18). Experimental evidence, however, do suggests that long alkylsilanes form very tightly packed monolayers that are only slightly less dense than alkanethiols on gold (14).

Surface Characterization

The products of organic synthesis in solution can easily be purified and subsequently analyzed with rapid, sensitive techniques such as NMR spectroscopy, mass spectrometry, elemental analysis and X-ray spectroscopy. Solid-phase synthesis greatly facilitates the purification of products and has become the backbone of modern combinatorial chemistry, but the characterization of products bound to the solid supports is more difficult. Solid-state NMR spectroscopy is one possible way to monitor a reaction, but in general, the products can only be fully analyzed after cleavage from the support. When working with reactions on monolayers, the problems of monitoring the reaction, determining the products and estimating the yield become quite significant. The extremely small quantities involved render most analytical tools useless, and very often, a combination of techniques is necessary to prove the structure on the surface. Monolayers on gold nanoparticles / colloids (20 nm size range) have been used as models for 2D SAMs (19) and their reactivity studied by NMR spectroscopy in solution. However as these small particles are highly curved, it is not always straightforward to extrapolate yields from nanoparticles to planar surfaces.

One of the advantages of SAMs on smooth, reflective surfaces, is that reactions on these monolayers can be studied by a wide range of techniques including infrared spectroscopy (20), scanning electron microscopy (21), contact angle measurements (22), atomic force microscopy (AFM) (23), surface plasmon resonance (24), ellipsometry (25), low-angle X-ray reflectometry (26), surface acoustic wave and acoustic plate mode devices (27), X-ray photoelectron spectroscopy (25), sum frequency spectroscopy (28), quartz crystal microbalance (29), electrochemical methods (30), confocal and optical microscopy (31), secondary ion mass spectrometry (SIMS) (32) and near-edge X-ray absorption fine structure (NEXAFS) (33). Details of these techniques are discussed elsewhere (14).

In practice, IR spectroscopy, ellipsometry and XPS are the techniques most widely used to study chemical transformations, whereas AFM is particularly useful to image-patterned surfaces. The introduction of fluorescent tags and their detection using (confocal scanning) fluorescence microscopy is widely used to study the attachment of labelled biomolecules to a substrate. The quantitative analysis can be quite difficult, however instead of determining the yield while the molecules are in the monolayer, it is also possible to cleave the products from the solid support and analyze the molecules "off-line". Using very sensitive analytical tools, even the tiny amounts of material cleaved from substrates can be characterized. Butler et al. (34) measured the efficiency of phosphoramidite based oligonucleotide synthesis on surface tension arrays using capillary electrophoresis of cleaved products.

Surface Modification of SAMs

The modification of surface properties through the selection of the appropriate terminal functional group in the monolayer has led to the development of an emerging research field "surface organic chemistry", where the aim is to control the physicochemical properties of man made surfaces by the functionalization of these surfaces, yielding formation of "tailor-made" surfaces. Chemical transformations on SAMs have been studied in detail and provide new mechanistic insights as well as routes to tailored surface properties (11,35,36). Methodologies of surface modification of SAMs focus on two strategies (a) chemical modification and attachment of organic molecules after formation of SAMs (b) attachment prior to assembly, i.e. the desired attachment on alkanethiols are carried separately in solution and the synthesized molecules are then assembled on gold surfaces. This review focuses on chemical modification of SAMs after assembly to a substrate.

Variation of the head group of the monolayer makes it possible to control wettability etc., and also allows the introduction of different chemical moieties with specific properties such as nonspecific binding of proteins to surfaces. For example the introduction of oligoethylene glycol functionality to the end of the alkyl chain results in protein-resistant properties (37). Thus instead of synthesizing different thiols/silanes with different head groups, it is more convenient to use a number of "standard" SAMs and subsequently perform reactions on SAMs to modify the surface chemistry. Performing reactions on

SAMs allows us to tune the properties of surfaces at the molecular level, but due to the nature of SAMs (tightly packed, movements of molecules within monolayers restricted) the choice of reaction is important. One must consider that steric effects are likely to be exacerbated for certain surface reactions, leading to an energy barrier higher than would be expected in solution chemistry. To successfully functionalize a SAM, reaction conditions must not cause destruction of the monolayer or damage the underlying substrate.

Over the last decade, a considerable number of reactions has been studied (11,35): (i) olefins: oxidation (38,39), hydroboration, and halogenation (40); (ii) amines: silylation (41,42), amidation (43), and imine formation (44); (iii) hydroxyl groups: reaction with anhydrides (45), isocyanates (46), epichloro-hydrin and chlorosilanes (47); (iv) carboxylic acids: formation of acid chlorides (48), mixed anhydrides (49) and activated esters (50); (v) carboxylic esters: reduction and hydrolysis (51); (vi) aldehydes: imine formation (52); (vii) epoxides: reactions with amines (53), glycols (54) and carboxyl-terminated polymers (55). A list of all the major classes of reactions on SAMs plus relevant examples are discussed comprehensively elsewhere (11). The following sections will provide a more detailed look at reactions with some of the common functional SAMs, i.e hydroxyl and carboxyl terminated SAMs.

Organic Reactions on SAMs

Reactions of Hydroxy-Terminated SAMs

The reaction between substrate-bound alcohols and fluoroacetic anhydrides (Scheme 1) has been studied and the reaction was estimated to give a 80-90% yield. The reaction was thought not to go to completion because the larger fluorine atoms lead to a sterically hindered environment (56). The reaction was studied in more detail by Leggett et al (45). XPS experiments showed that trifluoroacetic anhydride completely reacted with the hydroxy monolayer, but longer chain anhydrides were found to give only an 80% conversion. Hydroxy-terminated SAMs on gold react with alkyltrichlorosilanes in an analogous manner to the chemisorption of trichlorosilane on hydroxysilicon substrates. This reaction represents the first example of the preparation of double layers using both the trichlorosilane and thiol methods of monolayer formation (47). The reaction between phenyl isocyanate and hydroxyl bearing SAMs has been described by Himmel et al (57). The resulting urethane linkage was obtained in an 87% yield in the condensed phase reaction. This urethane linkage was found to be thermally unstable at temperatures above room temperature, as shown by the strong decrease of the nitrogen signal in XPS. The resulting monolayer bound phenyl groups were found to be ordered with respect to the original monolayer.

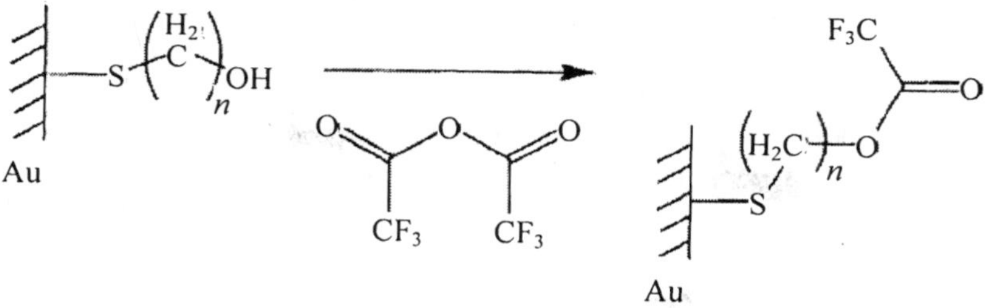

Scheme 1. Reaction of Hydroxyl terminated SAM

Reactions of Carboxylic Acid Terminated Monolayers

Carboxylic acid terminated monolayers self-assembled onto Au substrates have been studied by Leggett and co-authors (*45*). The reaction between a carboxylic acid functionalized SAM and trifluoroethanol in the presence of di-*tert*-butylcarbodiimide, an activator added to make the carboxylic acid monolayer more susceptible towards nucleophilic attack (*50*), was found to proceed slowly with only a 60% rate of conversion after several days. These sluggish reaction rates are in agreement with comparably slow reactions on poly (methacrylic acid) where steric interactions are believed to be responsible for the long reaction times (*58*). The authors concluded that the slow reaction on the monolayer was due to a combination of a) bulky *tert*-butyl groups on the diimide combined with the lack of space within the carboxylic acid SAM preventing attack; b) the sterically hindered nature of backside attack from the approaching alcohol directed towards the carbonyl group, and c) the adsorption of alcohol contaminants due to hydrogen bonding between the carboxylic acid and the ethanol used in preparation of the monolayer.

Terminal carboxyl groups in monolayers can also be activated by treatment with carbodiimides such as dicyclohexylcarbodiimide (DCC) or 1-ethyl-3-(3-dimethylaminopropyl)carbodiimide (EDC) (*59*). Alternatively, conversion to a mixed anhydride can be effected by reaction of a carboxyl-terminated film with ethyl chloroformate (*49*). Exposure of the surface to gaseous $SOCl_2$ has been reported to produce carboxyl chloride groups (*48,60*). These activated acid derivatives then react smoothly with alcohols or amines to form esters or amides (Scheme 2).

Other organic reactions on SAMs

Studies have been carried out to investigate the similarities and differences between chemical reactions in solution (three-dimensional reactions) and interfacial surface chemical reactions (two dimensional reactions). A variety of

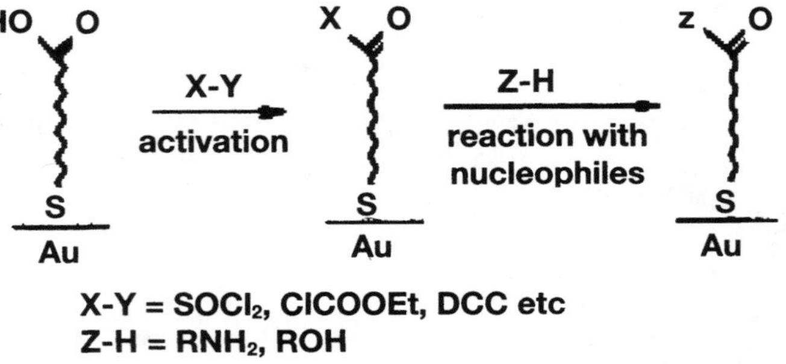

X-Y = SOCl$_2$, ClCOOEt, DCC etc
Z-H = RNH$_2$, ROH

Scheme 2. Reaction of Carboxyl terminated SAM

terminal functional groups and their chemical transformations on SAMs after their assembly have been examined (*44,45,47-52,57,61-69*). These studies have shown that many organic reactions that work well in solution are difficult to apply at surfaces because of steric hindrance. In such a hindered environment, backside reactions (e.g. S_N2 reaction) and reactions with large transition state (e.g., esterification, saponification, Diels-Alder reaction and others) often proceed slowly (*70*). Also, all these methodologies use organic solvents and involve tedious multi step protection deprotection chemistries (*71,72-74,48*). Few gas phase methods have been reported for surface modification of SAMs, however they do not provide the versatility and control over surface chemistry that chemical reaction provide, possibly due to vapor pressure requirements and the absence of catalysts (*75,76*). These studies show that the rules that govern chemical reactions in solution would be different from the rules that govern chemical reactions at interfaces. The intimate study of reactions and interactions within such films and with external reagents is sure to widen our understanding of the molecular behavior of such surfaces: an area that has not received sufficient attention for organic chemists.

Biocatalysis on Surfaces

Relatively very few reports exist on use of biocatalytic methodologies for carrying our surface modification of SAMs. Use of enzymes in organic synthesis (*77*) and polymer science (*78*) is well established and has been discussed elsewhere in comprehensive reviews. The rapidly increasing interest in *in vitro* enzyme-catalyzed organic and polymeric reactions has been due to the fact that several families of enzyme utilize and transform not only their natural substrates but also a wide range of unnatural compounds to yield a variety of useful products. Recent advances in non-aqueous enzymology have significantly expanded the potential conditions under which these reactions can be performed. Use of enzyme for surface modification of SAMs on metal surfaces could offer distinct advantages such as (1) development of methodologies of attaching organic moieties on SAMs after their assembly on metal surfaces, which due to steric hindrance are difficult to achieve via chemical means (2) avoidance of multiple protection/deprotection steps due to their high selectivity for a given organic transformation. (3) possible avoidance of organic solvents by carrying out these reactions in bulk (solvent-less), or aqueous medium (4) use of mild reaction conditions (room temperature to 70°C), thus ensuring structural integrity of the SAMs formed, and (5) reported selectivity of enzyme reactions that may provide spatial and topographical ordering of the surface.

There are numerous reports of hydrolysis of lipid monolayers using different lipases (*79,80*). Relatively few reports demonstrate lipase catalyzed esterification synthesis on air / water monolayers (*81,82*). Specifically Singh, et al., have reported the use of lipase lipozyme for the synthesis of glycerol and fatty acid on stearic acid monolayers (*81*). Singh, et al., have also reported lipase catalyzed esterification of oleic acid with glycerol in monolayers (*82*). Turner, et al. (*83*) have reported the hydrolysis of a phospholipid film which was covalently attached via chemical methods to a silica surface. However the rigid structural ordering of the SAMs on the metal surface offers significant bulk steric hindrance which may not be the case in the flexible lipid and air/water monolayers as reported above.

Enzymatic, surface-initiated polymerizations of aliphatic polyesters was reported for wider clinical use of aliphatic polyesters (*84*). The hydroxyl terminated SAM acted as an initiation site for lipase B catalyzed ROP of aliphatic polyesters, such as poly(ε-caprolactone) and poly (*p*-dioxanone) (Scheme 3). Another example of enzymatic SIP is the polymerization of poly (3-hydroxybutyrate) (PHB), where PHB synthase, fused with a His-tag at the N-terminus, was immobilized onto solid substrates through transition-metal complexes, Ni (II)-NTA, and the immobilized PHB synthase catalyzed the polymerization of 3-R-hydroxybutyryl-coenzyme A (3HB-CoA) to PHB (*85*).

Recently Mahapatro, et al., demonstrated the surface modification of functional self assembled monolayers on 316L stainless steel via lipase (Novozyme-435) catalysis (Scheme 4) (*86*). SAMs of 16-mercaptohexadecanoic acid (-COOH SAM) and 11-mercapto-1-undecanol (-OHSAM) were formed on 316L SS, and lipase catalysis was used to attach therapeutic drugs, perphenazine and ibuprofen, respectively, on these SAMs. The reaction was carried out in toluene at 60 °C for 5 h using Novozyme-435 as the biocatalyst. The FTIR, XPS and contact angle measurements collectively confirmed the biocatalytic surface modification of SAMs.

Biocatalysis could thus provide a viable alternate methodology for surface modification of SAMs with inherent advantages of enzyme catalysis and it also

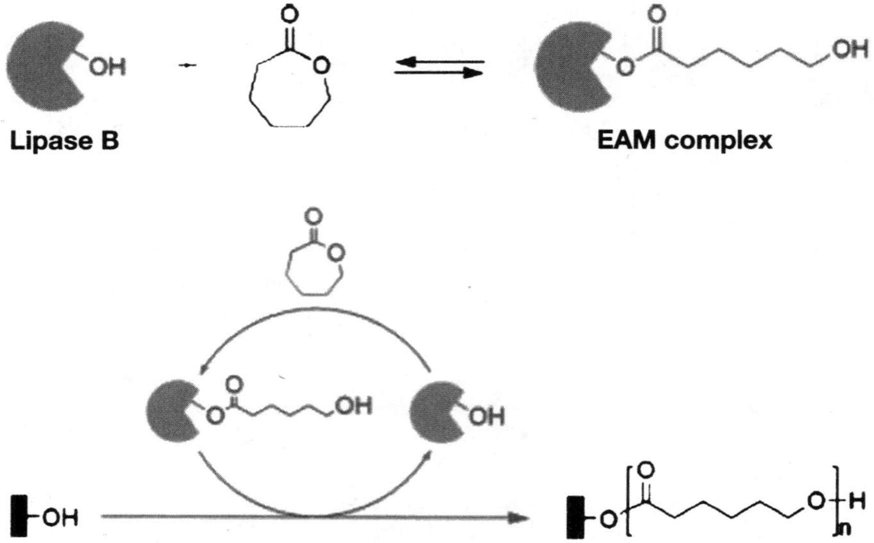

Scheme 3. Enzymatic surface initiated polymerizations

provides the possibility to overcome some of the limitations of organic methodologies mentioned in previous sections. Much work needs to be carried out to evaluate the viability of biocatalytic techniques for use in interfacial reactions such as surface modification of SAMs.

Summary and Outlook

We reviewed some recent findings in the field of biosurface organic chemistry, by focussing on interfacial chemical reactions on self assembled monolayers (SAMs). Although most type of organic reactions can be performed on SAMs, steric hindrance and diffusion barriers can hamper the yield or rate of reactions at the surface. Very limited reports exist on biocatalytic methodologies for surface modification reactions on SAMs. However advantages of mild reaction conditions and other inherent advantages of biocatalytic techniques offers promise in this growing field of biosurface organic chemistry. Also enforced positioning of functional groups in SAMs has great potential for selective rate enhancement and inhibition, which may provide links to better understanding of enzymatic reactions.

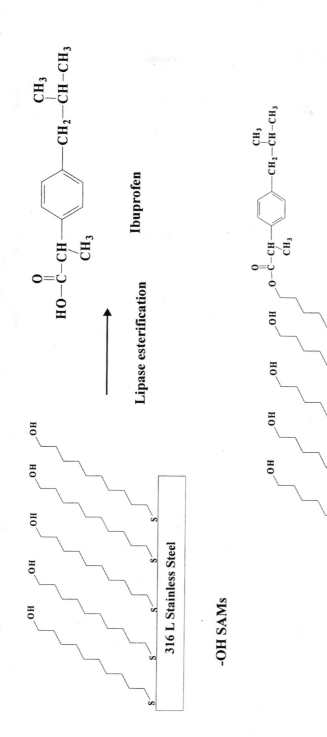

Scheme 4. Enzymatic modification of hydroxyl terminated SAMs

Acknowledgement

The author would like to acknowledge the NIH centre of Biotechnology and Biomedical Science at Norfolk State University for its financial support.

References

1. Clark, J. J.; Macquarrie, D. J. *Chemical Society Reviews* **1996**, *25*, 303-310.
2. Mirkin, C. A.; Ratner, M. A. *Ann. Rev. Phys. Chem.* **1992**, *43*, 719-54.
3. Ricco, A. J.; Crooks, R. M. *Accounts Chem. Res.* **1998**, *31*, 200.
4. Crooks, R. M.; Ricco, A. J. *Accounts Chem. Res.* **1998**, *31*, 219-227.
5. Ulman, A. *Chemical Reviews* **1996**, *96*, 1533-1554.
6. Allara, D. L. *Biosensors & Bioelectronics* **1995**, *10*, 771-83.
7. Mrksich, M. *Current Opinion in Colloid & Interface Sci.* **1997**, *2*, 83-88.
8. Mrksich, M.; Whitesides, G. M. *Annu Rev Biophys Biomol Struct* **1996**, *25*, 55-78.
9. Ulman, A.; Editor *Self-Assembled Monolayers of Thiols. [In: Thin Films (San Diego), 1998; 24]*, 1998.
10. Hostetler, M. J.; Murray, R. W. *Current Opinion in Colloid & Interface Sci.* **1997**, *2*, 42-50.
11. Terena P. Sullivan, W. T. S. H., 2003; Vol. 2003.
12. Tosatti, S.; Michel, R.; Textor, M.; Spencer, N. D. *Langmuir* **2002**, *18*, 3537-3548.
13. Bain, C. D.; Troughton, E. B.; Tao, Y. T.; Evall, J.; Whitesides, G. M.; Nuzzo, R. G. *J. Amer. Chem. Soc.* **1989**, *111*, 321-335.
14. Ulman, A. *An Introduction to Ultra thin Organic Films: from Langmuir-Blodgett to Self-Assembly*; Academic Press: London, 1991.
15. Bain, C. D.; Evall, J.; Whitesides, G. M. *J. Amer. Chem. Soc.* **1989**, *111*, 7155-7164.
16. Tsao, M. W.; Hoffmann, C. L.; Rabolt, J. F.; Johnson, H. E.; Castner, D. G.; Erdelen, C.; Ringsdorf, H. *Langmuir* **1997**, *13*, 4317-4322.
17. Chen, W.; Fadeev, A. Y.; Hsieh, M. C.; Oener, D.; Youngblood, J.; McCarthy, T. J. *Langmuir* **1999**, *15*, 3395-3399.
18. Fadeev, A. Y.; McCarthy, T. J. *Langmuir* **2000**, *16*, 7268-7274.
19. Ingram, R. S.; Hostetler, M. J.; Murray, R. W.; Schaaff, T. G.; Khoury, J.; Whetten, R. L.; Bigioni, T. P.; Guthrie, D. K.; First, P. N. *J. Amer. Chem. Soc.* **1997**, *119*, 9279-9280.
20. Liley, M.; Keller, T. A.; Duschl, C.; Vogel, H. *Langmuir* **1997**, *13*, 4190-4192.
21. Lopez, G. P.; Biebuyck, H. A.; Whitesides, G. M. *Langmuir* **1993**, *9*, 1513-1516.

22. Sigal, G. B.; Mrksich, M.; Whitesides, G. M. *J. Amer. Chem. Soc.* **1998**, *120*, 3464-3473.
23. O'Brien, J. C.; Stickney, J. T.; Porter, M. D. *Langmuir* **2000**, *16*, 9559-9567.
24. Mrksich, M.; Sigal, G. B.; Whitesides, G. M. *Langmuir* **1995**, *11*, 4383-5.
25. Ruan, C. M.; Bayer, T.; Meth, S.; Sukenik, C. N. *Thin Solid Films* **2002**, *419*, 95-104.
26. Wasserman, S. R.; Whitesides, G. M.; Tidswell, I. M.; Ocko, B. M.; Pershan, P. S.; Axe, J. D. *J. Amer. Chem. Soc.* **1989**, *111*, 5852-5861.
27. Dahint, R.; Grunze, M.; Josse, F.; Renken, J. *Analytical Chemistry* **1994**, *66*, 2888-2892.
28. Petralli-Mallow, T. P.; Plant, A. L.; Lewis, M. L.; Hicks, J. M. *Langmuir* **2000**, *16*, 5960-5966.
29. Hook, F.; Rodahl, M.; Kasemo, B.; Brzezinski, P. *Proc. National Acad. Sci. USA* **1998**, *95*, 12271-12276.
30. Ostuni, E.; Yan, L.; Whitesides, G. M. *Colloids and Surfaces, B: Biointerfaces* **1999**, *15*, 3-30.
31. Chen, C. S.; Mrksich, M.; Huang, S.; Whitesides, G. M.; Ingber, D. E. *Science (Washington, D. C.)* **1997**, *276*, 1425-1428.
32. Makohliso, S. A.; Leonard, D.; Giovangrandi, L.; Mathieu, H. J.; Ilegems, M.; Aebischer, P. *Langmuir* **1999**, *15*, 2940-2946.
33. Genzer, J.; Sivaniah, E.; Kramer, E. J.; Wang, J.; Koerner, H.; Char, K.; Ober, C. K.; DeKoven, B. M.; Bubeck, R. A.; Fischer, D. A.; Sambasivan, S. *Langmuir* **2000**, *16*, 1993-1997.
34. Butler, J. H.; Cronin, M.; Anderson, K. M.; Biddison, G. M.; Chatelain, F.; Cummer, M.; Davi, D. J.; Fisher, L.; Frauendorf, A. W.; Frueh, F. W.; Gjerstad, C.; Harper, T. F.; Kernahan, S. D.; Long, D. Q.; Pho, M.; Walker, J. A., 2nd; Brennan, T. M. *J. Amer. Chem. Soc.* **2001**, *123*, 8887-8894.
35. V. Chechik, R. M. C. C. J. M. S., 2000; Vol. 12.
36. Chi, Y. S.; Lee, J. K.; Lee, K.-B.; Kim, D. J.; Choi, I. S. *Bulletin of the Korean Chemical Society* **2005**, *26*, 361-370.
37. Chapman, R. G.; Ostuni, E.; Yan, L.; Whitesides, G. M. *Langmuir* **2000**, *16*, 6927-6936.
38. Maoz, R.; Sagiv, J. *Langmuir* **1987**, *3*, 1045-1051.
39. Maoz, R.; Sagiv, J. *Langmuir* **1987**, *3*, 1034-1044.
40. Haller, I. *J. Amer. Chem. Soc.* **1978**, *100*, 8050-8055.
41. Kurth, D. G.; Bein, T. *Journal of Physical Chemistry* **1992**, *96*, 6707-6712.
42. Kurth, D. G.; Bein, T. *Angew. Chem.* **1992**, *104*, 323-325 (See also Angew Chem , Int Ed Engl , 1992, 31(3), 336-338).
43. Kurth, D. G.; Bein, T. *Langmuir* **1993**, *9*, 2965-2973.
44. Moon, J. H.; Shin, J. W.; Kim, S. Y.; Park, J. W. *Langmuir* **1996**, *12*, 4621-4624.
45. Hutt, D. A.; Leggett, G. J. *Langmuir* **1997**, *13*, 2740-2748.

46. Persson, H. H. J.; Caseri, W. R.; Suter, U. W. *Langmuir* **2001**, *17*, 3643-3650.
47. Ulman, A.; Tillman, N. *Langmuir* **1989**, *5*, 1418-1420.
48. Duevel, R. V.; Corn, R. M. *Analytical Chemistry* **1992**, *64*, 337-342.
49. Wells, M.; Crooks, R. M. *J. Amer. Chem. Soc.* **1996**, *118*, 3988-3989.
50. Leggett, G. J.; Roberts, C. J.; Williams, P. M.; Davies, M. C.; Jackson, D. E.; Tendler, S. J. B. *Langmuir* **1993**, *9*, 2356-2362.
51. Wang, J. H.; Kenseth, J. R.; Jones, V. W.; Green, J. B. D.; McDermott, M. T.; Porter, M. D. *J. Amer. Chem. Soc.* **1997**, *119*, 12796-12799.
52. Horton, R. C.; Herne, T. M.; Myles, D. C. *J. Amer. Chem. Soc.* **1997**, *119*, 12980-12981.
53. Pirrung, M. C.; Davis, J. D.; Odenbaugh, A. L. *Langmuir* **2000**, *16*, 2185-2191.
54. Maskos, U.; Southern, E. M. *Nucleic Acids Research* **1992**, *20*, 1679-1684.
55. Luzinov, I.; Julthongpiput, D.; Liebmann-Vinson, A.; Cregger, T.; Foster, M. D.; Tsukruk, V. V. *Langmuir* **2000**, *16*, 504-516.
56. Bertilsson, L.; Liedberg, B. *Langmuir* **1993**, *9*, 141-149.
57. Himmel, H. J.; Weiss, K.; Jager, B.; Dannenberger, O.; Grunze, M.; Woll, C. *Langmuir* **1997**, *13*, 4943-4947.
58. Alexander, M. R.; Wright, P. V.; Ratner, B. D. *Surface and Interface Analysis* **1996**, *24*, 217-220.
59. Duan, C.; Meyerhoff, M. E. *Mikrochimica Acta* **1995**, *117*, 195-206.
60. Baker, M. V.; Landau, J. *Australian J. Chem.* **1995**, *48*, 1201-1211.
61. Sagiv, J. *J. Amer. Chem. Soc.* **1980**, *102*, 92-98.
62. Maoz, R.; Sagiv, J. *Langmuir* **1987**, *3*, 1034-1044.
63. Maoz, R.; Sagiv, J. *Langmuir* **1987**, *3*, 1045-1051.
64. Kurth, D. G.; Bein, T. *Angew. Chem. Internl Ed. Engl.* **1992**, *31*, 336-338.
65. Kurth, D. G.; Bein, T. *Langmuir* **1993**, *9*, 2965-2973.
66. Lofas, S.; Johnsson, B. *J. Chem. Soc.-Chem. Communications* **1990**, 1526-1528.
67. Lee, Y. W.; Reedmundell, J.; Sukenik, C. N.; Zull, J. E. *Langmuir* **1993**, *9*, 3009-3014.
68. Kohli, P.; Taylor, K. K.; Harris, J. J.; Blanchard, G. J. *J. Amer. Chem. Soc.* **1998**, *120*, 11962-11968.
69. Tillman, N.; Ulman, A.; Elman, J. F. *Langmuir* **1989**, *5*, 1020-1026.
70. Yan, L.; Huck, W. T. S.; Whitesides, G. M. *Journal of Macromolecular Science-Polymer Reviews* **2004**, *C44*, 175-206.
71. Yan, L.; Marzolin, C.; Terfort, A.; Whitesides, G. M. *Langmuir* **1997**, *13*, 6704-6712.
72. Smith, E. A.; Wanat, M. J.; Cheng, Y. F.; Barreira, S. V. P.; Frutos, A. G.; Corn, R. M. *Langmuir* **2001**, *17*, 2502-2507.
73. Frutos, A. G.; Brockman, J. M.; Corn, R. M. *Langmuir* **2000**, *16*, 2192-2197.

74. Brockman, J. M.; Frutos, A. G.; Corn, R. M. *J. Amer. Chem. Soc.* **1999**, *121*, 8044-8051.
75. Jun Hu, Y. L., Chalermchai Khemtong, Jouliana M. El Khoury, Timothy J. McAfoos, and Ian S. Taschner *Langmuir* **2004**, *20*, 4933-4938.
76. Wade, N.; Gologan, B.; Vincze, A.; Cooks, R. G.; Sullivan, D. M.; Bruening, M. L. *Langmuir* **2002**, *18*, 4799-4808.
77. Roberts, S. M. *Journal of the Chemical Society-Perkin Transactions 1* **2001**, 1475-1499.
78. Gross, R. A.; Kumar, A.; Kalra, B. *Chemical Reviews* **2001**, *101*, 2097-2124.
79. Tanaka, K.; Yu, H. *Langmuir* **2002**, *18*, 797-804.
80. Laboda, H. M.; Glick, J. M.; Phillips, M. C. *Biochemistry* **1988**, *27*, 2313-2319.
81. Singh, C. P.; Shah, D. O. *Colloids and Surfaces a-Physicochemical and Engineering Aspects* **1993**, *77*, 219-224.
82. Singh, C. P.; Skagerlind, P.; Holmberg, K.; Shah, D. O. *Journal of the American Oil Chemists Society* **1994**, *71*, 1405-1409.
83. Turner, D. C.; Peek, B. M.; Wertz, T. E.; Archibald, D. D.; Geer, R. E.; Gaber, B. P. *Langmuir* **1996**, *12*, 4411-4416.
84. K.R. Yoon, K. B. L. Y. S. C. W. S. Y. S. W. J. I. S. C. *Advanced Materials* **2003**, *15*, 2063-2066.
85. Kim, Y.-R.; Paik, H.-J.; Ober Christopher, K.; Coates Geoffrey, W.; Batt Carl, A. *Biomacromolecules* **2004**, *5*, 889-894.
86. Mahapatro, A.; Johnson, D. M.; Patel, D. N.; Feldman, M. D.; Ayon, A. A.; Agrawal, C. M. *Langmuir* **2006**, *22*, 901-905.

Chapter 13

Design and Synthesis of Biorelated Polymers by Combinatorial Bioengineering

Yoshihiro Ito[*], Hiroshi Abe, Akira Wada, and Mingzhe Liu

Nano Medical Engineering Laboratory, RIKEN (Institute of Physical and Chemical Research), 2–1 Hirosawa, Wako-shi, Saitama, 351–0198, Japan

Combinatorial bioengineering, in which targeting molecules are selected from a random molecular library, has been used to prepare functional polymers that have a molecular recognition and catalysis capability, and biodevice polymers based on these functionalities. Although the polymer content is limited to natural nucleic acids and amino acids, this has been extended to nonnatural compounds. The possibility of developing new functional polymers using combinatorial bioengineering is discussed.

1. Introduction

Combinatorial bioengineering, in which targeting molecules are selected from a random molecular library, has been developing since the 1990s. The current molecular library has been extended from natural biooligomers, such as oligonucleotides and peptides, to nonnatural compounds. Various methodologies for selecting targeting molecules have been devised. Ellington and Szostak (*1*) and Tuerk and Gold (*2*) were the first to develop *in vitro* selection and Systematic Evolution of Ligands by EXponential Enrichment (SELEX) methods, which are useful for the selection of functional oligonucleotides (*3–7*). Recently, the library has been extended to contain nonnatural oligomers or polymers (Figure 1). From this molecular library, we can select functional polymers, which include molecular recognition polymers, such as antibodies, and catalysts, such as enzymes. In this article, we review the recent progress in

194

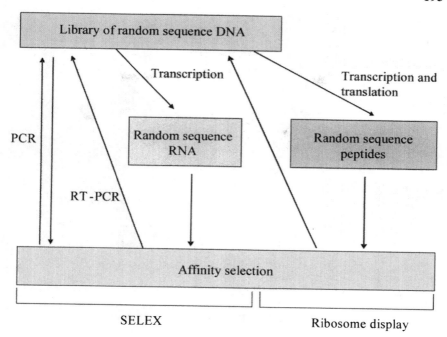

Figure 1. In vitro selections.

functional polymers that have a molecular recognition and catalysis capability, and biodevice polymers based on these functionalities.

2. *In Vitro* Selection

The principle of *in vitro* selection is governed by a number of the same principles that apply to the Darwinian theory of evolution, as shown in Figure 2. First, the random sequence DNA is prepared by automated solid-phase synthesis. A mixture of four types of nucleotide is added in a stepwise condensation reaction process. When necessary, this DNA library may be converted to an RNA library by *in vitro* transcription or to a peptide library by *in vitro* translation. Second, the prepared DNA, RNA, or peptide library is subjected to affinity selection, and the molecules that bind to a target molecule are selected. Because only a very small part of the library is selected in each selection, the selected fraction is then amplified by a polymerase chain reaction (PCR) or a reverse transcription PCR (RT-PCR) technique. Successive selection and amplification cycles bring about an exponential increase in the abundance of the targeting DNA, RNA, or peptide until it dominates the population.

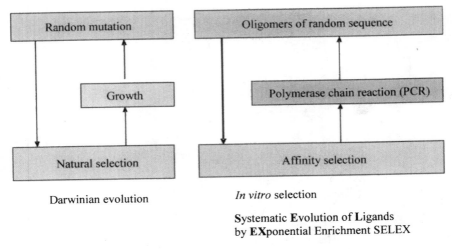

Figure 2. A comparison of Darwinian evolution theory and in vitro selection.

3. *In vitro* selection of molecular recognition polymers

The oligonucleotide obtained by the method described above is known as an "aptamer," a word derived from the Latin "*aptus*" meaning, "fit", and the Greek suffix -*mer*. An aptamer works in the manner of an artificial antibody and recognizes various types of "antigen." The composition of aptamers is extended from oligonucleotides to peptides as a "peptide aptamer". Aptamers have the following potential advantages over traditional antibody-based reagents: (i) they are not derived from living organisms and can be synthesized rapidly, reproducibly, and accurately by automated processes; (ii) they can be selected to bind to a wide range of targets, including those that are toxic or not inherently immunogenic; (iii) the labeling of oligonucleotides with various reporter molecules is simple and highly site specific; and (vi) the molecular weight of aptamers is usually lower than that of antibodies. It is possible to produce aptamers having a molecular weight of a few thousand.

3.1 Oligonucleotide aptamers

A large number of oligonucleotide aptamers for small-molecule targets, including organic dyes, porphyrins, other nucleotides and free nucleobases, amino acids, cofactors, basic antibiotics, and transition-state analogs, have been isolated over the previous decade. These targets include both planar and nonplanar compounds having overall negative or positive charges. In addition to these small molecules, proteins are also recognized by oligonucleotide aptamers. The first identification of an aptamer towards a protein that does not normally interact with RNA or DNA was the ssDNA aptamer to thrombin. A high number of molecules bind to thrombin with nanomolar affinity. In addition, oligonucleotide aptamers for various proteins have been identified, and information is available on the Internet (8). So far, we have isolated aptamers for folic acid (9), methotrexate (10), thyroxine (T_4) (11, 12), and chitin (13). As described above, one of the advantages of aptamers is that they are not derived from animals or living organisms, and therefore, sacrificing animals is not necessary. An automated process has been devised by some research groups, including our research group (14–17). If aptamers can be conveniently produced by machines in the future, then aptamers will be used instead of antibodies.

Recently, Cell-SELEX, which deals with living cells, was developed (18, 19). Using this method, aptamers that specifically recognize a cell's surface were obtained. To improve the selection process, some researchers have used surface plasmon resonance (SPR) or capillary electrophoresis (CE) (20–22). By developing CE-SELEX, Krylov's group obtained a DNA aptamer by a non-SELEX technique that requires no cyclical processes (23).

3.2   Peptide aptamers

Recently, peptide aptamers have been selected *in vitro* (24). In this selection, a yeast two-hybrid system (25, 26) and a phage display method (27 – 29) were employed. Using these methods, many types of peptide aptamer have been reported (28 –30). However, these methods used living cells, including yeast and *E. coli*, and so the peptide library was limited to nontoxic cases. In addition to the limited library population, a conditioning process of the cell culture and a highly sensitive sensor needed to detect the selected products were required. Therefore, to redress these disadvantages, ribosome display (31, 32), mRNA display (33, 34), and *in vitro* virus (35), have been developed (Figure 3) (24, 36).

In these display systems, molecular assemblies consisting of genotype molecules (mRNA) and phenotype molecules (peptide) are prepared. Ribosome and mRNA displays have peptide-ribosome-mRNA and peptide-puromycin-mRNA complexes, respectively. The former case is formed by inhibition of the attack of a releasing factor to a ribosome by removing the stop codons in the mRNA coding peptide library. As a result, the ribosome contains the mRNA and translated peptides. In the latter case, the 3' end of an mRNA coding peptide library is modified with puromycin. After translation, the puromycin binds to the peptide chain at the P site of the ribosome, and thus, the mRNA is linked to the peptide. After the formation of these complexes, the selection can be performed using the same method as the *in vitro* selection of oligonucleotides.

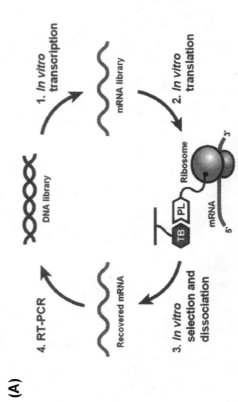

(A)

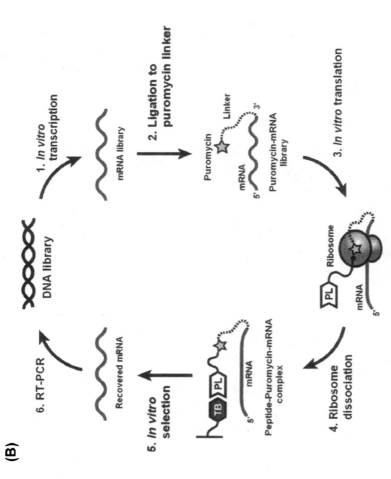

Binding of the puromycin to C-terminal of peptide

Figure 3. (A) Ribosome, and (B) mRNA displays.

4. *In Vitro* Selection of Biocatalysts

4.1 Oligonucleotide catalysts

Over the past few years, a variety of methods have been developed to allow the *in vitro* evolution of a range of biomolecules, including novel and improved biocatalysts (*37*), and ribozyme and deoxyribozyme (DNAzyme) have been discovered (*38*). When considering a catalytic antibody, the synthesis of an oligonucleotide catalyst can be naturally expected. When an oligonucleotide recognizing a transition-state analogue (TSA) is selected, then the oligonucleotide can act as a potential catalyst. The stabilization of the transition state of a reaction by the orchestration of functional groups attached to the backbone is considered to reduce the activation energy and thus to be essential for the synthesis of an efficient catalyst. For example, Cochran and Schultz (*39*) used a distorted porphyrin, *N*-methylmesoporphyrin IX, as a TSA to obtain antibodies that catalyze the metalation of mesoporphyrin IX. Similarly, Conn et al. (*40*) and Li and Sen (*41*) synthesized catalytic RNA and DNA, respectively, for porphyrin metalation using *N*-methylmesoporphyrin IX as the TSA.

Another method for selecting a catalyst has also been devised. This method is called the direct method, and the method scheme is shown in Figure 4. Usually, a biotin-labeled substrate is added to the oligonucleotide library pool, and some oligonucleotides self-catalyze the substrate to bind it. The biotin-labeled substrate-carrying oligonucleotides are purified using a streptavidin-immobilized gel. The bound nucleotides are collected and amplified by PCR. The same process is then repeated, and with an increase in the number of process rounds, oligonucleotides having self-catalytic activity can be selected. Using this method, ribozyme activity has been improved and deoxyribozyme (DNAzyme) has been discovered. The selected DNA or RNA catalyzed the ligation of RNAs and DNAs, and also catalyzed the ligation of DNA and RNA. DNAs and RNAs catalyzing the formation of an amide bond, an ester bond, and a Diels–Alder bond have also been selected using this method (*42*). Recently, it was demonstrated that a selected ribozyme catalyzed the bond formation between free substrates in addition to performing self-catalysis (*43*).

The advantage of natural catalysts (enzymes) is their high substrate specificity under mild conditions. However, the disadvantage is their deactivation under extreme conditions, including extremely low or high pH or temperature. The *in vitro* selection method is useful for the adaptation of ribozyme activity under extreme conditions. We have succeeded in the conversion of ligase ribozyme active at pH = 7 into a ribozyme that was active at pH = 4 using *in vitro* adaptation. A ribozyme active at pH = 7 that is not active at pH = 4 was chosen as the original ribozyme (*44*). The ribozyme was randomly mutated, and the resultant RNA library was mixed with biotin-labeled substrates at pH = 4. The reacted RNA was selected using a streptavidin-immobilized gel and eluted. The eluted RNA was amplified by reverse-

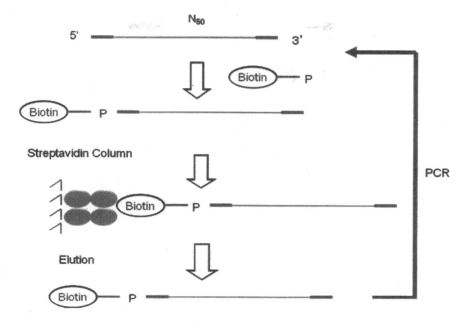

Figure 4. The direct selection method. If the substrate carrying the biotin binds to an oligonucleotide by the self-ligation reaction of an oligonucleotide, then the reacted oligonucleotide can be trapped by the streptavidin. The trapped oligonucleotide can be amplified by PCR for the next selection process.

transcription PCR. The amplified DNA was then transcribed, and the resultant RNA was mixed with biotin-labeled oligonucleotides at pH = 4. By repetition of this process the adapted ribozyme became 8,000 times more active than the original ribozyme at pH = 4 (*45*).

4.2 Peptide catalysts (enzyme evolution)

In the above-mentioned systems, the catalysts are basically limited to a self-catalytic activity, although an intermolecular catalyst has been found by accident. Therefore, a new methodology was developed, called "*in vitro* compart-mentalization" (IVC) (Figure 5) (*46–48*). In this system, a DNA random library and an *in vitro* translation system are enclosed in an emulsion droplet (w/o/w). Each droplet has a diameter of several micrometers and contains one DNA molecule and a corresponding peptide. Griffiths' group reported on various IVC systems containing enzymes that produce fluorescent molecules via chemical reactions, e.g., DNA methyl transferase and β-galactosidase. After the reaction in each emulsion occurred, the emulsion capsules exhibiting fluorescence were

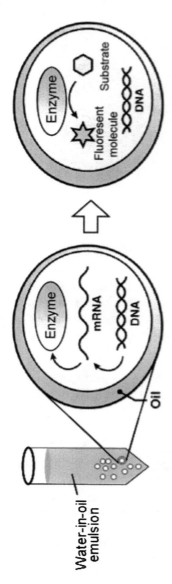

Figure 5. In vitro compartmentalization utilizes the aqueous core of water-in oil (w/o) or water/oil-in water (w/o/w) emulsions for in vitro translation. Proteins are expresses on the basis of DNA in each emulsion and the catalytic activity of protein is evaluated. Here, the fluorescence intensity of each emulsion is correlated with the enzymatic activity. Therefore, the emulsions with high fluorescence intensity are sorted by FACS, thus obtaining a pool of DNAs encoding enzymes with high activity.

isolated by fluorescence activated cell sorter (FACS), and from the isolated capsules, the catalytic peptide was isolated. Other systems have also been reported for screening and selection of peptide catalysts. These systems are based on approaches such as cell (49, 50) or phage (51) displays, or they are based on monitoring enzyme activity using fluorescent reporter proteins (52).

5. Extension to Non-Natural Oligomers

The type and number of the functional groups of a natural nucleic acid or amino acid are limited. If a nonnatural nucleic or amino acid can be applied to *in vitro* selection, then the scope of the functional oligomers or polymers will be extended (53, 54).

5.1 Oligonucleotide aptamers containing nonnatural components

Recently, natural DNA and RNA and also nonnatural DNAs and RNAs have been utilized in selection systems. Some research groups, including ours, have found that modified nucleotides can be proliferated by PCR or reverse-transcription PCR (RT-PCR) (55–64). DNA polymerase or RNA polymerase polymerizes modified DNA or RNA, respectively. Therefore, using the selection method, some modified oligonucleotides that specifically recognize molecules, such as antibodies, or those that catalyze chemical reactions have been synthesized.

Gold's group, a pioneer of the SELEX technique, has devised the PhotoSELEX method using photoreactive nucleotides as a nonnatural nucleotide (17). In the PhotoSELEX technique, a large number of random oligonucleotides carrying photoreactive deoxyuridine instead of deoxyuridine are exposed to the target of interest. Jhaveri et al. (65) have developed aptamers with a signal transduction ability. Anti-adenosine aptamers were selected from a pool that was skewed to contain very few fluoresceinated uridines. The primary family of aptamers showed a doubling of their relative fluorescence intensity under saturating concentrations of a cognate analyte, adenosine 5'-triphosphate (ATP), and they could sense ATP concentrations as low as 25 μM.

Chelliserrykattil and Ellington (66) have reported on the discovery of variant T7 RNA polymerases that will accept 2'-OMe A, C, and U (but not G). Burmeister et al. (67) reported on conditions under which the Y639F and Y639F/H784A polymerases will accept all four 2'-O-methyl nucleotides (68, 69). Using these methods, Burmeister et al. were able to discover a fully 2'-O-methyl aptamer for vascular endothelial growth factor (VEGF).

We have succeeded in the selection of RNA aptamers carrying multiple biotin groups in their side chains (70). Cytidine triphosphate (CTP) carrying the biotinyl group at the N^4-position was used for *in vitro* selection. A pool of random sequence RNAs containing biotinyl groups in the side chains was prepared, and the RNAs binding to ATP were selected. Because they carried multiple biotin groups on their side chains, the sensitivity was high.

In 2004, an aptamer, Macugen™, was approved by the US FDA for the treatment of age-related macular degeneration (71, 72). Macugen binds to and inhibits VEGF with a K_D = 49 pM (73). Of its 27 nucleotides, all but two bear either 2'-fluoro or 2'-O-methyl modifications. Macugen was discovered using the SELEX method, starting with a 2'-ribo purine, 2'-fluoro pyrimidine transcript library. Following the isolation of a VEGF binding sequence and minimization to remove extraneous nucleotides not required for binding, all but two of the 2'-ribonucleotides were substituted for 2'-O-methyl nucleotides to increase its stability to endogenous nucleases. The optimized molecule exhibited a long intraocular half-life of 10 d.

On the other hand, Kato et al. (62) reported the synthesis of 4'-thiopyrimidines (U and C) and their incorporation into transcripts by T7 RNA polymerase. The 4'-thio modification increased the stability of the transcripts 50-fold relative to the natural RNA transcripts. Using these libraries, Kato et al. (62) successfully performed a SELEX modification, and they were able to discover an aptamer to thrombin. Vaught et al. (56) showed that several UTPs substituted at the 5-position can be synthesized and subsequently transcribed using T7 RNA polymerase. Kuwahara et al. (59) have synthesized similar 5-modified dUTP compounds and have shown that a DNA polymerase efficiently catalyzed their template-dependent polymerization. Jaeger et al. (58) described a similar system for introducing a variety of modified bases into oligonucleotides using other DNA polymerases.

An alternative approach to the discovery of nuclease-resistant aptamers relies upon the fact that nucleases are highly enantioselective. While aptamers are created from the natural D-nucleotides, which are recognized by the nucleic-acid-degrading enzymes, a molecule synthesized as the mirror image L-oligonucleotide will not be degraded by any nucleases, because there are no such enzymes in the body capable of interacting with these nonnatural molecules. As such, the amplification methods used to create large aptamer libraries cannot be employed to synthesize the L-nucleic-acid-containing molecules. Based on the simple concept that if an aptamer binds to its natural target, the mirror image of the aptamer will identically bind to the mirror image of the natural target, then if the SELEX process is carried out using the mirror image of a target, an aptamer against this nonnatural mirror image will be obtained. The corresponding mirror image nucleic acid (L-oligonucleotide) of this aptamer is called a "Spiegelmer", from the German "Spiegel" or mirror (74, 75). Recent progress in preclinical developments has been reported by some researchers (76–78).

New strategies for post-SELEX optimization continue to be developed. Schmidt et al. (79) have demonstrated that LNA residues are substituted into a tenascin-C aptamer to increase the thermal stability and nuclease resistance of the aptamer. In common with other aptamer substitution strategies, some substitutions have led to a loss of target binding.

5.2 Peptide aptamers containing nonnatural components

Combinatorial peptide and protein libraries have been developed to accommodate nonnatural amino acids in a genetically encoded format via *in vitro* nonsense and sense suppression (*80–86*). General translation features and specific regioselective and stereoselective properties of a ribosome endow these libraries with a broad chemical diversity. Alternatively, amino acid residues can be chemically derivatized posttranslationally to add a preferred functionality to the encoded peptide.

Recently, Sisido's group has reported an *in vitro* selection of a peptide aptamer containing a nonnatural amino acid (*87*). They expanded the mRNA display method to include multiple nonnatural amino acids by introducing three different four-base codons at randomly selected single positions on the mRNA. Another nonnatural amino acid may be introduced by suppressing the amber codon that can appear from an $(NNK)_{(n)}$ nucleotide sequence on the mRNA. The mRNA display was expressed in Escherichia coli in an *in vitro* translation system in the presence of three types of tRNAs carrying different four-base anticodons and a tRNA carrying an amber anticodon, with the tRNAs being chemically aminoacylated with different nonnatural amino acids. The complexity of the starting mRNA-displayed peptide library was estimated to be 1.1×10^{12} molecules. The effectiveness of this four-base codon mediated mRNA display method was demonstrated in the selection of biocytin-containing peptides on streptavidin-coated beads. Moreover, a novel streptavidin-binding nonnatural peptide containing benzoylphenylalanine was obtained from the nonnatural peptide library. The nonnatural peptide library from the four-base codon mediated mRNA display provided a much wider functional and structural diversity than conventional peptide libraries that are constituted from 20 naturally occurring amino acids.

5.3 Biocatalysts containing nonnatural components

Nonnatural nucleotide-containing oligonucleotides have also been selected to improve or extend the functionality of ribozymes and DNAzymes, and nonnatural nucleotides carrying amino groups or fluoro groups have been used to make nuclease-resistant ribozymes (*88*). Our group has isolated alkylamino groups containing ribozymes using *in vitro* selection (*89, 90*). A random-sequence RNA library including N^6-aminohexyl-modified adenine residues in place of natural adenine residues was prepared, and ligase ribozyme was selected. The ribozyme mediated the formation of a phosphodiester bond with a DNA oligonucleotide through condensation with 5′-triphosphate moiety on the ribozyme. Among the clones isolated from this selection, one was shown to accelerate ligation by about 250 times more than the original random-sequence RNA library did. Almost zero acceleration was observed when N^6-aminohexyl-groups on adenine residues were omitted.

206

Eaton's group have incorporated imidazole onto oligonucleotides and selected Diels–Alderase from the random-sequence library (*91, 92*). The group of Joyce and Barbas III also applied this technique to the development of a DNAzyme that contained three catalytically essential imidazole groups to catalyze the cleavage of RNA substrates (*93*). The nucleic acid libraries for selection were constructed by polymerase-catalyzed incorporation of C5-imidazole-functionalized deoxyuridine in place of thymidine. Chemical synthesis was used to define a minimized catalytic domain composed of only 12 residues. The catalytic domain formed a compact hairpin structure that displayed three imidazole-containing residues. The imidazole-containing DNAzyme, which is one of the smallest known nucleic acid enzymes, combined the substrate-recognition properties of nucleic acid enzymes and the chemical functionality of protein enzymes in a molecule that was small, versatile, and catalytically efficient.

Two synthetically modified nucleoside triphosphate analogues (adenosine modified with an imidazole group and uridine modified with a cationic amine group) were enzymatically polymerized in tandem along a degenerate DNA library for the combinatorial selection of an RNAse A mimic by Perrin et al. (*94, 95*). Interestingly, the selected activity was consistent with both electrostatic and general acid/base catalysis at a physiological pH in the absence of divalent metal cations. This simultaneous use of two modified nucleosides to enrich the catalytic repertoire of DNA-based catalysts had not been demonstrated before, and evidence of a general acid/base catalysis at pH = 7.4 for DNAzyme had not been observed previously in the absence of a divalent metal cation and cofactor. This work illustrates how the incorporation of protein-like functionalities in nucleic acids can bridge the gap between proteins and oligonucleotides, underscoring its potential use in nucleic acid scaffolds in the development of new materials and improved catalysts for use in chemistry and medicine.

On the other hand, for the first time, we have synthesized a nonnatural ribozyme using an indirect method (*96, 97*). A pool of RNAs consisting of random sequences was obtained in the presence of 2'-aminocytidine triphosphate instead of CTP. The pool was incubated with an *N*-methylmesoporphyrin-immobilized gel, and the bound nonnatural RNAs were then collected and amplified by RT-PCR. The selected nonnatural RNA catalyzed the metalation of porphyrin using the same mechanism. In addition, the selected RNA exhibited a peroxidase activity by complex formation with hemin.

6. Polymer Device

Various modifications of functional oligonucleotides have been performed. The addition of allosteric functions was performed by Breaker's group (*98, 99*), Famulok's group (*100*), and Ellington's group (*101, 102*) using *in vitro* selection. Breaker's group (*98*) devised a new combinatorial strategy termed "allosteric selection", which favors the emergence of ribozymes that rapidly self-cleave only when incubated with their corresponding effector compounds. Using this method, RNA transcripts containing the hammerhead ribozyme were engineered to self-destruct in the presence of a specific nucleoside, 3',5'-cyclic monophosphate compounds. Breaker's group (*99*) showed that ribozymes were arrayed, and they used them to detect analytes as diverse as specific metal ions, nucleotides, cofactors, and drugs, all within a complex mixture. The RNA switch array has also been used to determine the phenotypes of *Escherichia coli* strains for adenylate cyclase function by detecting naturally produced 3',5'-cyclic adenosine monophosphate (cAMP) in bacterial culture media. Famulok's group (*100*) have applied the *in vitro* selection strategy based on an allosteric inhibition of a hammerhead ribozyme fused to a randomized RNA library using low concentrations of the antibiotic, doxycycline.

On the other hand, Ellington et al. (*102*) denoted such effector-responsive oligonucleotides as "aptazymes". They developed a technique for the selection of protein-dependent ribozyme ligases. After randomizing a previously selected ribozyme ligase, they selected variants that required one of two protein cofactors, a tyrosyl transfer RNA synthetase or hen egg white lysozyme. The resulting nucleoprotein enzymes were activated several thousandfold by their cognate protein effectors, and they could specifically recognize the structures of the native proteins.

Li and Lu developed a lead ion-dependent DNAzyme (*103*) for use as a sensor, based on the fluorescence of a fluorescence-labeled oligonucleotide released from DNAzyme activated by the lead ions. The DNAzyme sensitized the lead ions 80 times more than any other ions. Stojanovic et al. developed fluorescent sensors based on aptamer self-assemblies (*104, 105*). They adapted a DNA-based aptamer in two stages to signal the recognition of cocaine. An instability was engineered in one stem of a three-way junction that forms the cocaine-binding pocket, and the resultant short stem was end labeled with a fluorophore and a quencher. In the absence of cocaine, the two stems were open, but in the presence of cocaine, they were closed, and a three-way junction was formed. This major structural change brought the fluorophore and quencher together, thereby signaling the presence and concentration of the ligand. The sensor was selective for cocaine over its metabolite, could operate in serum, and was shown to be useful for the screening of cocaine hydrolases.

In addition to these sensing molecules, oligonucleotide-sensing deoxy ribozyme-based logic gates, their cascades, and automata have been developed by Stojanovic's group (*106–109*). Recently, Penchovsky and Breaker (*110*) have reported on a precise protocol for constructing and testing a complete set of ribozyme-based logic gates, whose self-cleavage activity was modulated by the presence of oligonucleotide inputs. To construct molecules that perform basic Boolean operations in solution, they generated allosteric ribozymes using a hammerhead ribozyme secondary structure motif. They significantly improved on previous approaches using a computational search method for rational design.

Some aptamer beacons using interactions between fluorescence and quenchers have been reviewed by Jayasena (*111*). On the other hand, Hartig et al. (*112*) developed a ribozyme-based detection scheme where an external substrate labeled with a fluorescent molecule at its 5' end and a fluorescence quencher at its 3' end was cleaved, resulting in a fluorescence dequenching, enabling real-time monitoring of ribozyme activity.

Fredriksson et al. (*113*) developed another technique for protein detection, in which the coordinated and proximal binding of a target protein by two DNA aptamers promoted the ligation of oligonucleotides linked to each aptamer affinity probe. The ligation of two such proximity probes gave rise to an amplifiable DNA sequence that reflected the identity and amount of the target protein. This proximity ligation assay detected a zeptomole amount of a cytokine platelet-derived growth factor without the need for washes or separations, and the mechanism can be generalized to other forms of protein analysis.

Recently Liu and Lu (*114*) have exploited the chameleon-like nature of gold particles to create a colorimetric indicator for the detection of specific biomolecules. First, each separated, and therefore red, gold nanoparticle had several single-stranded DNA sequences attached to it by covalent bonds. Two strands attached to neighboring nanoparticles were then bound to one another through a parallel DNA linker consisting of a series of nucleotides complementary to both strands. As several identical DNA strands were attached to each nanoparticle, in this way many nanoparticles were glued together. Aggregates formed with a characteristic blue-purple color. The innovation was that the DNA linker is itself an extension of a DNA aptamer that binds specifically to adenosine, representative small molecules. In the absence of adenosine, the aptamer behaved essentially as any short, linear DNA sequence and did not adopt a well-defined three-dimensional structure. Ellington and colleagues (*115*) used an aptamer-labeled quantum dot, the fluorescence of which was reduced by nonannealed fluorescent sequence. This sequence was then released in the presence of the analyte molecule to which the aptamer was linked.

As an application of an oligonucleotide catalyst, we immobilized DNAzyme carrying peroxidase activity with hemin on gold particles using a thiol modification of the DNAzyme's end to detect peroxide (*116*). Usually, horseradish peroxidase is immobilized on a solid support. Compared with protein enzymes, the DNAzyme is advantageous because of its thermal stability and convenience of preparation.

As an example of a combination of functional oligonucleotides, we have synthesized a DNA nanodevice that exhibited molecular recognition and enzymatic functions (*117*). In an enzyme-linked immunoassay (EIA), an

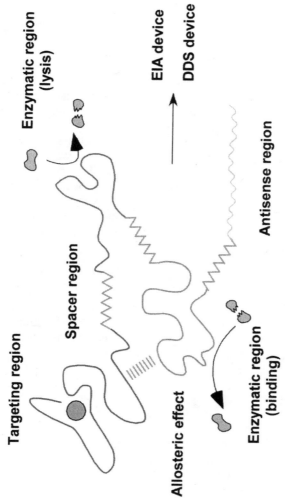

Figure 6. Various functions in polymer device.

enzyme-magnifying signal was attached to an antibody to enhance its sensitivity as a radioisotope. Instead of a bioconjugate, we designed DNA that carried both molecular recognition and catalytic regions. We synthesized DNA carrying both regions to bind to thyroxine as the analyte, and to have peroxidase for its enzyme activity. In the future, oligonucleotides integrating various functions will be designed and synthesized, as shown in Figure 6.

7. Future Research

DNA and proteins are recognized nanomaterials (*118*), and in particular, various types of DNA-based electronic wires, DNA-based computer elements, DNA molecular motors, DNA machines, and DNA nanorobotics have been reported (*119–122*). Current designs based on an extensive search of potential sequences are only possible for nucleic acid-based systems, for which the principles of Watson–Crick base pairing are well understood, the thermodynamic parameters are well studied, and the catalytic activity can be reliably predicted through a knowledge of the secondary structure. However, cells are adept at assembling protein-based molecular circuits from individual proteins. Recent progress in understanding the underlying principles of cellular signaling network organization—in particular, the ability to construct new protein gates using the principles of modular allostery, and to rewire networks using engineered scaffold proteins—suggests that proteins may be equally suited to the construction of artificial protein systems. In addition, the incorporation of nonnatural nucleic acids or amino acids into oligonucleotides and peptides will extend the scope of condensed polymer systems.

In the design of these polymer systems, combinatorial engineering is useful for creating functionalities. In the future, the functional polymers discussed in this article will contribute to the production of nanodevices by combining nanotechnology with combinatorial bioengineering.

References

1. Ellington, A. D.; Szostak, J. W. *Nature* **1990**, *346*, 818–822.
2. Tuerk, C.; Gold, L. *Science* **1990**, *249*, 505–510.
3. Wilson, D. S.; Szostak, J. W. *Ann Rev Biochem.* **1999**, *68*, 611–647.
4. Ito, Y.; Kawazoe, N.; Imanishi, Y. *Methods* **2000**, *22*, 107–114.
5. Perrin, D. M. *Comb. Chem. High Throughput Screen* **2000**, *3*, 243–269.
6. Brody, E. N.; Gold, L. *Rev. Mol. Biotechnol.* **2000**, *74*, 5–13.
7. Hesselberth, J.; Robertson, M. P.; Jhaveri, S.; Ellington, A. D. *Rev. Mol. Biotechnol.* **2000**, *74*, 15–25.
8. http://aptamer.icmb.utexas.edu.
9. Kawazoe, N.; Ito, Y.; Imanishi, Y. *Anal. Chem.* **1996**, *68*, 4309–4311.

10. Ito, Y.; Suzuki, A.; Kawazoe, N.; Imanishi, Y. *Bull. Chem. Soc. Jpn.* **1997**, *70*, 695–698.
11. Ito, Y.; Fujita, S.; Kawazoe, N.; Imanishi, Y. *Anal. Chem.* **1998**, *70*, 3510–3512.
12. Kawazoe, N.; Ito, Y.; Shirakawa, M.; Imanishi, Y. *Bull. Chem. Soc. Jpn.* **1998**, *71*, 1699–1703.
13. Fukusaki, E.; Kato, T.; Maeda, H.; Kawazoe, N.; Ito, Y.; Okazawa, A.; Kajiyama, S.; Kobayashi, A. *Bioorg. Med. Chem. Lett.* **2000**, *10*, 423–425.
14. Cox, J. C.; Rudolph, P.; Ellington, A. D. *Biotechnol. Prog.* **1998**, *14*, 845–850.
15. Zhang, H.; Hamasaki, A.; Toshiro, E.; Aoyama, Y.; Ito, Y. *Nucleic Acids Symp. Ser.* **2000**, *2000*, 219–220.
16. Cox, J. C.; Ellington, A. D. *Bioorg. Med. Chem.* **2001**, *9*, 2525–2531.
17. http://www.somalogic.com/tech/photoselex.html.
18. Blank, M.; Weinschenk, T.; Priemer, M.; Schluesener, H. *J. Biol. Chem.* **2001**, *276*, 16464–16468.
19. Shangguan, D.; Li, Y.; Tang, Z.; Cao, Z. C.; Chen, H. W.; Mallikaratchy, P.; Sefah, K.; Yang, C. J.; Tan, W. *Proc. Natl. Acad. Sci. U.S.A.* **2006**, *103*, 11838–11843.
20. Misono, T. S.; Kumar, P. K. R. *Anal. Biochem.* **2005**, *342*, 312–317.
21. Mendonsa, S. D.; Bowser, M. T. *J. Am. Chem. Soc.* **2005**, *127*, 9382–9383.
22. Mendonsa, S. D.; Bowser, M. T. *Anal. Chem.* **2004**, *76*, 5387–5392.
23. Berezovski, M.; Musheev, M.; Drabovich, A.; Krylov, S. N. *J. Am. Chem. Soc.* **2006**, *128*, 1410–1411.
24. Rothe, A.; Hosse, R. J.; Power, B. E. *FASEB J.* **2006**, *20*, 1599–1610.
25. Bickle, M. B. T.; Dusserre, E.; Moncorge, O.; Bottin, H.; Colas, P. *Nat. Protoc.* **2006**, *1*, 1066–1091.
26. Buerger, C.; Groner, B. *J. Cancer Res. Clin. Oncol.* **2003**, *129*, 669–675.
27. Hoess, R. H. *Chem. Rev.* **2001**, *101*, 3205–3218.
28. Ladner, R. C.; Sato, A. K.; Gorzelany, J.; de Souza, M. *Drug Discov. Today* **2004**, *9*, 525–529.
29. Krag, D. N.; Shukla, G. S.; Shen, G.-P.; Pero, S.; Ashikaga, T.; Fuller, S.; Weaver, D. L.; Burdette-Radoux, S.; Thomas, C. *Cancer Res.* **2006**, *66*, 7724–7733.
30. Hoppe-Seyler, F.; Butz, K. *J. Mol. Med.* **2000**, *78*, 426–430.
31. Yan, X.; Xu, Z. *Drug Discov. Today* **2006**, *11*, 911–916.
32. Rothe, A.; Hosse, R. J.; Power, B. E. *Expert Opin. Biol. Ther.* **2006**, *6*, 177–187.
33. Keefe, A. D.; Szostak, J. W. *Nature* **2001**, *410*, 715–718.
34. Wilson, D. S.; Keefe, A. D.; Szostak, J. W. *Proc. Natl. Acad. Sci. U.S.A.* **2001**, *98*, 3750–3755.
35. Miyamoto-Sato, E.; Yanagawa, H. *J. Drug Target.* **2006**, *14*, 505–511.
36. Lipovsek, D.; Pluckthun, A. *J. Immunol. Methods* **2004**, *290*, 51–67.
37. Griffiths, A. D.; Tawfik, D. S. *Curr. Opin. Biotechnol.* **2000**, *11*, 338–353.

212

38. Li, Y.; Breaker, R. R. *Curr. Opin. Struct. Biol.* **1999**, *9*, 315–323.
39. Cochran; A. G.; Schultz; P. G. *J. Am. Chem. Soc.* **1990**, *112*, 9414–9415.
40. Conn, M. M.; Prudent, J. R.; Schultz, P. G. *J. Am. Chem. Soc.* **1996**, *118*, 7012–7013.
41. Li, Y.; Sen, D. *Nat. Struct. Biol.* **1996**, *3*, 743–747.
42. Pan, T. *Curr. Opin. Chem. Biol.* **1997**, *1*, 17–25.
43. Seelig, B.; Keiper, S.; Stuhlmann, F.; Jaschke, A. *Angew. Chem. Int. Ed. Engl.* **2000**, *39*, 4576–4579.
44. Miyamoto, Y.; Teramoto, N.; Imanishi, Y.; Ito, Y. *Biotechnol. Bioeng.* **2001**, *75*, 590–596.
45. Miyamoto, Y.; Teramoto, N.; Imanishi, Y.; Ito, Y. *Biotechnol. Bioeng.* **2005**, *90*, 36–45.
46. Miller, O. J.; Bernath, K.; Agresti, J. J.; Amitai, G.; Kelly, B. T.; Mastrobattista, E.; Taly, V.; Magdassi, S.; Tawfik, D. S.; Griffiths, A. D. *Nat. Methods* **2006**, *3*, 561–570.
47. Mastrobattista, E.; Taly, V.; Chanudet, E.; Treacy, P.; Kelly, B. T.; Griffiths, A. D. *Chem. Biol.* **2005**, *12*, 1291–1300.
48. Matsuura, T.; Yomo, T. *J. Biosci. Bioeng.* **2006**, *101*, 449–456.
49. Olsen, M. J.; Stephens, D.; Griffiths, D.; Daugherty, P.; Georgiou, G.; Iverson, B. L. *Nat. Biotechnol.* **2000**, *18*, 1071–1074.
50. Kawarasaki, Y.; Griswold, K. E.; Stevenson, J. D.; Selzer, T.; Benkovic, S. J.; Iverson, B. L.; Georgiou, G. *Nucleic Acids Res.* **2003**, *31*, e126.
51. Cesaro-Tadic, S.; Lagos, D.; Honegger, A.; Rickard, J. H.; Partridge, L. J.; Blackburn, G. M.; Pluckthun, A. *Nat. Biotechnol.* **2003**, *21*, 679–685.
52. Santoro, S. W.; Wang, L.; Herberich, B.; King, D. S.; Schultz, P. G. *Nat. Biotechnol.* **2002**, *20*, 1044–1048.
53. Eaton, B. E. *Curr. Opin. Chem. Biol.* **1997**, *1*, 10–16.
54. Kusser, W. *J. Biotechnol.* **2000**, *74*, 27–38.
55. Wilson, C.; Keefe, A. D. *Curr. Opin. Chem. Biol.* **2006**, *10*, 607–614.
56. Vaught, J. D.; Dewey, T.; Eaton, B. E. *J. Am. Chem. Soc.* **2004**, *126*, 11231–11237.
57. Ahle, J. D.; Barr, S.; Chin, A. M.; Battersby, T. R. *Nucleic Acid Res.* **2005**, *33*, 3176–3184.
58. Jager, S.; Rasched, G.; Kornreich-Leshem, H.; Engeser, M.; Thum, O.; Famulok, M. *J. Am. Chem. Soc.* **2005**, *127*, 15071–15082.
59. Kuwahara, M.; Nagashima, J.; Hasegawa, M.; Tamura, T.; Kitagata, R.; Hanawa, K.; Hososhima, S.; Kasamatsu, T.; Ozaki, H.; Sawai, H. *Nucleic Acid Res.* **2006**, *34*, 5383–5394.
60. Hirao, I.; Kimoto, M.; Mitsui, T.; Fujiwara, T.; Kawai, R.; Sato, A.; Harada, Y.; Yokoyama, S. *Nat. Method* **2006**, *3*, 729–735.
61. Matsui, M.; Nishiyama, Y.; Ueji, S.; Ebara, Y. *Bioorg. Med. Chem. Lett.* **2007**, *17*, 456–460.
62. Kato, Y.; Minakawa, N.; Komatsu, Y.; Kamiya, H.; Ogawa, N.; Harashima, H.; Matsuda, A. *Nucleic Acid Res.* **2005**, *33*, 2942–2951.

63. Bugaut, A.; Toulme, J. J.; Rayner, B. *Org. Biomol. Chem.* **2006**, *4*, 4082–4088.

64. Shoji, A.; Kuwahara, M.; Ozaki, H.; Sawai, H. *J. Am. Chem. Soc.* **2007**, *129*, 1456–1464.

65. Jhaveri, S.; Rajendran, M.; Ellington, A. D. *Nat. Biotechnol.* **2000**, *18*, 1293–1297.

66. Chelliserrykattil, J.; Ellington, A. D. *Nat. Biotechnol.* **2004**, *22*, 1155–1160.

67. Burmeister, P. E.; Lewis, S. D.; Silva, R. F.; Preiss, J. R.; Horwitz, L. R.; Pendergrast, P. S.; McCauley, T. G.; Kurz, J. C.; Epstein, D. M.; Wilson, Keefe, A. D. *Chem. Biol.* **2005**, *12*, 25–33.

68. Padilla, R.; Sousa, R. *Nucleic Acid Res.* **1999**, *27*, 1561–1563.

69. Padilla, R.; Sousa, R. *Nucleic Acid Res.* **2002**, *30*, e138.

70. Ito, Y.; Suzuki, A.; Kawazoe, N.; Imanishi, Y. *Bioconjug. Chem.* **2001**, *12*, 850–854.

71. Gragoudas, E. S.; Adamis, A. P.; Cunningham, E. T. Jr.; Feinsod, M.; Guyer, D. R. *N. Engl. J. Med.* **2004**, *351*, 2805–2816.

72. Chakravarthy, U.; Adamis, A. P.; Cunningham, E. T. Jr.; Goldbaum, M.; Guyer, D. R.; Katz, B.; Patel, M. *Ophthalmology* **2006**, *113*, 1508e1–25.

73. Ruckman, J. Green, L. S.; Beeson, J.; Waugh, S.; Gillette, W. L.; Henninger, D. D.; Claesson-Welsh, L.; Janjic, N. *J. Biol. Chem.* **1998**, *273*, 20556–20567.

74. Wlotzka, B.; Leva, S.; Eschgfaller, B.; Burmeister J.; Kleiniung, F.; Kaduk, C.; Muhn, P.; Hess-Stumpp, H.; Klussmann, S. *Proc. Natl. Acad. Sci., U.S.A.* **2002**, *99*, 8898–8902.

75. Vater, A.; Klussmann, S. *Curr. Opin. Drug Discov. Devel.* **2003**, *6*, 253–261.

76. Shearman, L. P.; Wang, S. P.; Helmling, S.; Stribling, D. S.; Mazur, P.; Ge, L.; Wang, L.; Klussmann, S.; Macintyre, D. E.; Howard, A. D.; Strack, A. M. *Endocrinology* **2006**, *147*, 1517–1526.

77. Denekas, T.; Troltzsch, M.; Vater, A.; Klussmann, S.; Messlinger, K. *Br. J. Pharmacol.* **2006**, *148*, 536–543.

78. Purschke, W. G.; Eulberg, D.; Buchner, K.; Vonhoff, S.; Klussmann, S. *Proc. Natl. Acad. Sci., U.S.A* **2006**, *103*, 5173–5178.

79. Schmidt, K. S.; Borkowski, S.; Kurreck, J.; Stephens, A. W.; Bald, R.; Hecht, M.; Friebe, M.; Dinkelborg, L.; Erdmann, V. A. *Nucleic Acid Res.* **2004**, *32*, 5757–5765.

80. Frankel, A.; Li, S.; Starck, S. R.; Roberts, R. W. *Curr. Opin. Struct. Biol.* **2003**, *13*, 506–512.

81. Li, S.; Roberts, R. W. *Chem. Biol.* **2003**, *10*, 233–239.

82. Frankel, A.; Roberts, R. W. *RNA* **2003**, *9*, 780–786.

83. Frankel, A.; Millward, S. W.; Roberts, R. W. *Chem. Biol.* **2003**, *10*, 1043–1050.

84. Ja, W. W.; Roberts, R. W. *Biochemistry* **2004**, *43*, 9265–9275.

85. Starck, S. R.; Green, H. M.; Alberola-Ila, J.; Roberts, R. W. *Chem. Biol.* **2004**, *11*, 999–1008.
86. Millward, S. W.; Takahashi, T. T.; Roberts, R. W. *J. Am. Chem. Soc.* **2005**, *127*, 14142–14143.
87. Muranaka, N.; Hohsaka, T.; Sisido, M. *Nucleic Acids Res.* **2006**, *34*, e7.
88. Green, L. S.; Jellinek, D.; Bell, C.; Beebe, L. A.; Feistner, B. D.; Gill, S. C.; Jucker, F. M.; Janjic, N. *Chem. Biol.* **1995**, *2*, 683–695.
89. Teramoto, N.; Imanishi, Y.; Ito, Y. *Bioconjug. Chem.* **2000**, *11*, 744–748.
90. Teramoto, N.; Imanishi, Y.; Ito, Y. *J. Bioact. Comp. Polym. 15,* **2000**, 297–308.
91. Tarasow, T. M.; Tarasow, S. L.; Eaton, B. E. *Nature* **1997**, *389*, 54–57.
92. Tarasow, T. M.; Kellogg, E.; Holley, B. L.; Nieuwlandt, D.; Tarasow, S. L.; Eaton, B. E. *J. Am. Chem. Soc.* **2004**, *126*, 11843–11851.
93. Santoro, S. W.; Joyce, G. F.; Sakthivel, K.; Gramatikova, S. *J. Am. Chem. Soc.* **2000**, *122*, 2433–2439.
94. Perrin, D. M.; Garestier, T.; Helene, C. *J. Am. Chem. Soc.* **2001**, *123*, 1556–1563.
95. Lermer, L.; Roupioz, Y.; Ting, R.; Perrin, D. M. *J. Am. Chem. Soc.* **2002**, *124*, 9960–9961.
96. Kawazoe, N.; Teramoto, N.; Ichinari, H.; Imanishi, Y.; Ito, Y. *Biomacromolecules* **2001**, *2*, 681–686.
97. Teramoto, N.; Ichinari, H.; Kawazoe, N.; Imanishi, Y.; Ito, Y. *Biotechnol. Bioeng.* **2001**, *75*, 463–468.
98. Koizumi, M.; Soukup, G. A.; Kerr, J. N.; Breaker, R. R. *Nat. Struct. Biol.* **1999**, *6*, 1062–1071.
99. Seetharaman, S.; Zivarts, M.; Sudarsan, N.; Breaker, R. R. *Nat. Biotechnol.* **2001**, *19*, 336–341.
100. Piganeau, N.; Jenne, A; Thuillier, V; Famulok, M. *Angew. Chem. Int. Ed. Engl.* **2000**, *39*, 4369–4373.
101. Robertson, M. P.; Ellington, A. D. *Nat. Biotechnol.* **2001**, *19*, 650–655.
102. Robertson, M. P.; Ellington, A. D. *Nat. Biotechnol.* **1999**, *17*, 62–66.
103. Li, J.; Lu, Y. *J. Am. Chem. Soc.* **2000**, *122*, 10466–10467.
104. Stojanovic, M. N.; de Prada, P.; Landry, D. W. *J. Am. Chem. Soc.* **2000**, *122*, 11547–11548.
105. Stojanovic, M. N.; de Prada, P.; Landry, D. W. *J. Am. Chem. Soc.* **2001**, *123*, 4928–4931.
106. Stojanovic, M. N.; Mitchell, T. E.; Stefanovic, D. *J. Am. Chem. Soc.* **2002**, *124*, 3555–3561.
107. Stojanovic, M. N.; Stefanovic, D. *J. Am. Chem. Soc.* **2005**, 127, 6914–6915.
108. Kolpashchikov, D. M.; Stojanovic, M. N. *J. Am. Chem. Soc.* **2005**, 127, 11348–11351.
109. Stojanovic, M. N.; Stefanovic, D. *Nat. Biotechnol.* **2003**, *21*, 1069–1074.

110. Penchovsky, R.; Breaker, R. R. *Nat. Biotechnol.* **2005**, *23*, 1424–1433.

111. Jayasena, S. D. *Clin. Chem.* **1999**, *45*, 1628–1650.

112. Hartig, J. S.; Najafi-Shoushtari, S. H.; Grune, I.; Yan, A.; Ellington, A. D.; Famulok, M. *Nat. Biotechnol.* **2002**, *20*, 717–722.

113. Fredriksson, S.; Gullberg, M.; Jarvius, J.; Olsson, C.; Pietras, K.; Gustafsdottir, S. M.; Ostman, A.; Landegren, U. *Nat. Biotechnol.* **2002**, *20*, 473–477.

114. Liu, J.; Lu, Y. *Angew. Chem. Int. Edn.* **2005**, *45*, 90–94.

115. Levy, M.; Cater, S. F.; Ellington, A. D. *ChemBioChem* **2005**, *6*, 2163–2166.

116. Ito, Y.; Hasuda, H. *Biotechnol. Bioeng.* **2004**, *86*, 72–77.

117. Ito, Y. *Polym. Adv. Technol.* **2004**, *15*, 3–14.

118. Ito, Y.; Fukusaki, E. *J. Mol. Catal. B* **2004**, *28*, 155–166.

119. LaBean, T. H.; Li, H. Y. *Nano Today* **2007**, *2*, 26–35.

120. Liedl, T.; Sobey, T. L.; Simmel, F. C. *Nano Today* **2007**, *2*, 36–41.

121. Niemeyer, C. M. *Nano Today* **2007**, *2*, 42–52.

122. Bath, J.; Turberfield, A. J. *Nat. Nanotechnol.* **2007**, *2*, 275–284.

Chapter 14

Chemoenzymatic Synthesis of Block Copolymers

Matthijs de Geus[1], Anja R. A. Palmans[2], Christopher J. Duxbury[3], Silvia Villarroya[4], Steven M. Howdle[4], and Andreas Heise[1,3]

[1]Departments of Polymer Chemistry and [2]Macromolecular and Organic Chemistry, Technische Universiteit Eindhoven, Den Dolech 2, P.O. Box 513, 5600 MB Eindhoven, The Netherlands
[3]DSM Research, P.O. Box 18, 6160 MD Geleen, The Netherlands
[4]School of Chemistry, University of Nottingham, University Park, Nottingham, NG7 2RD, United Kingdom

The chemoenzymatic cascade synthesis of block copolymers combining enzymatic ring opening polymerization (eROP) and atom transfer radical polymerization (ATRP) is reviewed. Factors like reaction condition and initiator structure were investigated and optimized prior to the polymerization. The synthesis of block copolymers was successful in two consecutive steps, i.e. eROP followed by ATRP as evident from SEC and GPEC analysis. While in the one-pot approach, block copolymers could be obtained by sequential addition of the ATRP catalyst, side reactions were observed when all components were present from the start of the reaction. A successful one-pot synthesis was achieved by conducting the reaction in supercritical carbon dioxide.

Introduction

The design of complex macromolecular architectures has become a particular focus in polymer science. Some of these architectures possess unique properties, which makes them interesting candidates for specialty applications in nanostructured and biomedical materials such as block and branched copolymers. In particular the advent of controlled radical polymerization techniques provided convenient synthetic pathways to realize polymer architectures *(1-3)*.

Biocatalytic approaches in polymer science are expected to further increase the diversity of polymeric materials. Major progress has been achieved in recent years in applying enzyme catalysis in polymer science *(4,5)*. The application of enzymes in polymer synthesis and transformation is attractive due to their ability to function under mild conditions with high enantio- and regioselectivity. The stability of lipases in organic media and their ability to promote transesterification and condensation reactions on a broad range of low and high molar mass substrates has been shown in many examples. In particular, immobilized Lipase B from *Candida antarctica* (Novozym 435) has shown exceptional activity for a range of polymer forming reactions, including the ring opening polymerization of cyclic monomers (e.g. lactones, carbonates) *(6)*. Biocatalysis can therefore be an interesting complimentary, and in some cases alternative synthetic method for the design of functional polymer architectures. However, the full exploitation of biocatalysis in polymer science will require the development of mutually compatible chemo- and biocatalytic methods. In our laboratories, we thus explore the integration of biocatalytic and traditional polymer synthesis. Our goal is the development of chemoenzymatic polymerization reactions. In particular the combination of the distinctive attributes of enzymatic polymerization with controlled radical polymerization techniques can allow the formation of polymer structures with new properties not available by chemical polymerization. In that context we investigated the chemoenzymatic synthesis of block copolymers by combining enzymatic ring opening polymerization (eROP) with atom transfer radical polymerization (ATRP). This paper reviews our results in combining both techniques.

Results and Discussion

The synthesis of block copolymers is particularly suited to investigate the combination of two fundamentally different synthetic techniques, since the marriage of two chemically different building blocks often requires a considerable synthetic effort. Block copolymers with building blocks based on two intrinsically different polymerization mechanisms, e.g. polyester and polymethacrylate, can either be obtained by chemically linking two preformed polymer blocks or, alternatively, in two consecutive polymerizations using macro-initiators. The majority of the latter examples require an intermediate transformation step in order to convert the end-group of the first block into an active initiator for the second polymerization. A very powerful and elegant synthetic pathway to block copolymers is the use of an initiator combining two fundamentally different initiating groups in one molecule (dual or bifunctional initiator). This allows two consecutive polymerizations without an intermediate transformation step. The feasibility of this approach has been successfully demonstrated for the combination of various chemical polymerization techniques *(7)*.

As outlined in Scheme 1, the combination of ATRP and eROP for the synthesis of block copolymers can either be conducted in two consecutive steps, or in a one-pot cascade approach. The latter provides, without doubt, the higher synthetic and analytical challenge. In order to obtain high block copolymer yields, it is paramount to fine tune the reaction kinetics and minimize all side reactions leading to homopolymer formation. We therefore carefully investigated the governing factors to minimize the extent of undesired side reactions in order to obtain a maximum level of control over the polymerization and hence the polymer structure. Only the use of a bifunctional initiator equipped with an activated bromide group for ATRP and a primary hydroxy-group for the eROP allows the synthesis of block copolymers without an intermediate transformation step. The key to obtain high block copolymer yields is (i) high initiation efficiency from both initiator groups and (ii) exclusive initiation from the bifunctional initiator. Low initiator efficiency in either one of the two polymerizations or initiation by a species other than the bifunctional initiator inevitably results in homopolymer formation. Since the efficiency of ATRP is generally known to be high, we primarily focused our effort on the optimization of the enzymatic ROP, i.e., the conditions that lead to a high initiator efficiency in the enzymatic reaction.

Initiator Structure

It is generally accepted that the eROP of CL is initiated by a nucleophilic attack of a hydroxy compound on the enzyme activated monomer *(8-10)*. Various hydroxy-functionalized (macro)initiators have been reported in the past for the enzymatic synthesis of functionalized polymers and block copolymers, respectively *(8-11)*. As outlined above, only a high initiation efficiency of the dual initiator in the eROP leads to high block copolymer yield. This efficiency mainly depends on two factors: side reactions caused by competitive water initiation and the initiator design. We achieved the reduction of the water initiation by developing a thorough drying protocol prior to the enzymatic reaction *(12)*. In order to study the influence of the initiator structure, several dual

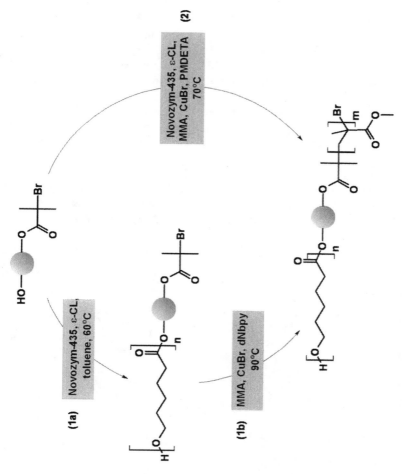

Scheme 1. Chemoenzymatic polymerization combining ATRP and eROP in two consecutive steps (1) and in a one-pot cascade approach (2).

initiators were synthesized and investigated. The goal was to identify the structure with the highest initiating efficiency that leads to a high degree of end-functionalized PCL in the eROP. All reactions were conducted at 60 °C for 1 hour in triplicate with Novozym 435 dried according to our optimized procedure *(12)*.

Figure 1 shows the consumption of two different dual initiators during the polymerization of CL. The figure illustrates the rapid conversion of initiator **2** in the initial phase of the reaction (> 90 % after 15 min.). Initiator **1**, on the other hand, is a poor initiator in this polymerization as evidenced from the slow initiator consumption and monomer conversion after one hour. This is surprising since we have applied this initiator in an earlier study for a similar synthesis of block copolymers without noticing significant amounts of non-functionalized PCL *(13)*. It can be speculated that the observed inhibition period in the polymerization using **1** implies that small amounts of water initiate the polymerization, causing very low monomer and initiator conversion. In addition, **1** is a mixture of enantiomers, which in conjunction with the enantioselectivity of the enzyme may influence the initiation kinetics. These results suggest that the incorporation of **2** into the growing polymer chain is predominantly through initiation. In contrast, **1** might very well be built into the chain by transesterification during the course of the reaction. As a consequence, high end-group functionalization with **1** can only be achieved at longer reaction times like those applied for the synthesis of ATRP initiator end-capped PCL in our earlier report.

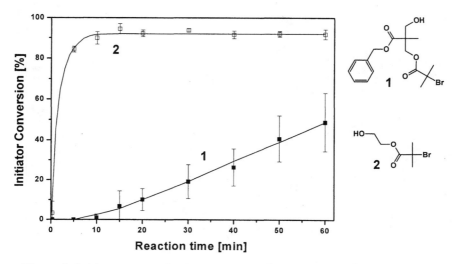

Figure 1. Initiator conversion in enzymatic polymerizations of CL, employing two different dual initiators. (Reproduced from reference 12. Copyright 2005 American Chemical Society.)

The kinetics of eROP was found to be first order with respect to monomer consumption, which agrees with previous literature reports *(8-10)*. Although the monomer consumption in eROP is supposed to be independent of the nucleophile (i.e., initiator), we clearly observe a relationship. The reason for this discrepancy might be the actual low amount of water present in our reaction whilst performing the kinetic experiments. If the water concentration is far higher than the amount of initiator used, differences in reaction kinetics between the initiators might not be observed.

Block copolymer synthesis

Consecutive polymerization (macro-initiation)

Based on the optimized reaction conditions a PCL macro-initiator was synthesized using initiator **2** yielding a polymer with a molecular weight of 6 800 g/mol and a polydispersity (PDI) of 2.0 after 120 min. ^{1}H-NMR analysis of the polymer confirms a low concentration of carboxylic acid end-capped polymers (< 2 %), suggesting a high degree of incorporation of **2** into the polymer. Further evidence for the incorporation of the initiator was obtained from MALDI-ToF-MS. Figure 2 shows a spectrum of a macro-initiator obtained from a typical enzymatic polymerization. Three distributions of peaks can be identified in this spectrum. The signals of every series are separated by 114 Da, which corresponds to the mass of the CL repeating unit. The distributions on the low molecular weight end of the spectrum (A in Figure 2) can be assigned to cyclic polymers, whereas the desired bromine-functional polymers are more prominent in the high molecular weight region of the spectrum (C in Figure 2). In agreement with the work of Hult, we observe that the relative amount of cyclic polymers depends on the reaction time and the monomer concentration *(10)*. The third distribution of peaks can be assigned to carboxylic acid end-functionalized polymer chains generated by side reactions caused by water (B in Figure 2).

The precipitated PCL macro-initiator was subsequently used for ATRP of MMA using CuBr/dNbpy as catalyst. An increase in the molecular weight from 13 400 to 36 900 g/mol was observed by Size Exclusion Chromatography (SEC). From ^{1}H-NMR, a block length ratio of PMMA to PCL of 4:1 was determined by comparison of the integrated peak area of the methoxy-protons of the MMA units (δ= 3.62 ppm) with the integrated peak area of the methylene ester protons in the PCL-backbone. This is in good agreement with the monomer feed ratio. In order to directly analyze the PMMA-block of the P(CL-*b*-MMA), the PCL-block was hydrolyzed. Comparison of the SEC trace of the remaining PMMA-block with that of P(CL-*b*-MMA) reveals a shift to lower molecular weight, namely to 24 400 g/mol, due to the removal of the PCL-block.

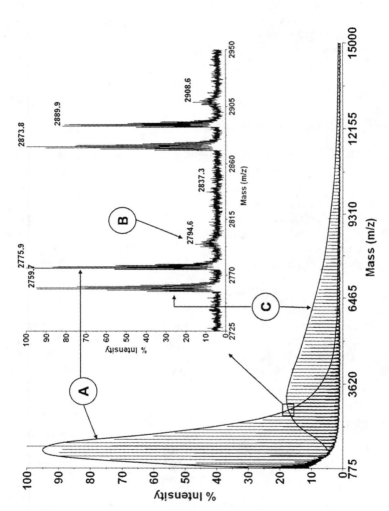

Figure 2. MALDI-ToF-MS spectra of PCL obtained from eROP using initiator A: cyclic polymers,
B: carboxylic acid end-functionalized polymers, C: polymers end-functionalized with 2.
(Reproduced from reference 12. Copyright 2005 American Chemical Society.)

Although this provides strong evidence for the block structure, the overlapping traces do not allow a final conclusion on the composition of the block copolymer, exclusively based on the SEC results.

In order to provide additional evidence for the block structure of the isolated polymers, these were further investigated by Gradient Polymer Elution Chromatography (GPEC) *(14)*. The separation in GPEC is based on both the interaction between the polymer and the stationary phase (adsorption chromatography) and on the solubility of the copolymer in the eluent mixture. Within certain molecular weight limits it allows the separation of block and homopolymers solely based on their chemical nature. In our case THF/heptane was used as solvent/non-solvent system. It was found that, by applying a gradient from 100 % heptane to 100 % THF in 10 minutes, a proper separation of a blend of the two homopolymers could be achieved. Whereas pure PCL elutes at a retention time of 8.3 minutes, pure PMMA elutes at 10.8 minutes (Figure 1). The GPEC-trace of the block copolymer, centered at a retention time of 10.4 min., is clearly separated from the traces of the two homopolymers. Further inspection of the chromatogram reveals only a small trace of PCL homopolymer present in the isolated block copolymer product, which most probably stems from the unfunctionalized and cyclic PCL.

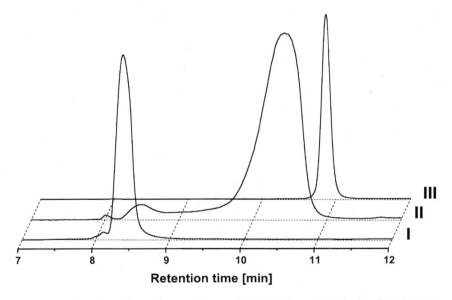

*Figure 1. GPEC-traces of homopolymer PCL (**I**; M_n = 13 000 g/mol), P(MMA-b-PCL) obtained in the chemoenzymatic reaction (**II**; M_n: 37 000 g/mol) and homopolymer PMMA (**III**; M_n = 15 000 g/mol). (Reproduced from reference **Error! Bookmark not defined.**. Copyright 2005 American Chemical Society.)*

A similar approach to block copolymers by combining eROP and ATRP was followed by Wang with a different bifunctional initiator *(15)*. Moreover, this approach was successfully applied for the synthesis of amphiphilic block copolymers in supercritical CO_2 ($scCO_2$) *(16)*. A hydrocarbon block and semifluorinated poly-(1*H*,1*H*,2*H*,2*H*-perfluorooctyl methacrylate) PFOMA block were synthesized using PCL macroinitiator containing chain ends with **2**. It has been demonstrated that PCL macroinitiators polymerize FOMA effectively, leading to a PFOMA block with controlled molecular weight.

One pot block copolymerization

Stimulated by the impact of one-pot cascade reactions in organic synthesis we investigated whether both polymerizations can be done in one-pot without intermediate work-up step *(17)*. Synthetically, one can distinguish two approaches: (i) both eROP and ATRP take place simultaneously, which requires that all components are present in the reaction flask from the beginning of the reaction. (ii) eROP and ATRP take place in one-pot but consecutively. In this case the second reaction has to be activated by an external stimulus on demand. This could be the addition of a reaction component required for the second reaction step, such as a reactant or catalyst. The success of a one-pot approach highly depends on the absence of mutual retardation by any present reaction component. The first successful reaction of this type was the block copolymer synthesis combining enzymatic ring opening polymerization (eROP) with nitroxide mediated controlled free radical polymerization (NMP) from a bifunctional initiator *(18)*. While the reaction was conducted in one pot without intermediate work-up, a compartmentalization of both techniques was achieved by the different activation temperatures of the individual polymerizations (eROP: 60 °C; NMP: > 90 °C). Hence, the enzymatic polymerization could be conducted undisturbed by the chemical process, which was subsequently activated by an increase of the reaction temperature.

A similar compartmentalization resulted in a successful one-pot process for the combination of ATRP and eROP, i.e., in a one-pot, two-step procedure *(17)*. For this process CL, *t*-butyl methacrylate (*t*-BMA), Novozym 435 and **2** were heated to 60 °C to initiate the eROP and obtain the PCL block end-capped with **2**. After 120 min. CuBr/dNbipy was added in order to activate the ATRP and thus the block copolymer formation. Figure 4 shows that during the first step of the consecutive process, i.e. the eROP, CL conversion reached 95 % while only negligible conversion of *t*-BMA was detected. Only upon addition of the ATRP catalyst, the radical polymerization started and reached ca. 43 % conversion within 180 min (300 min. total). Both reaction kinetics are comparable with the kinetics observed from the homopolymerization under similar conditions, suggesting that both reactions run undisturbed by each other (Figure 4). A clear

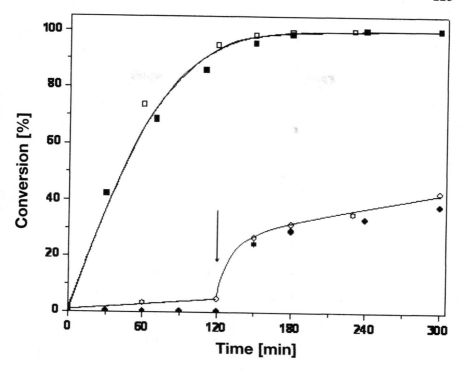

Figure 4. Conversion of CL (□) and t-BMA (○) in a consecutive one-pot cascade polymerization (arrow marks addition of ATRP catalyst) in comparison with the conversion of CL (■) and t-BMA (●) in a homopolymerization. Lines are added to guide the eye for data set of the cascade polymerization. (Reproduced with permission from reference 17. Copyright 2006 John Wiley and Sons, Ltd..)

increase of the molecular weight from 4 200 g/mol (first step) to 6 800 g/mol after the second step was seen in SEC, suggesting the formation of block copolymers. This result was confirmed by GPEC measurements.

We applied a similar approach under kinetic resolution conditions (Figure 5) *(19)*. First a mixture of **2** and racemic 4-methylcaprolactone (4-MeCL) was reacted until all *(S)*-4-MeCL had been converted, i.e., to about 50 % total monomer conversion. Subsequently, oxygen was removed from the reaction medium by several consecutive freeze-pump-thaw cycles. The ATRP was initiated by adding MMA and $Ni(PPh_3)_2Br_2$ and raising the reaction temperature to 80 °C. In this case the nickel complex acts as both ATRP initiator and enzyme inhibitor thereby preventing any side reactions caused by the enzyme. Chiral block copolymers were obtained by this approach as evident from SEC analysis.

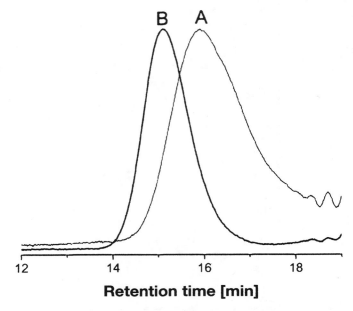

B A

Retention time [min]

*Figure 5. SEC traces of unprecipitated P(4-MeCL) [trace A, M$_n$: 3 200 g/mol,
PDI: 1.8] and P(4-MeCL-b-MMA) (trace B, M$_n$:10 900 g/mol, PDI: 1.3).
(Reproduced with permission from reference 19. Copyright 2004.)*

Compartmentalization by sequential monomer addition was also successfully demonstrated in the chemoenzymatic synthesis of block copolymers in scCO$_2$ using 1H,1H,2H,2H-perfluorooctyl methacrylate (FOMA) as an ATRP monomer. Detailed analysis of the obtained polymer P(FOMA-b-PCL) confirmed the presence of predominantly block copolymer structures *(16)*. The clear advantage of the scCO$_2$ in this approach is that unlike conventional solvents it solubilises the fluorinated monomer.

The combination of eROP with ATRP in a one-pot cascade reaction without intermediate activation proved more challenging *(17)*. This stems partly from the fact that ATRP needs metal catalyst, which adds an additional component and thus potential inhibitor to the reaction. For example when CuBr, a common ATRP catalyst, was added to an eROP of CL the reaction stopped at a monomer conversion of ca. 20 %, after which total enzyme inhibition was observed. Interestingly, an immediate inhibition was observed with CuBr$_2$. This suggests that the actual inhibiting species is Cu^{2+}, which in the case of the Cu$^+$ addition might slowly be generated by oxidation. This would explain the induction period before inhibition. The exact mechanism of the inhibition is not clear but one can speculate that complexation of the cation by the histidine of the amino acid triad

(Ser-His-Asp) in the active site plays a significant role. In the presence of typical ATRP multivalent ligands like PMDETA or dNbipy the enzyme inhibition is much less pronounced. This is not unexpected since the concentration of metal ions available for inhibition of the enzyme is reduced due to complexation by the ligand. Interestingly, the reaction in the presence of the tridentate PMDETA shows a faster CL conversion than dNbipy, which is probably due to stronger coordination of the Cu^+ by the former ligand. The results show that both ligand systems allow enzymatic polymerization in the tested systems.

Another important parameter in the cascade polymerization is the monomer. While we did not observe any influence of the CL on the ATRP of MMA, MALDI-ToF analysis confirmed that the enzymatic polymerization in the presence of MMA inevitably results in transesterification, and thus in the incorporation of methacrylate groups to a high extent. Moreover, even the methanol released in this transesterification process acts as an active component in the reaction, as evident from the presence of PCL molecules with methyl ester end-groups. Consequently, MMA cannot be used for a one-pot cascade reaction in organic solvents. This problem can be solved by using *tert*-butylmethacrylate (t-BMA) as the vinyl component, because of the inertness of tertiary alcohols/esters in the enzymatic reaction. Side-reactions similar to those seen with MMA have not been observed in MALDI-ToF-MS with this monomer. While block copolymers were successfully synthesized in this cascade approach a significant trace of t-BMA homopolymer can be detected in GPEC. Moreover, comparison of the conversion plots show that in this reaction the CL conversion stops at 30 % after 60 min reaction time, suggesting an enzyme inhibition, while the conversion of t-BMA proceeds *(17)*.

A successful, simultaneous one-pot synthesis of block copolymers by combining eROP of CL with ATRP of methyl methacrylate (MMA) was achieved in $scCO_2$ *(20)*. We demonstrate that both enzyme- and metal-based ATRP catalysts function concurrently in that solvent. Use of CL as a cosolvent allows the ATRP-catalyzed growth of the PMMA block to proceed with good control, despite the fact that the CL cosolvent is being consumed by the enzymatic ROP. The key to success is to ensure that the ATRP process remains homogeneous in $scCO_2$. This is not normally the case (vide infra), but in this process CL acts as a very effective cosolvent, while also supplying the monomer for consumption by the enzymatic polymerization. Intriguingly, control over the ATRP reaction was preserved, despite the consumption of the CL cosolvent as it polymerized to form the second block. A key advantage of the use of $scCO_2$ is that the ATRP can occur in the $scCO_2$ plasticized PCL that is formed, as well as in the CL monomer/$scCO_2$ solution. The two catalysts do not appear to have any adverse effect on each other; both PCL and PMMA peaks were clearly observed in the ^1H NMR spectrum of the product. From the NMR analysis, the block copolymers were found to be approximately 15 mol % PMMA. To demonstrate block copolymer formation, the product was hydrolyzed. SEC analysis of the product clearly shows a shift to lower molecular weight, implying the removal of the PCL from the block copolymer (Figure 6). Recently also the successful combination of eROP with RAFT in scCO$_2$ was reported *(21)*.

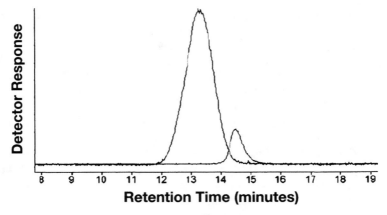

Figure 6. SEC traces for the block copolymer before (M_w: 41 000, PDI: 2.11) and after (M_w: 10 000, PDI: 1.07) hydrolysis of the PCL block. Concentration of the hydrolyzed sample is approximately 10 mol % relative to that of the block copolymer. (Reproduced from reference 20. Copyright 2005 American Chemical Society.)

Conclusions

We investigated the chemoenzymatic synthesis of block copolymers combining eROP and ATRP using a bifunctional initiator. A detailed analysis of the reaction conditions revealed that a high block copolymer yield can be realized under optimized reaction conditions. Side reactions, such as the formation of PCL homopolymer, in the enzymatic polymerization of CL could be minimized to < 5 % by an optimized enzyme drying procedure. Moreover, the structure of the bifunctional initiator was found to play a major role in the initiation behavior and hence, the yield of PCL macroinitiator. Block copolymers were obtained in a consecutive ATRP. Detailed analysis of the obtained polymer confirmed the presence of predominantly block copolymer structures. Optimization of the one-pot procedure proved more difficult. While the eROP was compatible with the ATRP catalyst, incompatibility with MMA as an ATRP monomer led to side-reactions. A successful one-pot synthesis could only be achieved by sequential addition of the ATRP components or partly with inert monomers such as *t*-butyl methacrylate. One-pot block copolymer synthesis was successful, however, in supercritical carbon dioxide. Side reactions such as those observed in organic solvents were not apparent.

Acknowledgments

MdG acknowledges financial support from the Dutch Polymer Institute (DPI). CJD and SV acknowledge financial support from a Marie Curie Research Training Network (BIOMADE, MRTN-CT-2004-505147). SMH is a Royal Scociety-Wolfson Research Merit Award Holder.

References

1. Matyjaszewski, K. Ed. Advances in Controlled/Living Radical Polymerization, ACS Symposium Series No. 854, Oxford University Press, **2003**.
2. Kamigaito, M.; Ando, T.; Sawamoto, M. *Chem. Rev.* **2001**, *101*, 3689
3. Hawker, C. J.; Bosman, A. W.; Harth, E. *Chem. Rev.* **2001**, *101*, 3661.
4. Kobayashi, S.; Uyama, H.; Kimura, S. *Chem. Rev.* **2001**, *101*, 3793.
5. Gross, R. A.; Kumar, A.; Kalra, B. *Chem. Rev.* **2001**, *101*, 2097.
6. Anderson, E. M.; Larsson, K. M.; Kirk, O. *Biocatal. Biotransform.* **1998**, *16*, 181.
7. Bernaerts, K. V.; Du Prez, F. E. *Prog. Polym. Sci.* **2006**, *31*, 671.
8. Henderson, L. A.; Svirkin, Y. Y.; Gross, R. A.; Kaplan, D. L.; Swift, G. *Macromolecules* **1996**, *29*, 7759.
9. Uyama, H.; Suda, S.; Kobayashi, S. *Acta Polym.* **1998**, *49*, 700.
10. Cordova, A.; Iversen, T.; Hult, K. *Polymer* **1999**, *40*, 6709.
11. Kumar, A.; Gross, R. A.; Wang, Y. B.; Hillmeyer, M. A. *Macromolecules* **2002**, *35*, 7606.
12. de Geus, M.; Peeters, J.; Wolffs, M.; Hermans, T.; Palmans, A. R. A.; Koning, C. E.; Heise, A. *Macromolecules* **2005**, *38*, 4220.
13. Meyer, U.; Palmans, A. R. A.; Loontjens, T.; Heise, A. *Macromolecules* **2002**, *35*, 2873.
14. Philipsen, H. J. A. *J. Chrom., A*, **2004**, *1037*, 329.
15. Sha, K.; Li, D.; Wang, S.; Qin, L.; Wang, J *Polym Bull* **2005**, *55*, 349.
16. Villarroya, S; Zhou, J.; Duxbury, C. J.; Heise, A.; Howdle, S. M. *Macromolecules* **2006**, *39*, 633.
17. de Geus, M.; Schormans, L; Palmans, A. R. A.; Koning, C. E.; Heise, A. *J. Polym. Sci. A; Polym.* **2006**, *44*, 4290.
18. As, B. van; Thomassen, P.; Palmans, A. R. A.; Kalra, B; Gross, R. A.; Heise, A., *Macromolecules* **2004**, *37*, 8973.
19. Peeters, J.; Palmans, A. R. A.; Veld, M.; Scheijen, F.; Heise, A.; Meijer, E. W. *Biomacromolecules* **2004**, *5*, 1862.
20. Duxbury, C. J.; Wang, W.; de Geus, M.; Heise, A.; Howdle, S. M. *J. Am. Chem. Soc.* **2005**, *127*, 2384.
21. Thurecht, K.J.; Gregory, A.M.; Villarroya, S.; Zhou, J.; Heise, A.; Howdle, S.M. *Chem. Commun.* **2006**, *42*, 4383.

Chapter 15

Ring-Opening of ω-Substituted Lactones by Novozym 435: Selectivity Issues and Application to Iterative Tandem Catalysis

Anja R. A. Palmans, Bart A. C. van As, Jeroen van Buijtenen, and E. W. Meijer

Laboratory of Macromolecular and Organic Chemistry, Technische Universiteit Eindhoven, Den Dolech 2, P.O. Box 513, 5600 MB Eindhoven, The Netherlands

The enantioselectivity of the Novozym® 435 catalysed ring-opening of lactones is related to the conformation of the ester: *cisoid* lactones show *S*-selectivity or no selectivity while *transoid* lactones show a pronounced *R*-selectivity. The inability of Novozym 435 to polymerise 6-MeCL stems from its opposing selectivities for the alcohol and acyl moiety. By combining Novozym 435 with a racemisation catalyst, unreactive terminal alcohols in the *S*-configuration can be turned into reactive terminal alcohols of the *R*-configuration which can propagate. Combining 2 different catalysts that work together to accomplish propagation, also referred to as iterative tandem catalysis, is an elegant approach to convert a racemic monomer quantitatively into a homochiral polymer.

Introduction

Synthetic polymers applied in everyday life rarely possess well-defined stereochemistries of their backbones. This sharply contrasts with the polymers made by Nature where perfect control is the norm. An exception is poly-L-lactide; this polyester is frequently used in a variety of biomedical applications.[1] By simply playing with the stereochemistry of the backbone, properties ranging from a semi-crystalline, high melting polymer (poly-L-lactide) to an amorphous high T_g polymer (poly-*meso*-lactide) can be achieved.[2] The *synthetic* synthesis of such chiral polymers typically starts from optically pure monomers obtained form the chiral pool. The fermentation product L-lactic acid, for example, is the starting material for the synthesis of poly(L-lactide).

The stereoselective polymerisation of lactides and β-butyrolactone using chiral metal based initiators or the kinetic resolution polymerisation of substituted lactones with Novozym 435 leads to enantio-enriched polymers starting from racemic monomer mixtures.[3,4] Converting a racemic mixture of monomers *quantitatively* into a homochiral polymer is less straightforward.[5] This is surprising considering the enormous potential of asymmetric catalysis and tandem catalysis developed in organic chemistry in the past decades.[6] Dynamic kinetic resolution (DKR) of *sec*-alcohols, for example, is a powerful and elegant method to quantitatively convert a racemic mixture of *sec*-alcohols into an enantiopure *R*-ester by combining lipase-catalysed transesterification with Ru-catalysed racemisation.[7]

The extension of tandem catalysis to polymer chemistry is, however, not trivial. In order to reach high molecular weight polymers, each reaction has to proceed with almost perfect selectivity and conversion. Obviously, combining different catalytic reactions limits the choice of suitable reactions since they must also be compatible with each other. We recently introduced Iterative Tandem Catalysis (ITC), a novel polymerisation method in which chain growth during polymerisation is effectuated by two or more intrinsically different catalytic processes that are both compatible and complementary. If the catalysts and monomers are carefully selected, ITC is able to produce chiral polymers from racemic monomers, as was shown by us for the ITC of 6-MeCL and the DKR polymerisation of *sec*-diols and diesters.[8,9]

Here, we will discuss some fundamental aspects of the selectivity of Novozym 435 —*Candida antarctica* Lipase B immobilised on an acrylic resin— for ω-substituted lactones. This selectivity determines the polymerisability of a ω-substituted lactone employing Novozym 435 as the catalyst. We will first address the influence of the conformation of the ester bond in the lactone (*cisoid* or *transoid*) on the selectivity of the ring-opening of a selection of lactones with ring sizes varying from a 4- to 13-membered ring. Then, we will address the influence of the size of the substituents at the ω-position on the selectivity and reactivity of lactones. Finally, we will show that ITC is successfully applied in the polymerisation of ω-methylated lactones such as 6-MeCL and 6-EtCL, but is less broadly applicable than expected.

Selectivity of Novozym 435 for ω-substituted lactones

Ring-opening of ω-substituted lactones by Novozym 435

Research by Huisgen and co-workers showed that the ester bond in unsubstituted lactones can exist in two conformations: the higher energy *cisoid* conformation and the lower energy *transoid* conformation (Figure 1).[10] Up to the 7-membered ring (ε-caprolactone) conformational strain only allows for a *cisoid* conformation of the ester bond. Starting from the 8-membered ring, the *transoid* conformation becomes possible and from the 10-membered ring on, the ester bond is exclusively in the *transoid* conformation.

transoid   cisoid

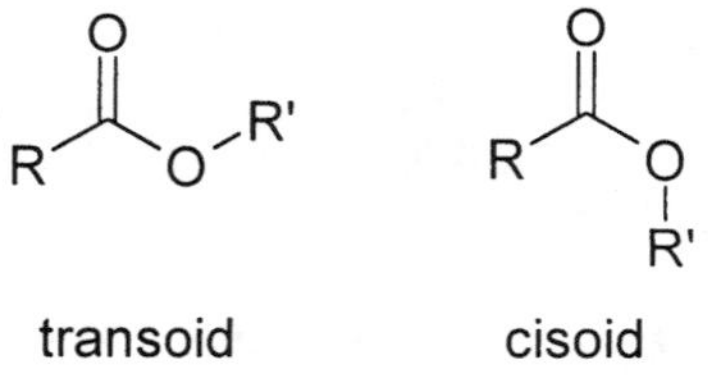

Figure 1. Cisoid and transoid conformation of the ester bond

We were interested in the selectivity of the ring-opening of ω-substituted lactones employing Novozym 435 as the catalyst. We assume that a methyl-substituent at the ω-position will not dramatically affect the conformational preferences (*cisoid* or *transoid*) of the ester bond in the substituted lactones. Several ω-methylated lactones from a 4- to a 13-membered ring were synthesized in their racemic form (Figure 2). The lactones were then subjected to Novozym 435 catalysed ring-opening employing benzyl alcohol (BA) as the nucleophile. The results are summarized in Table 1.

Fascinating differences in the selectivities and reaction rates of the ring-opening of the different lactones with Novozym 435 were observed (Table 1). While ring-opening of the small lactones that are exclusively in a *cisoid* conformation is S-selective (βBL and 6-MeCL) or non-selective (5-MeVL), ring-opening of the lactones that can adopt a *transoid* conformation (7-MeHL, 8-MeOL, 12-MeDDL) is exclusively R-selective. The change in selectivity is abrupt: while Novozym 435 is moderately selective for *(S)*-6-MeCL (E ratio =

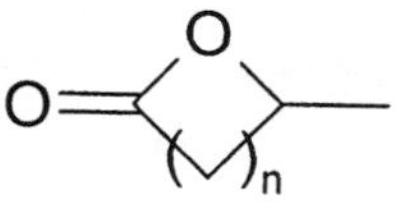

n = 1: β-butyrolactone (βBL)
n = 3: ω-methyl-δ-valerolactone (5-MeVL)
n = 4: ω-methyl-ε-caprolactone (6-MeCL)
n = 5: ω-methyl-7-heptanolactone (7-MeHL)
n = 6: ω-methyl-8-octanolactone (8-MeOL)
n = 10: ω-methyl-12-dodecanolactone (12-MeDDL)

Figure 2. Racemic methylated lactones employed in this study

12 at 60°C, *vide infra*), almost complete enantioselectivity is observed for the *R*-enantiomer in 7-MeHL. The reactivity of the faster reacting enantiomers of the different lactones also varies quite significantly. The relative reactivities are similar to the relative reactivities found previously for the ring-opening polymerisation of unsubstituted lactones.[11]

Table 1. Selectivity of Novozym 435 for ω-Methylated Lactones[a]

Lactone	Selectivity	k_{cat} [1/s]	
		S-enantiomer	*R*-enantiomer
βBL	*S*	45.7	n.d.[b]
5-MeVL[c]	-[d]	7.9	7.6
6-MeCL	*S*	49.3	8.5
7-MeHL	*R*	0.01[e]	204.4
8-MeOL	*R*	n.d.[b]	10.3
12-MeDDL[f]	*R*	n.d.[b]	23.3

[a] Reaction conditions: lactone (4 mmol), BA (1 mmol), Novozym 435 (27 mg), 1,3,5-tri-t-butylbenzene (0.3 mmol, internal standard) in toluene (2 mL); reaction at 70 °C. [b] Could not be determined. [c] Experiment performed with excess of 1-octanol (8 mmol) as initiator because of ring-chain equilibrium; enzyme dried overnight at 50 °C over P_2O_5. [d] No significant enantioselectivity was observed for the reaction. [e] Determined in a separate experiment using 2 mmol of isolated (S)-7-MeHL. [f] 2 mmol of 12-MeDDL.
SOURCE: Reproduced from *J. Am. Chem. Soc.* **2004**, *129*, 7393. Copyright 2004 American Chemical Society.

Selectivity of Novozym 435 for acyl donor and alcohol nucleophile

The action of *Candida antarctica* Lipase B (CALB) relies on a two-step mechanism with an acylation step and a deacylation step, involving a covalent acyl-enzyme intermediate (Scheme 1).[12] After binding of an acyl donor (typically an ester) in the active site, an acyl-enzyme intermediate (acylation of the enzyme) is formed. When a nucleophile (typically an alcohol) attacks, the product leaves the active site and the free enzyme remains (deacylation of the enzyme). In case of a lactone as the acyl donor, the product formed after ring-opening of the ring can act as nucleophile, explaining the propensity of lactones to be polymerised by Novozym 435.

In the Novozym 435 catalysed ring-opening of a (chiral) substituted lacton, both acylation and deacylation can be enantioselective. For example, it is well known that CALB shows pronounced selectivity for R-secondary alcohols in the deacylation step.[13] Since the forward and backward reaction exhibits by definition the same selectivity, esters comprising a substituent at the alcohol side are expected to show pronounced R-selectivity in the acylation step.[14] This is indeed observed for 7-MeHL, 8-MeOL and 12-MeDDL (Table 1). However, the selectivity for acyl donor in the case of βBL, 5-MeVL and 6-MeCL –lactones in which the ester bond is exclusively in a *cisoid* conformation– is low or for the S-enantiomer. We can speculate that lactones in a *cisoid* conformation must attain a different orientation in the active site in order to be activated.[15]

We investigated the ring-opening of 6-MeCL in more detail using benzyl alcohol as the nucleophile and Novozym 435 as the catalyst in a 1/4 BA/lactone ratio (Figure 3A). The conversion of both enantiomers of 6-MeCL and BA as a function of time is shown in Figure 3B. Ring-opening is S-selective, as evidenced by the faster *(S)*-6-MeCL conversion compared to *(R)*-6-MeCL. Since 6-MeCL is an ω-substituted lactone, ring-opening results in a terminal *sec*-alcohol, in this case with the S-configuration. These S-alcohols are not accepted as a nucleophile by Novozym 435 since the lipase-catalysed transesterification is highly R-selective. Figure 3B indeed shows that 1 equivalent of *(S)*-6-MeCL is consumed. Consequently, the S-configuration of the terminal secondary alcohol prevents polymerisation of 6-MeCL from taking place on a realistic timescale. This was observed by us previously.[4b-c]

For the higher membered ring lactones, on the other hand, the R-lactone is the preferred substrate (Table 1). As a consequence, ring-opening affords an R-terminal alcohol which is accepted as the nucleophile. Indeed, Novozym 435 catalyses the ring-opening polymerisation of 7-MeHL, 8-MeOL and 12-MeDDL, furnishing the corresponding *(R)*-polyesters and the unreacted *(S)*-lactones.[15]

Ring-opening of other ω-substituted ε-caprolactones by Novozym 435

Because of the fascinating opposite selectivity of Novozym 435 for acyl donor and alcohol nucleophile in the case of 6-MeCL, we extended this investigation to ε-caprolactones with larger substituents at the ω-position. For this purpose, 6-ethyl-ε-caprolactone (6-EtCL), 6-*n*-propyl-ε-caprolactone (6-

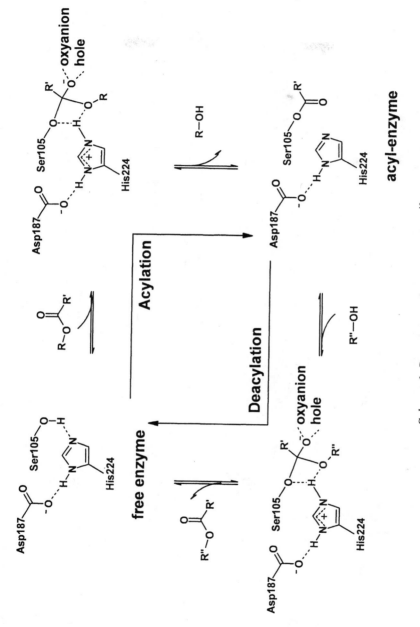

Scheme 1. Reaction mechanism of CALB[12]

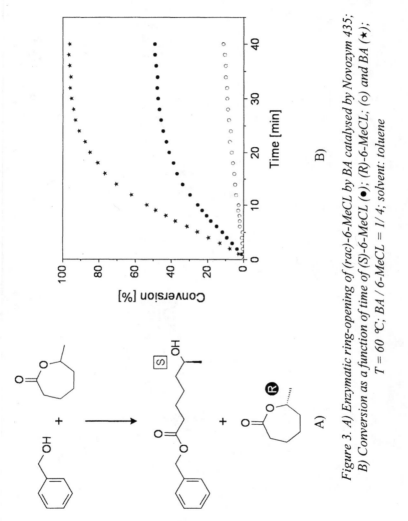

Figure 3. A) Enzymatic ring-opening of (rac)-6-MeCL by BA catalysed by Novozym 435;
B) Conversion as a function of time of (S)-6-MeCL (●); (R)-6-MeCL; (○) and BA (★);
T = 60 °C; BA / 6-MeCL = 1/4; solvent: toluene

PrCL) and 6-n-butyl-ε-caprolactone (6-BuCL) were synthesised and subjected to a Novozym 435 catalysed ring-opening. The turnover frequencies (TOF)[16] and the enantiomeric ratio E, which is a measure of the enantioselectivity of the enzyme for the lactone,[17] were derived from the kinetic plots.

Table 2 shows the kinetic and enantioselectivity data of the enzymatic ring-opening of 6-MeCL to 6-BuCL. In all cases, the S-enantiomer is the preferred substrate. The reactivity of 6-MeCL and 6-EtCL is similar (entries 1 and 2) with TOF of 340 and 132 min⁻¹, respectively. The reactivity reduces significantly when the substituent size increases to propyl and butyl (entries 3 and 4) where the TOF is 14 and 35 min⁻¹, respectively. Only 6-MeCL shows an appreciable enantioselectivity (E = 12), while the enantioselectivity for the lactones with larger substituent is significant, but low (2.8 to 4.8). This makes both 6-MeCL and 6-EtCL interesting substrates for ITC since they are readily accepted by Novozym 435 as a substrate (high TOF), and show moderate selectivities.

Table 2. Kinetic and Enantioselectivity Data for the Enzymatic Ring-Opening of ω-Substituted ε-Caprolactones.[a]

entry	substrate	TOF [1/min]	E ratio	Preferred enantiomer
1	6-MeCL	340	12 +/- 0.5	S
2	6-EtCL	132	4.8 +/- 0.1	S
3	6-PrCL	14.0	2.8 +/- 0.1	S
4	6-BuCL	35.0	4.1 +/- 0.3	S

[a] Reaction conditions: BA/Lactone = 1/4; T=60°C; solvent = toluene; [lactone] = 3.4 M

Chiral polymers from racemic ω–substituted lactones by ITC

6-MeCL and 6-EtCL are good substrates for Novozym 435 as evidenced by the high activity in the ring-opening reactions. However, they cannot be polymerised as a result of the opposite selectivity of Novozym 435 for the nucleophile (R-selectivity) and the lacton (S-selectivity). By combining Novozym 435 catalysed ring-opening of 6-MeCL with racemisation of the terminal alcohol, it is theoretically possible to polymerise 6-MeCL. This concept is an example of iterative tandem catalysis (ITC). Scheme 2 shows the different steps that are required to enable a successful polymerisation. First, ring-opening of (S)-6-MeCL, the preferred enantiomer of Novozym 435, will lead to a ring-opened product with an S-alcohol chain end. Since Novozym 435 is highly enantioselective towards R-secondary alcohols, this chain end is not accepted as a nucleophile. Therefore, the product is virtually unreactive and propagation does not occur. To enable polymerisation of (S)-6-MeCL, racemisation of the

S-alcohol that is formed upon ring-opening is required, furnishing a reactive *R*-chain end. These can subsequently react with another molecule of *(S)*-6-MeCL.

Racemisation can be achieved with a variety of homogeneous catalysts. We selected the well-known Shvo catalyst (Scheme 2).[18] Since many DKR processes have been successfully developed with the combination Novozym 435 and this Ru-based catalyst, the two catalysts are both compatible and complementary.[19] *In-situ* racemisation of the terminal secondary alcohol of the propagating polymer chain should provide reactive chain ends, theoretically resulting in an enantiopure polymer in a 100% yield starting from the racemic monomer (Scheme 2). If the less reactive *(R)*-6-MeCL is incorporated (which will occur since the selectivity for lactones is moderate with an E value of 12, see Table 2) a reactive *R*-chain end is obtained and propagation occurs instantly. In this way, both enantiomers of the monomer are consumed.

In order to keep the system as simple as possible, we conducted the initial experiments with *(S)*-6-MeCL as the substrate.[20] To compensate for the effect of dehydrogenation of the end-groups, 2,4-dimethyl-3-pentanol was added as a hydrogen donor.[21] Figure 4A shows the conversion of *(S)*-6-MeCL as a function of time in a typical ITC experiment. Although relatively slow, the conversion of lactone increases steadily and after 318 hr full conversion is reached. After 5, 8, and 13 days of reaction, samples were evaluated with size exclusion chromatography (SEC) to assess the formation of polymers. As is clear from Figure 4B, the molecular weight increases over time, confirming that polymerisation indeed takes place. Poly-*(R)*-6-MeCL with a promising $ee_{polymer}$ of 86% and M_p of 8.2 kDa was obtained after work-up. The low rate of reaction compared to DKR (typically complete after 48 h with the Shvo catalyst) is attributed to the low concentration of the terminal alcohol as well as to the iterative nature of the system.

We then switched to *(rac)*-6-MeCL as the substrate. Figure 5 shows the consumption of both *(R)*- and *(S)*-6-MeCL as a function of time in a typical ITC experiment. Coincidentally, comparable rates of reaction for both enantiomers were observed. *(rac)*-6-MeCL was polymerised within 220 h with complete conversion of both enantiomers, yielding a high $ee_{polymer}$ of 92% and a M_p of 9.4 kDa.

In order to evaluate if polymers of higher molecular weight are accessible, the reaction was repeated with M/I = 206 and 1,6-heptanediol as the initiator. This diol was selected since hydrogenolysis of the lactone, a minor side reaction catalysed by the Shvo catalyst, produces the diol in small amounts. By controlling the amount of initiator in this way, a polymer with an M_p of 20.8 kDa was obtained. The polymer was precipitated from methanol to remove the catalyst (yield 34%) and the number-averaged molecular weight obtained from SEC analysis was 25.0 kDa with a PDI of 1.23 (Figure 6). Interestingly, poly-*(R)*-6-MeCL shows a melting point of 28°C while its racemic analogue is an oil.

239

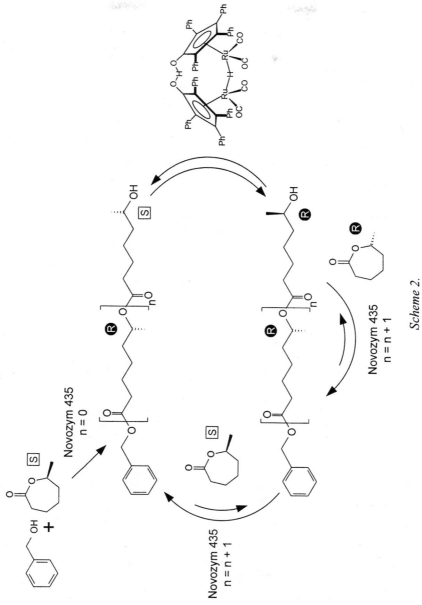

Scheme 2.

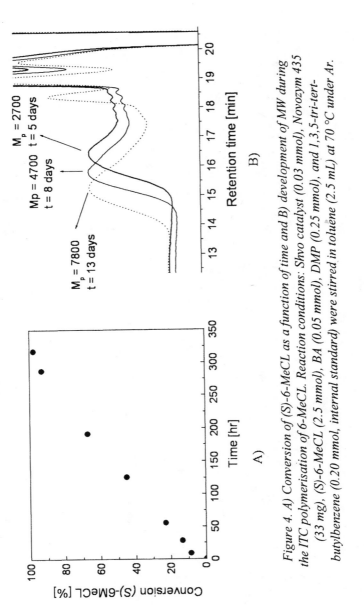

Figure 4. A) Conversion of (S)-6-MeCL as a function of time and B) development of MW during the ITC polymerisation of 6-MeCL. Reaction conditions: Shvo catalyst (0.03 mmol), Novozym 435 (33 mg), (S)-6-MeCL (2.5 mmol), BA (0.05 mmol), DMP (0.25 mmol), and 1,3,5-tri-tert-butylbenzene (0.20 mmol, internal standard) were stirred in toluene (2.5 mL) at 70 °C under Ar.

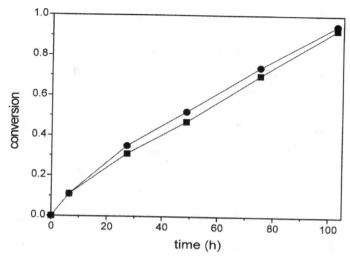

Figure 5. Conversion of (S)-6-MeCL (■) and (R)-6-MeCL (●) vs. time in a typical ITC experiment with 40 eq of 6-MeCL with respect to BA, 1.2 mol% of 2 and 14 mg Novozym 435 / mmol 6-MeCL. (Reproduced by permission of the Royal Society of Chemistry from ref 8b. Copyright 2006.)

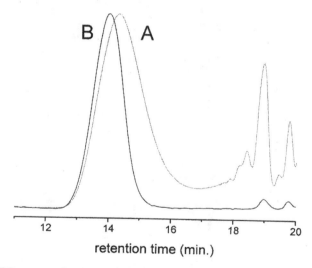

Figure 6. SEC traces of polymer obtained by ITC of (rac)-6-MeCL before (A) and after (B) precipitation from methanol ($M_n = 25.0$ kDa, PDI = 1.23). (Reproduced by permission of the Royal Society of Chemistry from ref 8b. Copyright 2006.)

Other monomers were also tested in the ITC polymerisation. Gratifyingly, ITC of 6-EtCL was successful. A chiral polymer with an $ee_{polymer}$ of 93% and a molecular weight M_p of 6.4 kDa was obtained. 6-PrCL and 6-BuCL are not suitable as a substrate since CALB can only accomodate secondary alcohols with methyl or ethyl substituents in its active site.[13b] Larger ring lactones such as 7-MeHL, 8-MeOL and 12-MeDDL are, unfortunately, unsuitable in ITC since the R-enantiomer will propagate rapidly and the racemisation catalyst will only delay the reaction. Moreover, S-enantiomers are not accepted as a substrate so quantitative conversion of the racemic mixture in reasonable time scales is not feasible. Smaller ring lactones such as 5-MeVL and βBL were also not suitable in ITC: 5-MeVL showed a high propensity for cyclisation and could not be polymerised while βBL showed a low reactivity and enantioselectivity for the transesterification of the terminal alcohols and no significant polymerization was observed within a reasonable amount of time.

Conclusions

The enantioselectivity of Novozym 435 catalysed ring-opening of lactones is related to the conformation of the ester: *cisoid* lactones show S-selectivity or no selectivity while *transoid* lactones show a pronounced R-selectivity. The inability of Novozym 435 to polymerise 6-MeCL stems from its opposing selectivities for the alcohol and acyl moiety. By combining Novozym 435 with a racemisation catalyst, unreactive alcohol chain ends in the S-configuration can be turned into reactive alcohol chain end of the R-configuration which can propagate. Combining 2 different catalysts that work together to accomplish propagation, also referred to as iterative tandem catalysis, is an elegant approach to convert a racemic monomer quantitatively into a homochiral polymer.

Acknowledgement

The authors would like to thank G. Verzijl, Rinus Broxterman and Andreas Heise of DSM and Emmo Meijer of Unilever for stimulating discussions. Marloes Verbruggen of the Junior Researchers Institute is gratefully acknowledged for lactone syntheses. NRSC-C, NWO and DSM are acknowledged for financial support.

References and Notes

1. Ikada, Y.; Tsuji, H. *Macromol. Rapid Commun.* **2000**, *21*, 117.
2. Lou, X.; Detrembleur, Ch.; Jérôme, R. *Macromol. Rapid. Commun.* **2003**, *24*, 161.

3. a) Dechy-Cabaret, O.; Martin-Vanca, B.; Bourissou, D. *Chem. Rev.* **2004**, *104*, 6147. b) Spassky, N.; Wisniewsky, M.; Pluta, Ch.; Le Borgne, A. *Macromol. Chem. Phys.* **1996**, *197*, 2627. c) Amgoune, A.; Thomas, C. M.; Ilinca, S.; Roisnel, T.; Carpentier, J.-F. *Angew. Chem Int. Ed.* **2006**, *45*, 2782. d) Zhong, Z.; Dijkstra, P. J.; Feijen, J. *J. Am. Chem. Soc.* **2003**, *125*, 11291. e) Ovitt, T. M.; Coates, G. W. *J. Am. Chem. Soc.* **1999**, *121*, 4072.

4. a) Al-Azemi, T. F.; Kondaveti, L.; Bisht, K. S. *Macromolecules* **2002**, *35*, 3380. b) Peeters, J.; Palmans A.R. A.; Veld, M.; Scheijen, F.; Heise, A.; Meijer, E. W. *Biomacromolecules* **2004**, *5*, 1862. c) Peeters, J. W.; van Leeuwen, O.; Palmans, A. R. A.; Meijer, E. W. *Macromolecules* **2005**, *38*, 5587.

5. Okamoto, Y.; Nakano, T. *Chem. Rev.* **1994**, *94*, 349.

6. a) Wasilke, J. C.; Obrey, S. J.; Baker, R. T.; Bazan, G. C. *Chem. Rev.* **2005**, *105*, 1001. b) Bruggink, A.; Schoevaart, R.; Kieboom, T. *Org. Process Res. Dev.* **2003**, *7*, 622. c) Jacobsen, E. N.; Pfaltz, A.; Yamamoto, H.; Eds. Comprehensive Asymmetric Catalysis; Springer: Berlin, **1999**; Vol. 1. d) Ojima, I.; Ed. Catalytic Asymmetric Synthesis; 2nd Ed.; **2000**. e) Blaser, H.-U.; Malan, C.; Pugin, B.; Spindler, F.; Steiner, H.; Studer, M. *Adv. Synth. Catal.* **2003**, *345*, 103. f) Blaser, H.-U. *Chem. Commun.* **2003**, 293.

7. a) Pamies, O.; Backvall, J.-E. *Chem. Rev.* **2003**, *103*, 3247. b) Pamies, O.; Backvall, J.-E. *Trends Biotechnol.* **2004**, *22*, 130.

8. a) van As, B. A. C.; van Buijtenen, J.; Heise, A.; Broxterman, Q. B.; Verzijl, G. K. M.; Palmans, A. R. A.; Meijer, E. W. *J. Am. Chem. Soc.* **2005**, *127*, 9964. b) van Buijtenen, J.; van As, B. A. C.; Meuldijk, J.; Palmans, A. R. A.; Vekemans, J.; Hulshof, L. A.; Meijer, E. W. *Chem. Commun.* **2006**, 3169.

9. a) Hilker, I.; Rabani, G.; Verzijl, G. K. M.; Palmans, A. R. A.; Heise, A. *Angew. Chem. Int. Ed.* **2006**, *45*, 2130. b) van As, B. A. C.; van Buijtenen, J.; Mes, T.; Palmans, A. R. A.; Meijer, E. W. *Chemistry*, **2007**, *accepted for publication.*

10. a) Huisgen, R. *Angew. Chem.* **1957**, *69*, 341. b) Huisgen, R.; Ott, H. *Angew. Chem.* **1958**, *70*, 312. c) Huisgen, R.; Ott, H. *Tetrahedron* **1959**, *6*, 253.

11. Van der Mee, L.; Helmich, F.; De Bruijn, R.; Vekemans, J. A. J. M.; Palmans, A. R. A.; Meijer, E. W. *Macromolecules* **2006**, *39*, 5021.

12. Anderson, E. M.; Larsson, K. M.; Kirk, O. *Biocatal. Biotransform.* **1998**, *16*, 181.

13. a) Janes, L. E.; Kazlauskas, R. J. *Tetrahedron Asymm.* **1997**, *8*, 3719. b) Rotticci, D. R; Haeffner, F.; Orrenius, C.; Norin, T.; Hult, K. *J. Mol. Catal. B - Enzymatic* **1998**, *5*, 267.

14. Ohtani, T; Nakatsukasa, H.; Kamezawa, M.; Tachibana, H.; Naoshima, Y. *J. Mol. Catal. B -Enzymatic-* **1998**, *4*, 53.

15. van Buijtenen, J.; van As, B. A. C.; Verbruggen, M.; Roumen, L.; Vekemans, J. A. J. M.; Pieterse, K.; Hilbers, P. A. J.; Hulshof, L. A.; Palmans, A. R. A.; Meijer, E. W. *J. Am. Chem. Soc.* **2007**, *129*, 7393.

16. The turnover frequency (TOF) is defined as the number of turnovers per enzyme molecule per second at the start of reaction. In order to calculate this number, an active protein content of 10 wt.% of the immobilized preparation is assumed. The TOF is calculated using the formule TOF = ki * (initial substrate concentration) / (total enzyme concentration). The initial rate constant (ki) is the slope of the ln(1-conversion) versus time plot.

17. Chen, C. S.; Fujimoto, Y.; Girdaukas, G.; Sih, C. J. *J. Am Chem. Soc.* **1982**, *104*, 7294.
18. Shvo, Y. ; Czarkie, D.; Rahamim, Y. *J. Am. Chem. Soc.* **1986**, *108*, 7400.
19. a) Larsson, A. L. E.; Persson, B. A.; Bäckvall, J.-E. *Angew. Chem. Int. Ed.* **1997**, *36*, 1211. b) Choi, J. H.; Choi, Y. K.; Kim, Y. H.; Park, E. S.; Kim, E. J.; Kim, M. J.; Park, J. W. *J. Org. Chem.* **2004**, *69*, 1972. c) Verzijl, G. K. M.; De Vries, J. G.; Broxterman, Q. B. *Tetrahedron Asymm.* **2005**, *16*, 1603.
20. van As, B. A. C.; Chan, D.-K.; Kivit, P.J.J.; Palmans, A. R. A.; Meijer, E. W. *Tetrahedron: Asymmetry*, **2007**, *18*, 787.
21. Pamiès, O.; Bäckvall, J.-E., *J. Org. Chem.* **2002**, *67*, 1261.

Polyesters and Polyamides

Chapter 16

Enzymatic Degradation of Diol–Diacid Type Polyesters into Cyclic Oligomers and Its Application for the Selective Chemical Recycling of PLLA-Based Polymer Blends

Asato Kondo, Satoko Sugihara, Kohei Okamoto, Yota Tsuneizumi, Kazunobu Toshima, and Shuichi Matsumura[*]

Faculty of Science and Technology, Keio University, 3–14–1, Hiyoshi, Kohoku-ku, Yokohama 223–8522, Japan

The lipase-catalyzed degradation of poly(butylene adipate) (PBA) and poly(butylene succinate) (PBS) into cyclic oligomers and their polymerizability were analyzed. The enzymatic polymerization of a cyclic oligomer was dependent on its ring-size. The highest molecular weight PBA with an M_w of 173000 was produced by the polymerization of the cyclic BA monomer The molecular weight decreased to around M_w 60000 by polymerization of the cyclic BA dimer and trimer. Similar results were obtained for PBS. For the efficient chemical recycling of PBA and PBS, the cyclic BA monomer and the cyclic BS dimer had the best oligomer sizes, respectively. As an application of the enzymatic degradation for the chemical recycling of PBS, the selective recycling of the PBS/poly(L-lactic acid)(PLLA) polymer blend was carried out using a lipase and a clay catalyst, montmorillonite K5.

Introduction

Diol-diacid type aliphatic polyesters, such as poly(butylene adipate) (PBA) and poly(butylene succinate) (PBS), have recently attracted attention as biodegradable plastics with a variety of physico-chemical properties. Furthermore, adipic acid and succinic acid can be produced by the fermentation method, and butane-1,4-diol is produced by the reduction of succinic acid, PBA and PBS are regarded as potentially bio-based plastics. However, these processes require a considerably high amount of production energies. Thus, the sustainable chemical recycling of the polymer will reduce the required production energies as well as the carbon resources. The chemical recycling of such polyesters may be realized using an enzyme catalyst. Enzyme-catalyzed polymer synthesis and degradation have received increased interest due to the environmentally friendliness (1-3). It is reported that the degradation of polycaprolactone took place using lipase in toluene to produce enzymatically repolymerizable oligomer (4,5). The chemical recycling of diol-diacid type polyesters was conventionally carried out by hydrolysis to the corresponding diol and diacid using an alkaline catalyst and a high temperature. These processes required neutralization and purification that use significant amounts of energy. Also, for the repolymerization, the precise stoichiometry of diol and diacid is needed. If diol-diacid type polyesters could be degraded into repolymerizable cyclic oligomers, which had the exact stoichiometry of the two monomers, the chemical recycling process might be significantly simplified. We previously reported that aliphatic polyesters were degraded into cyclic oligomers in an organic solvent using a lipase (5-7). However, the polymerization of the cyclic oligomer has not been extensively studied with respect to the cyclic oligomer size, i.e., degree of oligomerization. The ring-size of the oligomer may be responsible for both the ring-opening polymerization and the molecular weight of the produced polymer. The quantitative evaluation of the lipase-catalyzed polymerization of small- to large-sized (4- to 16-membered) lactones has been made using the Michaelis-Menten constants (8,9). However, at present no study on the polymerizability of cyclic oligomers with different oligomerization degree with lipase.

In this report, the lipase-catalyzed degradation of PBA and PBS into cyclic oligomers and their polymerizabilities were analyzed using molecularly pure cyclic oligomers with the definite oligomerization degree of 1 to 3. Also, as an application of the lipase-catalyzed degradation of PBS into cyclic oligomers, the selective chemical recycling of PBS/poly(L-lactic acid) (PLLA) polymer blend using a lipase and a clay catalyst was reported. The concepts of the production and sustainable chemical recycling of the potentially bio-based polyesters are summarized in Figure 1.

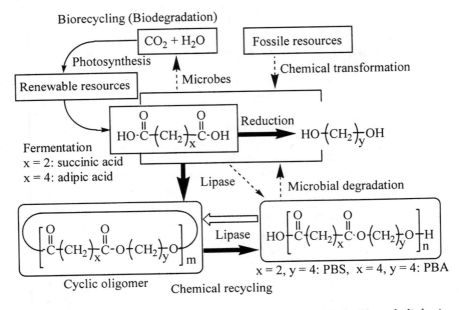

Figure 1. Production and chemical recycling of potentially bio-based aliphatic polyesters via cyclic oligomers using lipase.

Experimental Part

Materials and Measurements

PBA having an M_w = 22000 and M_w/M_n = 1.5 was purchased from Aldrich Chemical Co. (Osaka, Japan). PBS having an M_w = 99000 and M_w/M_n = 1.7 was supplied by Showa High Polymer Co. Ltd. (Osaka, Japan). PLA with an M_w = 151000 was purchased from Sigma-Aldrich Corp. (St. Louis, MO, USA). The PLLA was purified by the reprecipitation method using chloroform as a solvent and reprecipitated by methanol. Montmorillonite K5 (MK5) and K10 (MK10) were purchased from Fluka Chemie GmbH (Germany). Immobilized lipase from *Candida antarctica* [lipase CA: Novozym 435 (triacylglycerol hydrolase + carboxylesterase) having 10000 PLU/g (propyl laurate units: lipase activity based on ester synthesis)] was kindly supplied by Novozymes Japan Ltd. (Chiba, Japan). The enzyme was dried under vacuum over P_2O_5 at 25 °C for 2 h before use.

The weight-average molecular weight (M_w), number-average molecular weight (M_n) and molecular weight distribution (M_w/M_n) for polymers were measured by a size exclusion chromatography (SEC) using SEC columns (Shodex K-804L + K-800D, Showa Denko Co., Ltd., Tokyo, Japan) with a refractive index detector. Chloroform was used as the eluent at 1.0 mL/min. The SEC system was calibrated with polystyrene standards having a narrow molecular weight distribution. The molecular weight was also measured by matrix-assisted laser desorption ionization time-of-flight mass spectrometry (MALDI-TOF MS). The MALDI-TOF MS was measured with a Bruker Ultraflex mass spectrometer. The supercritical carbon dioxide fluid chromatography (SFC) for the analysis and fractionation of the oligomer mixture was performed on a JASCO SFC-201 chromatograph equipped with a 4.6 mm i.d. x 250 mm preparative column packed with silica gel (SFCpak SIL-5, JASCO Ltd. Tokyo, Japan) using supercritical fluid of CO_2 (scCO$_2$) as the mobile phase, and ethanol as the modifier. The fluid pressure was controlled at 22 MPa and column temperature was kept at 70 °C. Chromatograms were recorded using a UV detector operating at a wavelength of 215 nm. The ^{1}H NMR spectra were recorded with a JEOL Model Lambda 300 (300 MHz) spectrometer (JEOL, Ltd., Tokyo, Japan).

General Enzymatic Degradation Procedure for Polyesters

The general procedure for the enzymatic degradation was carried out as follows. The polyester was dissolved in toluene and the immobilized lipase was added. It was then stirred under a nitrogen atmosphere in an oil bath. After the reaction, the insoluble enzyme was removed by filtration. The solvent was then evaporated under slightly reduced pressure to quantitatively obtain the oligomer as the degradation product. The oligomer was analyzed by ^{1}H NMR, SEC and MALDI-TOF MS. The crude oligomer was further fractionated using the preparative SFC column with scCO$_2$ and ethanol according to the oligomerization degree. The molecular structure of the isolated cyclic oligomer was analyzed by ^{1}H NMR and MALDI-TOF MS. The spectral data of the cyclic butylene adipate (BA) dimer and cyclic butylene succinate (BS) dimer were found to be representative. ^{1}H NMR (300 MHz, CDCl$_3$): cyclic BA dimer δ = 1.68 (m, 4H, -OCH$_2$CH$_2$CH$_2$CH$_2$O-), 1.73 (m, 4H, -CO-CH$_2$CH$_2$CH$_2$CH$_2$-CO-), 2.34 (m, 4H, -CH$_2$-COO-), 4.12 (m, 4H, -CH$_2$-O-CO-); cyclic BS dimer δ = 1.71 (m, 4H, -OCH$_2$CH$_2$CH$_2$CH$_2$O-), 2.63 (m, 4H, -CH$_2$-COO-), 4.11 (m, 4H, -CH$_2$OCO-).

General Enzymatic Polymerization Procedure for Cyclic Oligomers

The general procedure for the polymerization of a cyclic oligomer was carried out in a screw-capped vial with molecular sieves 4A placed at the top of the vial. A mixture of the cyclic oligomer and immobilized lipase CA with/without an organic solvent was stirred under a nitrogen atmosphere in an oil bath. The reaction mixture was then dissolved in chloroform, and the insoluble enzyme was removed by filtration. The solvent was evaporated under reduced pressure to quantitatively obtain the polymer. The polymer structure was analyzed by ^{1}H NMR and SEC.

Results and Discussion

Enzymatic Degradation into the Cyclic Oligomer

Typical diol-diacid type biodegradable aliphatic polyesters, PBA and PBS, were degraded in a diluted toluene solution by the immobilized lipase CA which exclusively formed the cyclic oligomers. Their composition was partially dependent on the degradation conditions, such as the enzyme concentration and temperature. On the other hand, in a more concentrated condition, the cyclic oligomer was readily repolymerized by the ring-opening polymerization.

PBA (300 mg) with an M_w of 22000 was degraded in toluene (150 mL) using immobilized lipase CA (150 mg, 50 wt% relative to PBA) at 60 °C for 24 h. The MALDI-TOF MS analysis of the degradation products indicated that the degradation products consisted of a homologous series of cyclic BA oligomers, and no linear oligomer, which had a molecular mass of 18 m/z higher than that of the corresponding cyclic homologues, was detected. Figure 2 shows the typical SFC chromatogram of the degradation products using $scCO_2$ fluid and ethanol. It was found that the oligomer peaks from the monomer to tetramer were clearly separated by the preparative SFC column. Therefore, the cyclic oligomer was fractionated according to the oligomerization degree using the preparative SFC column for further studies. The complete isolation of each cyclic BA oligomer was confirmed by SEC, MALDI-TOF MS and ^{1}H NMR. The composition of the cyclic BA monomer to tetramer was calculated based on the isolated yields of each oligomer. The isolated yields of the cyclic BA monomer to tetramer, which are shown in Figure 2, were 6, 47, 17 and 5%. The remaining amount was mainly due to the isolation loss.

Though the BA oligomer mixture, which was produced by the enzymatic degradation of PBA, was repolymerized by lipase to produce PBA, the polymerizability of the isolated oligomer by lipase CA, particularly the molecular weight of the produced polymer, might vary according to the oligomerization degree. This information is important for the efficient chemical recycling of polyesters using lipase via cyclic oligomers.

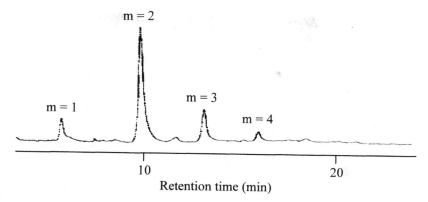

Figure 2. Typical SFC chromatogram of the degradation products using scCO$_2$ fluid and ethanol. m: oligomerization degree of cyclic BA oligomer.

Ring-Opening Polymerization of Molecularly Pure Cyclic Oligomer

The ring-opening polymerization of the molecularly pure cyclic BA oligomer was basically carried out using lipase CA in the more concentrated toluene solution of 300 mg/mL at 60 °C for 24 h, and the M_ws of the produced polymer are listed in Table I. It was found that a cyclic oligomer was polymerized by the lipase CA; however, the M_w of the produced polymer was dependent on the oligomerization degree for the BA and BS oligomers. Surprisingly, the cyclic BA monomer showed the highest M_w of 173000 and the M_ws produced from the BA dimer and trimer clearly decreased. These results indicated that the cyclic BA monomer was the best for the production of a high-molecular weight PBA when using lipase CA with respect to both the chemical recycling of PBA and the polymer production.

Cyclic BS oligomers mainly consisting of dimers and trimers were produced by the degradation of PBS by lipase CA in toluene. Each cyclic oligomer was isolated according to the oligomerization degree using the preparative SFC with scCO$_2$ fluid and ethanol. Similar repolymerization tendencies were observed for the cyclic BS oligomers, such that the molecular weight of the produced PBS varied according to the oligomerization degree (*10*). The cyclic BS dimer was polymerized by lipase CA to produce a high-molecular weight PBS with an M_w of 172000. This value could not be attained by the direct polycondensation of succinic acid and butane-1,4-diol using the lipase CA. On the other hand, the M_w of PBS produced by the ring-opening polymerization of the cyclic BS trimer decreased to an M_w of 128000. The cyclic BS monomer was scarcely produced both by the lipase-catalyzed degradation of PBS and the lipase-catalyzed direct condensation of the diol and dimethyl succinate probably due to the ring-strain

Table I. Molecular weight of PBA by the polymerization of cyclic oligomer and apparent kinetic parameters for ring-opening reaction.

Cyclic oligomer	Apparent kinetic parameters of lipase CA for ring-opening reaction			M_w of the polymer produced by the ring-opening polymerization
	K_m (mol/L)	V_{max} (mol/L min)	V_{max}/K_m (/min)	
BA 1-mer	0.75	0.023	0.031	173000
BA 2-mer	0.68	0.47	0.69	63800
BA 3-mer	0.54	0.40	0.74	55100

of the cyclic monomer. Therefore, the cyclic BS dimer might be the best oligomer size for the rapid and efficient repolymerization by lipase for the chemical recycling of PBS.

Kinetic Constants for Ring-Opening Reaction with Lipase

The proposed mechanism for the lipase-catalyzed polymerization of a lactone via an enzyme-activated monomer (EAM) was presented by Kobayashi and coworkers (*11,12*) The key step is the reaction of the lactone with lipase involving the ring-opening reaction of the lactone to produce the EAM. The initiation is the nucleophilic attack of water, which is contained in the enzyme, on the acyl carbon of the EAM, yielding ω-hydroxy acid, which is regarded as the initiator of the propagation reaction. The propagation reaction occurs by the successive nucleophilic reaction of the EAM with the ω-hydroxy acid. This mechanism is applicable for the lipase-catalyzed polymerization of cyclic oligomers as shown in Figure 3. That is, the cyclic oligomer reacts with the hydroxy group of the serine residue at the active site of lipase CA forming EAM. Initiation is the nucleophilic attack of the water on the acyl carbon of the EAM forming the ω-hydroxy acid as the initiator (Figure 3a). The propagation reaction starts by the successive nucleophilic reaction of the ω-hydroxy group on the initiator-derived growing oligomer chain with EAM (Figure 3b).

The apparent kinetic constants of the molecularly pure cyclic BA monomer to trimer for lipase CA were determined according to the literature using Lineweaver-Burk plots for the enzymatic ring-opening reaction of the cyclic oligomers (*13,14*). The apparent K_m, V_{max} and V_{max}/K_m of lipase CA for the cyclic oligomers are shown in Table I. Similar apparent K_m values for the cyclic BA monomer to trimer were obtained, indicating that these cyclic oligomers and

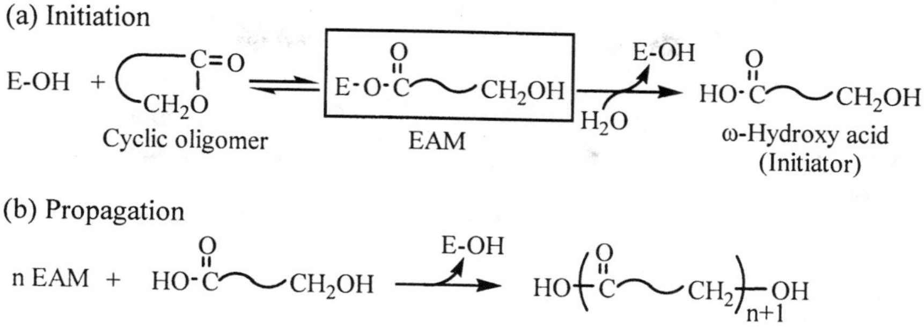

(a) Initiation

$$E\text{-}OH \;+\; \underset{\text{Cyclic oligomer}}{\overset{\displaystyle C{=}O}{\underset{CH_2O}{\Big\backslash}}} \;\rightleftharpoons\; \underset{\text{EAM}}{\boxed{E\text{-}O\text{-}\overset{O}{\overset{\|}{C}}\!\!\sim\!\! CH_2OH}} \;\xrightarrow[H_2O]{E\text{-}OH}\; \underset{\substack{\omega\text{-Hydroxy acid}\\ (\text{Initiator})}}{HO\text{-}\overset{O}{\overset{\|}{C}}\!\!\sim\!\! CH_2OH}$$

(b) Propagation

$$n\,EAM \;+\; HO\text{-}\overset{O}{\overset{\|}{C}}\!\!\sim\!\! CH_2OH \;\xrightarrow{E\text{-}OH}\; HO\!\!\left(\!\overset{O}{\overset{\|}{C}}\!\!\sim\!\! CH_2\!\right)_{\!\!n+1}\!\!\!OH$$

Figure 3. Proposed mechanism of the lipase-catalyzed polymerization of cyclic oligomers.

lipase CA could form EAM with similar affinities. That is, the rate of the EAM formation was similar for all the cyclic oligomers. However, V_{max} for the cyclic BA monomer was significantly lower than those of the cyclic dimer and trimer. This means that the rate of the initiator BA monomer formation was slow when compared to those of the BA dimer and trimer. If the polymerization exclusively follows the mechanism shown in Figure 3, the molecular weight of the produced polymer is dependent on the molar ratio of the initiator hydroxy acid to EAM. The low V_{max} indicates the slow formation of the initiator hydroxy acid by the ring-opening reaction of the cyclic BA monomer when compared to the cyclic dimer and trimer, thus the molecular weight of the produced polymer was greater than those produced from the BA dimer and trimer. No significant differences in K_m and V_{max} values between cyclic dimer and trimer were observed. Both the cyclic BA dimer and cyclic BA trimer, which were polymerized by lipase to produce PBA with similar molecular weights, were observed.

Lipase-Catalyzed Degradation into Cyclic Oligomers having the Desired Oligomer Size

In order to reproduce a high-molecular weight PBA, the selective degradation of PBA into the cyclic BA monomer will be effective. Therefore, the degradation conditions were analyzed with respect to the selective formation of the cyclic BA monomer by the enzymatic degradation of PBA using lipase CA in toluene. It was found that the cyclic BA monomer gradually increased with the increasing degradation temperature from 40 °C to 80 °C. At 80 °C, degradation of the cyclic BA monomer increased to 30%. This is due to the increase in the lipase activity at 80 °C. Figure 4 shows the cyclic oligomer composition as a function of the lipase concentration. The composition of the cyclic BA monomer gradually increased from 8% to 38% with the increasing lipase concentration from 50 to 300%. On the other hand, the cyclic BA dimer decreased with the increasing cyclic BA monomer composition. Based on these results, the relatively high reaction temperature of 80 °C and the high ratio of the polymer to lipase might have significant effects on increasing the cyclic BA monomer content. The high reaction temperature may reduce the durability of the immobilized lipase CA. Therefore, the actual ratio of lipase to polymer was increased by passing it through an enzyme column at a relatively low temperature of 40 °C.

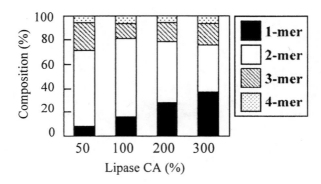

Figure 4. Composition of the cyclic BA monomer to tetramer as a function of lipase concentration.

The enzymatic degradation was carried out by passage through the enzyme-packed column (*15*). The reaction system consisted of an enzyme-packed column placed in a column oven at 40 °C and an HPLC pump. The enzyme-packed column was prepared by packing the immobilized lipase CA (6.8 g) in a 7.8 mm i.d. × 300 mm column. The degradation of the polymer was performed by the batch injection of a 1.0 mL polymer solution containing 0.1 – 0.5 % polymer through a 1.0 mL loop injector with elution by toluene at a flow rate of 0.2 – 0.5 mL/min. The eluents containing the degradation products as measured by UV detection at 215 nm were collected at the outlet of the UV detector. Figure 5 shows the composition of the cyclic oligomer at various flow rates and polymer concentrations through the enzyme column. It was found that both a low flow rate and diluted polymer concentration promoted the formation of the cyclic BA monomer. At the polymer concentration of 1 mg/mL and the flow rate of 0.2 mL/min, the composition of the cyclic BA monomer increased to about 48%.

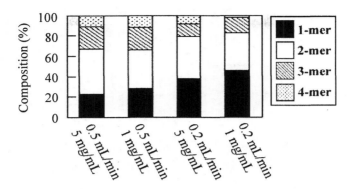

*Figure 5. The composition of cyclic BA oligomer at various flow rate (mL/min)
and polymer concentration (mg/mL).*

Chemical Recycling of PBS/PLLA Polymer Blend

PLLA was blended with PBS in order to improve the mechanical properties.
The chemical recycling of PLLA has been extensively studied with respect to the
thermal degradation to lactide (*16,17*), alkaline hydrolysis and hydrothermal
depolymerization into the lactic acid monomer (*18*). Thermal degradation and
hydrothermal depolymerization generally require a high temperature above 200
°C, thus consuming a considerable amount of energies and often accompany the
isomerization. We previously reported the PLLA degradation into repolymerizable oligomers using an environmentally benign solid acid, MK10 (*19*). As
an application for the degradation of PBS by lipase, typical PLLA-based
polymer blend, PBS/PLLA, was evaluated with respect to the selective chemical
recycling. PBS is blended with PLLA in order to provide flexibility to the rigid
PLLA for the practical application of PLLA in various fields. Therefore,
selective chemical recycling of these two components will be needed to establish
a sustainable chemical recycling of such polymer blends. A conceptual diagram
of the chemical recycling of the PBS/PLLA polymer blend is shown in Figure 6.
PBS in the PBS/PLLA polymer blend was first degraded by lipase CA into the
repolymerizable cyclic BS oligomer without degradation of PLLA. The cyclic
BS oligomer was isolated as a soluble fraction by the reprecipitation of PLLA.
The isolated PLLA was then degraded into the LA oligomer by MK5. Thus the
produced BS oligomer and LA oligomer were repolymerized to produce the
corresponding polymers.

The MK5 could be repeatedly used at least five times without any
significant decrease in the degradation activities. That is, after the reaction, the
MK5 was recovered by filtration, washing with water, and dried under vacuum at

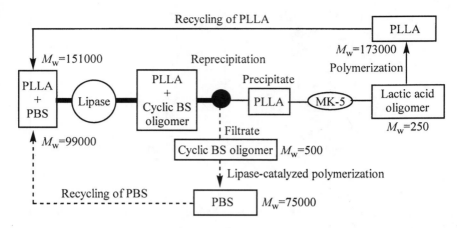

Figure 6. A conceptual diagram of the chemical recycling of the PBS/PLLA polymer blend

room temperature for 1 d. Similarly, the immobilized lipase CA could be repeatedly used for the degradation of PBS in toluene.

Enzymatic Degradation of PBS in PBS/PLLA Polymer Blend

The typical enzymatic degradation of PBS in the PBS/PLLA polymer blend was carried out as follows. The PBS/PLLA polymer blend (1:1 wt%, 2.0 g) was dissolved in toluene (440 mL) at 100 °C and then immobilized lipase CA (3.0 g) was added and stirred using a magnetic stirring bar under a nitrogen atmosphere in an oil bath at 100 °C for 24 h. After the reaction, the reaction mixture was diluted with chloroform (400 mL) and the insoluble lipase CA was removed by filtration for recycling use. The solvent was then evaporated under slightly reduced pressure to almost quantitatively obtain the degradation mixture of the BS oligomer and unreacted PLLA as a white and brittle film. The two components of the degradation mixture were further separated by two reprecipitations for the purification. That is, the degradation mixture (2.0 g) was dissolved in chloroform (10 mL) and then slowly added to methanol (1000 mL) with stirring for the precipitation of PLLA. The precipitated PLLA was collected by filtration and the filtrate was evaporated under slightly reduced pressure to obtain the BS oligomer. The precipitated PLLA was again dissolved in chloroform (6 mL) and then slowly added to methanol (600 mL) with stirring for the reprecipitation of PLLA. The precipitated PLLA was collected by filtration and dried under slightly reduced pressure to obtain the isolated PLLA in 92% yield and 99 % purity as analyzed by ^{1}H NMR. On the other hand, the filtrate containing the BS oligomer was evaporated and the oligomer was dissolved in chloroform (2 mL) and reprecipitated by hexane (200 mL) with stirring. The precipitated PLLA containing a small amount of the BS oligomer was removed by filtration and the filtrate was evaporated under slightly reduced pressure to obtain the BS oligomer in 78% yield and 99 % purity as analyzed by ^{1}H NMR. The oligomer structure was analyzed by ^{1}H NMR, SEC and MALDI-TOF MS. No PLLA derived oligomeric fraction was detected in the BS oligomer fraction as analyzed by both ^{1}H NMR and MALDI-TOF MS. Cyclic BS oligomer: ^{1}H-NMR (300 MHz, CDCl$_3$): δ = 1.58 - 1.80 (m, -OCH$_2$CH$_2$), 2.55 - 2.71 (m, -OOCCH$_2$CH$_2$COO-), 4.04 - 4.24 (m, -CH$_2$O-).

Figure 7 shows the ^{1}H NMR of the selective degradation products of PLLA/PBS polymer blend using lipase CA. Figure 7a shows the ^{1}H NMR of the degradation mixture using lipase CA. It was found that the degradation products mainly consisted of PLLA and the BS oligomer. The degradation products were isolated by the reprecipitation method. By the reprecipitation of PLLA using methanol followed by hexane, the BS oligomer was isolated as a soluble fraction as shown in Figure 7b. From Figure 7b, no peak ascribed to PLLA was detected by ^{1}H NMR. It was also confirmed that the methylene protons at δ = 3.68 ppm due to the terminal hydroxy methyl group of the BS oligomer denoted by the arrow in Figure 7b were scarcely observed. These results indicated the cyclic structure of the BS oligomer. Figure 7c shows the ^{1}H NMR of the isolated PLLA after the reprecipitation of the enzymatic degradation products. It was found that PLLA was exclusively isolated by the reprecipitation using methanol.

Figure 8 shows the MALDI-TOF MS spectra of the BS oligomer obtained by the reprecipitation of the degradation products as a soluble fraction. From the MALDI-TOF MS of the BS oligomer, the cyclic dimer and trimer were the main components followed by the lower amount of the cyclic tetramer. No significant peak due to the linear type-BS oligomer having 18 m/z greater than that of the corresponding cyclic BS oligomer was observed. The small peak of 16 m/z greater than that of the corresponding cyclic oligomer was the peak with the potassium ion instead of the sodium ion. It was also confirmed by the ^{1}H NMR that no PLLA degradation occurred with lipase CA.

Repolymerization of BS Oligomer

In order to evaluate the polymerizability of the BS oligomer obtained by the selective degradation of the PBS/PLLA polymer blend by lipase CA, the isolated BS oligomer, mainly consisting of the cyclic dimer to tetramer, was subjected to the ring-opening polymerization by lipase CA according to the method described in a previous report (10). The repolymerization was carried out in a screw-capped vial with molecular sieves 4A placed at the top of the vial. A mixture of the cyclic oligomer with an M_w of 300 (30 mg), immobilized lipase CA (12 mg)

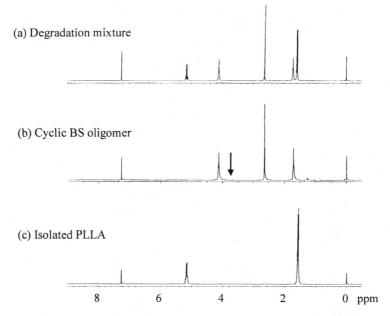

Figure 7. ^{1}H *NMR of the selective degradation products of PBS/PLLA polymer blend by lipase CA. (a) Enzymatic degradation mixture by lipase CA; (b) isolated BS oligomer as a soluble fraction. The arrow indicates the terminal methylene protons of the BS oligomer at $\delta = 3.68$ ppm; (c) isolated PLLA by reprecipitation after the degradation of PBS/PLLA by lipase CA.*

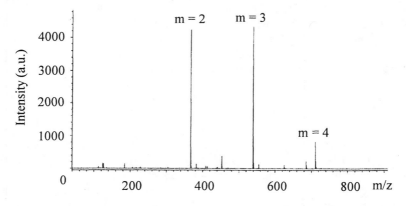

Figure 8. MALDI-TOF MS spectrum of the isolated cyclic BS oligomer produced by the degradation of PBS by lipase CA at 100 °C for 24 h. m: Oligomerization degree of cyclic BS oligomer (+Na^{+})

and toluene (0.6 mL) as the solvent was stirred under a nitrogen atmosphere in an oil bath at 100 °C for 24 h. After the reaction, the reaction mixture was dissolved in chloroform, and the insoluble enzyme was removed by filtration. The solvent was then evaporated under reduced pressure to quantitatively obtain PBS with an M_w of 75000. PBS: ^{1}H-NMR (300 MHz, CDCl$_3$): δ = 1.67 - 1.80 (m, -OCH$_2$CH_2), 2.55 - 2.71 (m, -OOCCH$_2$CH$_2$COO-), 4.04 - 4.24 (m, -CH$_2$O-).

Degradation of PLLA Isolated from the PBS/PLLA Blend using MK5

The PLLA obtained by the reprecipitation after enzymatic degradation of the PBS/PLLA polymer blend was subjected to degradation by MK5 as follows. The isolated PLLA (800 mg) dissolved in toluene (10 mL) at 100 °C and MK5 (3.2 g) was added and then stirred using a magnetic stirring bar under a nitrogen atmosphere in an oil bath at 100 °C for 1 h. After the reaction, the reaction mixture was diluted with chloroform and the insoluble MK5 was removed by filtration. The filtrate was then evaporated under slightly reduced pressure to almost quantitatively obtain the lactic acid (LA) oligomer as the degradation product. The LA oligomer was analyzed by ^{1}H NMR, SEC and MALDI-TOF MS. LA oligomer: ^{1}H NMR (300 MHz, CDCl$_3$): δ = 1.40 - 1.66 (m, CH$_3$), 4.38 (q, J = 6.6, CHOH), 5.10 - 5.30 (m, inner CH$_3$CH). The M_n was 250 as determined by the ratios of the two ^{1}H NMR peak areas of the terminal hydroxy methyne (δ = 4.4) and inner methyne (δ = 5.2) groups. It was found that a linear LA oligomer with an average oligomerization degree of 3.8 was the main component (78 wt%), and the other was the free monomeric lactic acid (22 wt%) as determined by ^{1}H NMR (20). The LA oligomer was also analyzed by MALDI-TOF MS as shown in Figure 9. It was confirmed that almost all fractions of the degradation products were of the linear form, and no cyclic oligomer having a molecular mass of 18 Da less than that of the corresponding linear oligomer was detected.

Repolymerization of the LA Oligomer from PBS/PLLA

The repolymerization of the LA oligomer was carried out basically according to the method for the polymerization of the monomeric lactic acid by Fukushima et al. (21). The LA oligomer produced by the degradation of PLLA by MK5 was readily repolymerized by a conventional chemical catalyst in a three-step gradual polymerization to produce a high-molecular weight PLLA. That is, 800 mg of the LA oligomer was polymerized in a melt using 0.5 % SnCl$_2$/p-TSA (molar ratio 1:1) at 175 °C for 6 h at 10 mmHg. The solidified reactants were then granulated and further reacted for crystallization at 105 °C

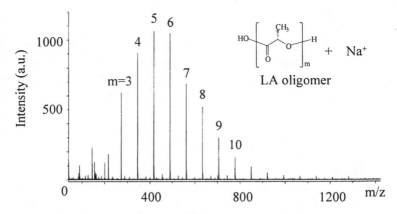

Figure 9. MALDI-TOF MS spectrum of the LA oligomer produced by the degradation of PLLA by MK5 at 100 °C for 1 h. m: Oligomerization degree of LA oligomer (+Na⁺)

for 2 h at 3 mmHg and finally at 150 °C for 20 h at 3 mmHg. After the reaction, the crude polymer was dissolved in chloroform (0.2 mL) and reprecipitated by methanol (30 mL) to give PLLA with an M_w of 173000 and M_w/M_n = 1.7 with a 67 % yield. Under these polymerization conditions, a considerable amount of LL-lactide was produced and deposited at the top of the reaction tube by the evaporation from the polymerization mixture due to the high temperature and reduced pressure during the repolymerization. The purity of the LL-lactide was at least 95% and contained a small amount of L-lactic acid oligomer as analyzed by ^1H NMR. This was collected and used in the next run for the polymerization. Thus, the total recovery reached almost a quantitative value. It was confirmed that no decreasing in molecular weight was observed by the addition of this recovered lactide for the repolymerization. PLLA: ^1H-NMR (300 MHz, CDCl₃): δ = 1.58 - 1.66 (3H, d, J = 7.00, CH₃), 5.16 (1H, q, J = 6.6, CH). ^{13}C-NMR (75 MHz, CDCl₃): δ = 16.6 (CH₃), 69.0 (CH=), 169.6 (CO).

Conclusions

It was found that the enzymatic polymerizability of cyclic oligomers, which were produced by the lipase-catalyzed degradation of diol-diacid type aliphatic polyesters, was dependent on the ring-size, or oligomerization degree, as analyzed using molecularly pure cyclic oligomers. The highest molecular weight PBA with an M_w of 173000 was produced by the ring-opening polymerization of the cyclic BA monomer. The molecular weight of PBA decreased to around M_w

60000 by the lipase-catalyzed polymerization of the cyclic BA dimer and trimer. Similarly, the cyclic BS dimer was polymerized by lipase to produce PBS with an M_w of 172000. On the other hand, the molecular weight of PBS decreased to M_w 128000 by the polymerization of the cyclic BS trimer. These results were due to the reactivity of the cyclic oligomer with the lipase. For efficient chemical recycling of PBA and PBS, cyclic BA monomer and the cyclic BS dimer were the best oligomer sizes, respectively. The content of the cyclic BA monomer could be significantly increased to about 50% by passing it through the lipase-packed column.

Typical PLLA-based polymer blend, PBS/PLLA, was evaluated with respect to the selective chemical recycling of bio-based plastics. The PBS/PLLA polymer blend was successfully recycled first by the selective degradation of PBS by lipase CA into the repolymerizable cyclic BS oligomer using lipase CA followed by the degradation of PLLA into the LA oligomer by MK5. No significant inhibition by the blending polymer for the selective degradation was observed.

Acknowledgements

This work was partially supported by a Grant-in-Aid for General Scientific Research and by a Grant-in-Aid for the 21st Century COE Program "KEIO LCC" from the Ministry of Education, Culture, Sports, Science, and Technology, Japan. Immobilized lipase from *Candida antarctica* (CA, Novozym 435) was kindly supplied by *Novozymes Japan Ltd.* (Chiba, Japan).

References

1. Kobayashi, S.; Uyama, H.; Kimura, S. *Chem. Rev.* **2001**, *101*, 3793.
2. Gross, R.; Kumar, A.; Kalra, B. *Chem. Rev.* **2001**, *101*, 2097.
3. Matsumura, S. *Macromol. Biosci.* **2002**, *2*, 105.
4. Kobayashi, S.; Uyama, H.; Takamoto, T. *Biomacromolecules* **2000**, *1*, 3.
5. Matsumura, S.; Ebata, H.; Kondo, R.; Toshima, K. *Macromol. Rapid Commun.* **2001**, *22*, 860.
6. Okajima, S.; Kondo, R.; Toshima, K.; Matsumura, S. *Biomacromolecules* **2003**, *4*, 1514.
7. Matsumura, S. *Adv. Polym. Sci.* **2006**, *194*, 95.
8. Uyama, H.; Namekawa, S.; Kobayashi, S. *Polym. J.* **1997**, *29*, 299.
9. Van der Mee, L.; Helmich, F.; De Bruijin, R.; Vekemans, J. A. J. M.; Palmans, A. R. A.; Meijer, E. W. *Macromolecules* **2006**, *39*, 5021.
10. Sugihara, S.; Toshima, K.; Matsumura, S. *Macromol. Rapid Commun.* **2006**, *27*, 203.

11. Uyama, H.; Takeya, K.; Kobayashi, S. *Bull. Chem. Soc. Jpn.* **1995**, *68*, 56.
12. Kobayashi, S.; Uyama, H.; Namekawa, S. *Polym. Degrad. Stab.* **1998**, *59*, 195.
13. Kobayashi, S.; Uyama, H. *Macromol. Symp.* **1999**, *144*, 237.
14. Namekawa, S.; Suda, S.; Uyama, H.; Kobayashi, S. *Int. J. Biol. Macromol.* **1999**, *25*, 145.
15. Osanai, Y.; Toshima, K.; Matsumura, S. *Green Chem.* **2003**, *5*, 567.
16. Nishida, H.; Mori, T.; Hoshihara, S.; Fan, Y.; Endo, T.; Shirai, Y. *Polym. Degrad. Stab.* **2003**, *81*, 515.
17. Fan Y.; Nishida, H.; Shirai, Y.; Endo, T. *Green Chem.* **2003**, *5*, 575.
18. Saeki, T.; Tsukegi, T.; Tsuji, H.; Daimon, H.; Fujie, K. *Kobunshi Ronbunshu* **2004**, *61*, 561.
19. Okamoto, K.; Toshima, K.; Matsumura, S. *Macromol. Biosci.* **2005**, *5*, 813.
20. Espartero, J. L.; Rashkov, I.; Li, S. M.; Manolova, N.; Vert, M. *Macromolecules* **1996**, *29*, 3535.
21. Fukushima, T.; Sumihito, Y.; Koyanagi, K.; Hashimoto, N.; Kimura, Y.; Sakai, T. *Intern. Polym. Process* **2000**, *15*, 38.

Chapter 17

Cutinase: A Powerful Biocatalyst for Polyester Synthesis by Polycondensation of Diols and Diacids and ROP of Lactones

Mo Hunsen[1], Abul Azim[2], Harald Mang[2], Sabine R. Wallner[2], Asa Ronkvist[2], Wenchun Xie[2], and Richard A. Gross[2,*]

[1]Department of Chemistry, Kenyon College, Gambier, OH 43022
[2]NSF I/UCRC for Biocatalysis and Bioprocessing of Macromolecules, Polytechnic University, Six Metrotech Center, Brooklyn, NY 11201

Humicola insolens Cutinase (HIC) catalyzed *polymerization reactions* were investigated to gain insight into the potential of Cutinase from *Humicola insolens* as a novel catalyst for polymerization reactions. It was found that the immobilized HiC is a powerful catalyst for both polycondensation of diols and diacids and ring opening polymerization reactions of lactones. The optimal activity of immobilized HIC was at 70 °C. The immobilized HiC was inactive for catalysis of L-lactide and (R,S)-β-butyrolactone ROP. HiC-catalyzed ROP of ε-CL in toluene and in bulk and of PDL in toluene was successfully performed giving high molecular weight polymers.

Introduction

New paradigms in polymer synthesis are needed to meet increasing demands for structural complexity without a concurrent increased environmental burden. This requires catalysts that are selective while operating under mild conditions. Some enzymes, in nonaqueous media, have proven to be surprisingly active for a wide range of polyester and polycarbonate synthetic reactions. Reactions leading to polymers include step-condensation, transesterification, and ring-opening polymerization.[1] Surprisingly, the majority of enzymes studied for polymerization reactions have been from the lipase family with Lipase B from *Candida antarctica* as the dominant enzyme. This paper reports for the first time that a cutinase from *Humicola insolens* (HiC) has been found to have this unusual characteristic of catalyzing polyester synthesis. Cutinases are extracellular fungal enzymes whose natural function is catalyzing the hydrolysis of ester bonds in cutin, a lipid-polyester found in the cuticle of higher plants.[2] With molecular weights of around 20 kDa, cutinases are the smallest members of the serine α/β hydrolase superfamily.[3] Thus far, the majority of published work on cutinase-catalyzed biotransformations have focused on degradation of polyesters[4] and on the esterification or transesterification of small molecules.[5] Herein, we report that the cutinase from *Humicola insolens* (HiC), obtained from Novozymes, has promising activity for lactone ring-opening and condensation polymerization reactions.

Results and Discussion

Previous work has shown that, for polymerization reactions, it is preferable to immobilize enzymes on high surface area supports.[6] This increases the enzymes accessibility to high molecular weight substrates that must diffuse to and from the catalyst active site. Furthermore, it is well-known that immobilization of enzymes on solid supports often increases its thermal stability.[7] Many literature reports on lipase-catalyzed polymerizations use Lipase B from *Candida antarctica* immobilized on Lewatit beads.[8] Hence, as a starting point for studies of HiC activity for polyester synthesis, this enzyme was similarly immobilized by physical adsorption onto Lewatit beads. A study was performed to determine HiC activity for polyester synthesis via condensation reactions. A series of diols and diacids were selected that differ in chain length (see Scheme 1). Diacids were used in the free acid form without activation by esterification with groups such as vinyl or halogenated alkyl esters. Indeed, many literature reports on lipase-catalyzed polyester synthesis have relied on activation of diacids with such leaving groups.[9] First, polymerizations were performed between adipic acid and diols of differing chain length (see Table 1, entries 1-3).

Table 1. HiC-Catalyzed Polycondensation Reactions between Various Alcohols and Diacids[a]

Entry	Alcohol	Diacid	Mn^b [Da]	Mw/Mn^b	DP[c]
1	1,4-butanediol	Adipic acid	2700	1.4	13
2	1,6-hexanediol	Adipic acid	7000	1.5	31
3	1,8-octanediol	Adipic acid	12000	1.6	47
4	1,4-CHDM	Succinic acid	900	2.4	4
5	1,4-CHDM	Adipic acid	4000	2.3	16
6	1,4-CHDM	Suberic acid	5000	2.8	18
7	1,4-CHDM	Suberic acid	19000	1.7	61

[a]Reaction conditions: 1% w/w enzyme, 48 h, vacuum (10 mm of Hg), 70 °C, bulk, quantitative conversion. [b]Determination of conversion and molecular weight by GPC in THF using polystyrene standards. [c]Degree of polymerization (DP) polymer Mn/molecular weight of repeating unit.

With an increase in diol chain length from C4 to C6 and C8, Mn increased from 2700 to 7000 and 12 000, respectively. The polydispersity (Mw/Mn) values were between 1.4 and 1.6, well below that for statistically random condensation polymerization reactions. In another series of reactions, 1,4-cyclohexanedimethanol (1,4-CHDM) was used as the diol, and the diacid chain length was varied from C4 to C10 (Table 1, entries 4-7, and Figure 1). Inspection of Table 1 shows that Mn values were 900, 4000, 5000, and 19 000 for the C4, C6, C8, and C10 diacids, respectively.

Hence, these experiments show that HiC has good activity for polycondensation polymerizations and that HiC activity is dependent on diol and diacid chain length. Increase in the diol and diacid chain length to C8 and C10, respectively, resulted in polyesters of increased molecular weight. Preference of HiC for longer chain length building blocks is reminiscent of substrate specificity observed for Candida antarctica Lipase B-catalyzed polycondensations.[10] Furthermore, cutin, the natural substrate hydrolyzed by cutinases, consists of long chain (e.g., C16, C18) hydroxyacids. 1,4-Cyclohexanedimethanol was found to be a good substrate for HiC-catalyzed polycondensation reactions. This is useful since the cyclohexyl ring can impart greater rigidity to the resultant polyesters. Low polydispersities for some polymerizations in Table 1 indicate HiC is "chain selective".[11] In other words, chain growth occurs where HiC has higher activity for formation and/or esterifications between oligomers of certain chain lengths. Interestingly, chain selectivity during polymerization reactions was similarly observed for Candida antarctica Lipase B-catalyzed polycondensation reactions.[11]

Initial investigations of immobilized HiC thermal stability were conducted by performing polycondensation reactions at different temperatures (see Figure 1).

Polymerizations between 1,4-CHDM and diacids with chain lengths C4 to C10 were performed at 50, 60, 70, 80, and 90 °C. Good activity was found over a large temperature range with an optimum at 70 °C. Large decreases in HiC activity occurred for reactions performed at 80 and 90 °C. Since polymerization reactions are often run in viscous reaction media, the ability to perform enzyme-catalyzed reactions at relatively higher temperatures, approaching 100 °C, is desirable.

Inspired by the finding that HiC is active for polycondensation polymerizations, studies were performed to assess HiC activity for lactone ring-opening polymerizations (Scheme 2). HiC-catalyzed ROP was carried out in either bulk or toluene. To determine the relationship between reaction temperature and enzyme activity for lactone ring-opening polymerizations, ε-caprolactone (1) in bulk was taken as the model system.

Inspection of Table 2 , entries 1-4, shows that optimum conversion and polyester molecular weight from **1** were at 70 °C. HiC activity dropped precipitously for polymerizations conducted at 80 °C. These results agree with those described above for condensation polymerizations. By performing poly-(ε-caprolactone) in toluene instead of in bulk at 70 °C, Mn increased from 16 000 to 24 900 and Mw/Mn decreased from 3.1 to 1.7. An increase in Mn is expected for solution polymerizations since the solvent decreases the viscosity of the reaction medium, thereby easing diffusion constraints between substrates and the enzyme. However, the decrease in polydispersity for the solution polymerizations is less easily explained. One possibility is that transesterification reactions leading to broader polydispersity occur more rapidly for reactions conducted in bulk.[10] HiC catalysis of ω-pentadecalactone (**2**) polymerization in

Table 2. HiC-Catalyzed Ring-Opening Polymerization of Various Lactones[a]

Entry	monomer	solvent	T [°C]	$conv^b$ [%]	Mn^c [Da]	Mw/Mn^c	DP^d
1	1	bulk	50	57	4300	4.2	38
2	1	bulk	60	83	6300	3.2	55
3	1	bulk	70	99	16000	3.1	140
4	1	bulk	80	14	2500	2.7	22
5	1	toluene	70	99	24900	1.7	218
6	2	toluene	70	99	44600	1.7	186

[a]Reaction conditions: 0.1% w/w enzyme, 24 h, no vacuum, in toluene or bulk. [b]Determination of conversion by 1H NMR (CDCl3). [c]Determination of molecular weight by GPC in THF (entries 1-5) and CHCl3 (entry 6), using polystyrene standards. [d]Degree of polymerization (DP) = polymer Mn/molecular weight of repeating unit.
SOURCE: Reproduced from reference 12. Copyright 2007 American Chemical Society.

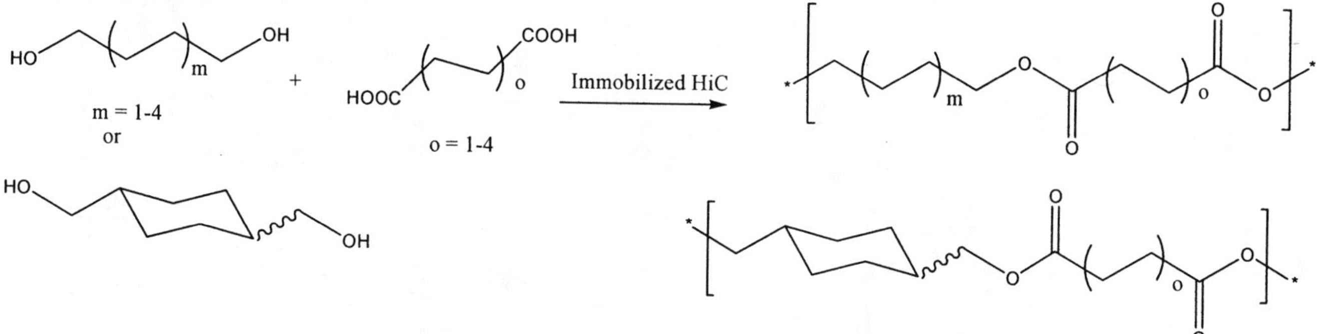

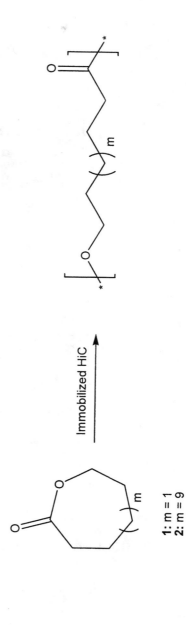

Scheme 1. HiC-Catalyzed Polycondensation Reactions between Diols and Diacids

1: m = 1
2: m = 9

Scheme 2. HiC-Catalyzed Ring-Opening Polymerization of Lactones

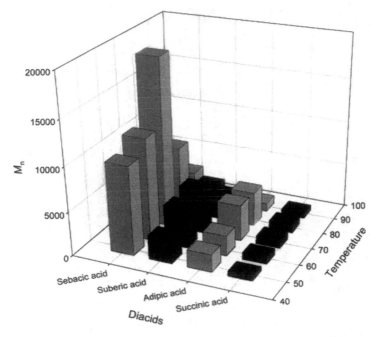

Figure 1. Polycondensation between 1,4-CHDM and various carboxylic diacids (C4, C6, C8, and C10, respectively) over the temperature range of 50-90 °C showing obtained molecular weights (M_n). (Reproduced from reference 12. Copyright 2007 American Chemical Society.)

toluene at 70 °C gave poly(ω-pentadecalactone) with M_n and M_w/M_n of 44 600 and 1.7, respectively. Thus, both 7- and 16-membered lactones **1** and **2** are excellent substrates for HiC-catalyzed polymerizations.

The thermal properties of the polyesters was investigated using differential scanning calorimetry (DSC). Polymers were dissolved in $CHCl^3$, precipitated in methanol and dried in vacuum before analysis by differential scanning calorimetry (DSC). Summarized data for polymerization products are given in Table 3.

In summary, the cutinase from *Humicola insolens* is a new, promising biocatalyst for polyester synthesis. The immobilized cutinase showed optimal activity at 70 °C and catalyzed a broad range of condensation and lactone ring-opening polymerizations in bulk and using toluene as solvent, yielding high molecular weight polyesters. On the basis of an extensive literature review and experience in our laboratory, we conclude that the activity of cutinase from *Humicola insolens* for the polyester synthesis reactions studied herein is rivaled only by Lipase B from *Candida antarctica*. Work is in progress to better define

Table 3. Thermal properties of polyesters.

Polymer	Tg[a] [°C]	Tm[b] [°C]	Tm[c] [°C]	ΔHm[d] [Jg⁻¹]	Scan rate [°C min⁻¹]	Range [°C]
Poly(butylene adipate)	-60.1	56.7	52.3	48.2	5	-85 to +80
Poly(hexylene adipate)	n.d.[e]	60.0	-	74.4	5	-85 to +80
Poly(octylene adipate)	n.d.[e]	68.6	-	88.4	5	-85 to +80
Poly(cyclohexane dimethylene adipate)	-23.4	107.4	100.1	53.2	5	-85 to +120
Poly(cyclohexane dimethylene suberate	-35.1	82.6	-	42.1	5	-85 to +100
Poly(cyclohexane dimethylene sebacate)	-40.7	62.1	55.7	36.5	5	-85 to +80
Poly(ε-caprolactone)	n.d.[e]	56.1	-	58.6	10	-85 to +80
Poly(ω-pentadecalactone)	n.d.[e]	96.1	-	125.6	10	-85 to +120

[a]Glass transition temperature; [b]Peak melting temperature of the higher melting transition; [c]Peak melting temperature of the lower melting transition; [d]Heat of fusion; [e] n.d.=not determined.

SOURCE: Reproduced from reference 12. Copyright 2007 American Chemical Society.

the activity, stability, and recyclability of *Humicola insolens* immobilized on other solid supports.

Experimental Procedures

1. Materials

Diacids (succinic acid, adipic acid, suberic acid, and sebacic acid), diols (cyclohexane-1,4-dimethanol, 1,8-octanediol, 1,6-hexanediol and 1,4-butanediol), and lactone (ω-pentadecalactone) were purchased from Aldrich Chemical Co. in the highest available purity and used as received ω-Caprolactone, a gift from Union Carbide, was dried over calcium hydride and distilled under reduced pressure in a nitrogen atmosphere. *Humicola insolens* cutinase was a gift from Novozymes (Bagsvaerd, Denmark). Lewatit OC VOC 1600 was received as a gift from Rohm and Haas.

2. Instrumental Methods

Nuclear Magnetic Resonance (NMR). NMR spectra were recorded in CDCl3 using a Bruker DPX 300 at 300 (^{1}H) and 75 (^{13}C) MHz. Chemical shifts are reported relative to TMS (δ 0.00) and coupling constants (J) are given in Hz. Polymerization of lactones was monitored by ^{1}H NMR to determine monomer conversion.

Gel Permeation Chromatography (GPC). The number and weight average molecular weights (Mn and Mw, respectively) were determined by size exclusion chromatography using a Waters 510 pump, a 717 plus autosampler, and a Wyatt Optilab DSP interferometeric refractometer coupled to 500, 10^3, 10^4, and 10^5 Å Ultrastyragel columns in series. Trisec GPC software version 3 was used for calculations. THF (CHCl$_3$ for PPDL) was used as eluent with a flow rate of 1.0 mL min-1 at 35 °C. Molecular weights were determined on the basis of a conventional calibration curve generated by narrow molecular weight polystyrene standards obtained from Aldrich Chemical Company.

Differential Scanning Calorimetry (DSC). Calorimetric analysis was performed on a TA Instrument DSC 2920 Differential Scanning Calorimeter. After a first heating run at 10 °C min-1 to 120 °C, the sample was cooled at 10 °C min^{-1} to -85 °C and then heated from -80 °C to 120 °C at 5 °C min^{-1} unless otherwise specified (Table 3).

3. Procedures

Immobilization of Humicola insolens cutinase

Humicola insolens cutinase was immobilized on Lewatit OC VOC 1600. The resin (700 mg) was first activated with ethanol (10 mL) and dried under vacuum for 60 min to remove traces of ethanol. The resin was further washed with phosphate buffer (pH 7.8, 0.1 M, 3 x 10 mL). Cutinase (used as received, 1.7 mg mL^{-1}, 25 mL) was added to the resin and incubated with gentle shaking (100 rpm) at 4 °C for 72 hrs. Remaining supernatant was removed by centrifugation (10 000 rpm, 5 min, 15 °C). The cutinase loaded resin was freeze dried and the dry weight determined (W). The protein contents of supernatants before (P1) and after immobilization (P2) were determined using the bicinchoninic acid method. The protein loading was calculated as mg protein g^{-1} of resin = (P1 – P2)/W. The enzyme loading was 60 mg protein g^{-1} or 6 % by wt.

Method A: General Procedure for Polycondensation between Diols and Diacids

Diol (1 mmol) and diacid (1 mmol) were heated to 100 °C to obtain a monophasic solution of reactants and the temperature was then reduced to the reaction temperature (50 °C to 90 °C). Enzyme beads were added to the reaction so that the ratio (w/w) of enzyme to total monomer was 1% and vacuum (10 mm of Hg) was applied 1 hr after addition of the immobilized cutinase. After 48 hrs the reaction was terminated by adding THF (2 mL) and filtration to remove enzyme beads. Solvent was stripped by rotoevaporation and remaining volatiles were eliminated by thorough drying in a vacuum oven. The isolated products were analyzed by NMR (^{1}H & ^{13}C), GPC, and DSC. Conversion was determined by GPC.

Method B: General Procedure for Ring-Opening Polymerization of Lactones

Lactone (1 g), toluene (2 mL, for reactions not run in bulk) and immobilized enzyme (0.1 % w/w enzyme to total weight of monomer) was added under nitrogen atmosphere in a parallel reactor (Argonaut Advantage Series 2050). The reactions were stirred magnetically while maintaining reactions at pre-determined temperatures (50 °C to 90 °C). After 24 hrs, reactions were terminated by adding CHCl$_3$ (20 mL), filtration to remove the catalyst beads and stripping the solvent by roto-evaporation. The product obtained was dried in a vacuum oven and analyzed by NMR (^{1}H & ^{13}C), GPC and DSC. GPC was recorded using THF as the eluent for poly(ω-caprolactone) and CHCl$_3$ as the eluent for poly(ω-pentadecalactone). Monomer conversion to polymer was measured by ^{1}H NMR.

4. Polymer characterization

Poly(butylene adipate)

Synthesis was by Method A using 1,4-butanediol (90 mg, 1 mmol) and adipic acid (146 mg, 1 mmol) as monomers.
^{1}H NMR (CDCl$_3$): δ 1.66-1.93 (8H, m), 2.33-2.54 (4H, m), 4.09 (4H, t, J = 6.6 Hz)
^{13}C NMR (CDCl$_3$): δ 24.8, 25.7, 34.2, 64.2, 173.7

Poly(hexylene adipate)

Synthesis was by Method A using 1,6-hexanediol (118 mg, 1 mmol) and adipic acid (146 mg, 1 mmol) as monomers.
^{1}H NMR (CDCl$_3$): δ 1.33-1.44 (4H, m), 1.50-1.77 (8H, m), 2.26-2.39 (4H, m), 4.06 (4H, t, J = 6.6 Hz)
^{13}C NMR (CDCl$_3$): δ 24.8, 26.0, 28.9, 34.3, 64.7, 173.8

272

Poly(octylene adipate)

Synthesis was by Method A using 1,8-octanediol (146 mg, 1 mmol) and adipic acid (146 mg, 1 mmol) as monomers.
^{1}H NMR (CDCl$_3$): δ 1.25-1.40 (8H, m), 1.52-1.72 (8H, m), 2.26-2.37 (4H, m), 4.05 (4H, t, J = 6.6 Hz)
^{13}C NMR (CDCl$_3$): δ 24.8, 26.2, 29.0, 29.5, 34.3, 64.9, 173.8

Poly(cyclohexane dimethylene adipate)

Synthesis was by Method A using cyclohexane-1,4-dimethanol (144 mg, 1 mmol) and adipic acid (146 mg, 1 mmol) as mononers.
^{1}H NMR (CDCl$_3$): δ 0.96-1.03 (4H, m), 1.35-1.81 (10H, m), 2.33 (4H, t, J = 6.3 Hz), 3.89 (3H, d, J = 6.5 Hz) *trans*, 3.99 (1H, d, J = 7.2 Hz) *cis*
^{13}C NMR (CDCl$_3$): δ 24.8, 29.2, 34.3, 37.4, 69.7, 173.8

Poly(cyclohexane dimethylene suberate)

Synthesis was by Method A using cyclohexane-1,4-dimethanol (144 mg, 1 mmol) and suberic acid (174 mg, 1 mmol) as monomers.
^{1}H NMR (CDCl$_3$): δ 0.96-1.04 (4H, m), 1.32-1.81 (14H, m), 2.30 (4H, t, J = 7.4 Hz), 3.89 (3H, d, J = 6.5 Hz) *trans*, 3.98 (1H, d, J = 7.2 Hz) *cis*
^{13}C NMR (CDCl$_3$): δ 25.2, 29.2, 29.3, 34.6, 37.4, 69.6, 174.2

Poly(cyclohexane dimethylene sebacate)

Synthesis was by Method A using cyclohexane-1,4-dimethanol (144 mg, 1 mmol) and sebacic acid (202 mg, 1 mmol) as monomers.
^{1}H NMR (CDCl$_3$): δ 0.97-1.04 (4H, m), 1.30-1.82 (18H, m), 2.29 (4H, t, J = 7.4 Hz), 3.89 (3H, d, J = 6.5 Hz) *trans*, 3.98 (1H, d, J = 7.2 Hz) *cis*
^{13}C NMR (CDCl$_3$): δ 25.4, 29.3, 29.5, 29.6, 34.7, 37.5, 69.6, 174.3

Poly(ε-caprolactone)

Synthesis was by Method B and the monomer was Σ-caprolactone (1 g, 8.76 mmol).
^{1}H NMR (CDCl$_3$): δ 1.35-1.43 (2H, m), 1.61-1.66 (4H, m), 2.30 (2H, t, J = 7.4 Hz), 4.06 (2H, t, J = 6.6 Hz),
^{13}C NMR (CDCl$_3$): δ 24.9, 25.9, 28.7, 34.5, 64.5, 173.9

The unique signals at 4.23 (t, OCH_2) and 2.64 (t, C(O)CH_2) were assigned to the protons of ε-caprolactone. Signals at 4.06 (t, OCH_2), 3.64 (t, HOCH_2), and 2.30 (t, C(O)CH_2) appeared after the onset of polymerization reactions and were assigned to poly(ε-caprolactone). The relative intensity of signals at 4.06 to 4.23 and 4.06 to 3.64 were used to calculate monomer conversion and Mn, respectively.

Poly(ω-pentadecalactone)

Synthesis was by Method B and the monomer was ω-pentadecalactone (1 g, 4.16 mmol).

^{1}H NMR (CDCl$_3$): δ 1.24-1.63 (24H, m), 2.28 (2H, t, J = 7.5 Hz), 4.06 (2H, t, J = 6.7 Hz),

^{13}C NMR (CDCl$_3$): δ 25.4, 26.3, 29.1, 29.6, 29.6, 29.7, 29.9, 29.9, 30.0, 30.0, 30.1, 34.8, 64.8, 174.3

Acknowledgment

The authors thank the NSF and the industrial members of the NSF-I/UCRC for Biocatalysis and Bioprocessing of Macromolecules at the Polytechnic University for their financial support, intellectual input, and encouragement during the course of this research.

References

1. (a) Gross, R. A.; Kumar, A.; Kalra, B. *Chem. ReV.* **2001**, *101*, 2097-2124. (b) Kobayashi, S.; Uyama, H.; Kimura, S. *Chem. ReV.* **2001**, *101*, 3793-3818.
2. (a) Kolattukudy, P. E. *Science* **1980**, *208*, 990-1000. (b) Kolattukudy, P. E. In *Lipases*; Borgstroem, B., Brockman, H. L., Eds.; Elsevier: Amsterdam, 1984; pp 471-504. (c) Ettinger, W. F.; Thukral, S. K.; Kolattukudy, P. E. *Biochemistry* **1987**, *26*, 7883-7892.
3. Longhi, S.; Cambillau, C. *Biochim. Biophys. Acta* **1999**, *1441*, 185-196.
4. Yoon, M. Y.; Kellis, J.; Poulose, A. J. *AATCC Rev.* **2002**, *2*, 33-36.
5. Carvalho, C. M. L.; Aires-Barros, M. R.; Cabral, J. M. S. *Electron. J. Biotechnol.* **1998**, *1*, 160-173.
6. Ison, A. P.; Macrae, A. R.; Smith, C. G.; Bosley, J. *Biotechnol. Bioeng.* **1994**, *43*, 122-130.
7. Epton, R.; Marr, G.; Shackley, A. T. *Polymer* **1981**, *22*, 553-557.

8. (a) Hu, J.; Gao, W.; Kulshrestha, A.; Gross, R. A. *Macromolecules* **2006**, *39*, 6789-6792. (b) van der Mee, L.; Helmich, F.; de Bruijn, R.; Vekemans, J. A. J. M.; Palmans, A. R. A.; Meijer, E. W. *Macromolecules* **2006**, *39*, 5021-5027. (c) Peeters, J.; Palmans, A. R. A.; Veld, M.; Scheijen, F.; Heise, A.; Meijer, E. W. *Biomacromolecules* **2004**, *5*, 1862-1868.

9. (a) Magolin, A. L.; Creene, J. Y.; Klibanov, A. M. *Tetrahedron Lett.* **1987**, *28*, 1607-1609. (b) Wallace, J. S.; Morrow, C. J. *J. Polym. Sci., Part A: Polym. Chem.* **1989**, *27*, 3271-3284. (c) Chaudhary, A. K.; Lopez, J.; Beckman, E. J.; Russell, A. J. *Biotechnol. Prog.* **1997**, *13*, 318-325.

10. Mahapatro, A.; Kalra, B.; Kumar, A.; Gross, R. A. *Biomacromolecules* **2003**, *4*, 544-551.

11. Mahapatro, A.; Kumar, A.; Kalra, B.; Gross, R. A. *Macromolecules* **2004**, *37*, 35-40.

12. For a communication on the work described here see: Hunsen, M.; Azim, A.; Mang, H.; Sabine R. Wallner, S. R.; Ronkvist, A.; Xie, W.; Gross, R. A. *Macromolecules* **2007**, *40*, 148-150.

Chapter 18

"Sweet Polyesters": Lipase-Catalyzed Condensation– Polymerizations of Alditols

Jun Hu, Wei Gao, Ankur Kulshrestha, and Richard A. Gross[*]

NSF/I/UCRC for Biocatalysis and Bioprocessing of Macromolecules, Polytechnic University, Six Metrotech Center, Brooklyn, NY 11201

A series of 4-, 5- and 6-carbon natural polyols were studied as monomers for immobilized *Candida antarctica* Lipase B (CALB) catalyzed polymerizations. Reactions were performed with a molar ratio of adipic acid to 1,8-octanediol to polyol of 1.0:0.8:0.2. Molecular weight increase as a function of reaction time was measured. The following is the relative order of M_w as a function of polyol structure for 46 h polymerizations: D-mannitol (73.0 ± 0.4 K) > erythritol (38.1 ± 4.4 K), xylitol (42.3 ± 2.2 K), ribitol (38.4 ± 2.9 K) > D-glucitol (27.7 ± 2.0 K) > galactitol (11.0 ± 0.9 K). Replicate experiments using a needle valve and a digital vacuum regulator showed that, reproducible results were obtained only with the digital vacuum regulator that more accurately controlled the reduced pressure in reaction vessels. From this limited set of polyols, no apparent correlation was found between polyol chain length and its polymerization by CALB catalysis. Plots of log[η] vs. logM_w were prepared from SEC-MALS-Viscosity analyses of poly(octamethylene adipate) and corresponding terpolymers with galactitol, D-glucitol and D-mannitol. Comparison of exponent *a* values from slopes of these plots showed copolymers from D-mannitol had the largest degree of branching and, therefore, greatest propensity for combined reactivity at both primary and secondary hydroxyl groups. Differences in alditol reactivity were analyzed with respect to their stereochemical configurations at polyol β-carbons, closest to terminal primary hydroxyl groups, and secondary hydroxyl groups, that upon reaction lead to branching.

Alditols polyols are readily renewable, inexpensive and harmless to the environment. By incorporation of polyols into aliphatic polyesters, functional linear or hyperbranched polymers can be prepared with specific biological activities and/or that respond to environmental stimuli.[1] Polyesters with carbohydrate or polyol repeat units in chains have been prepared by chemical methods.[2a-d,4a-c,8] In some cases, the reaction conditions led to hyperbranched polymers (HBPs).[4a-c,8] The highly branched architecture of HBPs leads to unusual mechanical, rheological and compatibility properties.[4-8] These distinguishing characteristics have garnered interest for their use in numerous industrial and biomedical fields.[8] Chemical routes to linear polyol-polyesters require elaborate protection-deprotection steps[2a-d]. Furthermore, condensation routes to hyperbranched polymers generally require harsh reaction conditions such as temperatures above 150 °C and highly acidic catalysts[4a-c,8].

Single-step chemical routes to HBPs from multifunctional monomers, without protection-deprotection chemistry, leads to randomly branched polymer topologies. To achieve perfectly branched polymers researchers have used stepwise synthetic methods to prepare dendrimers. A need exists for new, simple synthetic methods that don't rely on protection-deprotection methods to prepare both functional linear polymers and polymers with improved control over branching. A promising approach to address these challenges is the use of isolated enzymes as catalysts for polymerization reactions. Lipases are already well-established catalysts for regioselective esterification of low molar mass substrates at mild temperatures (30 to 70 °C).[12] Early work assumed that activation of carboxylic acids by electron withdrawing groups was needed to perform enzyme-catalyzed copolymerizations of polyols[13a-j]. Furthermore, since polyols (e.g. glucitol) are generally insoluble in non-polar organic media, polar solvents were used.[13f-j] Unfortunately, these solvents cause large reductions in enzyme activity.[13f-j]

Recently, our laboratory reported copolymerizations without activation of the diacid or adding solvent [14a,b]. The monomers were combined so they formed monophasic mixtures, *Candida antarctica* Lipase B (CALB), physically immobilized on Lewatit beads (N435), was then added. For example, a hyperbranched copolyester with 18 mol % glycerol-adipate units was formed in 90% yield, with M_w 75 600 (by SEC-MALLS), M_w/M_n 3.1, and 27 mol% of glycerol units that are branch sites[14a]. Also, N435 catalyzed the polymerization of glucitol and adipic acid, in-bulk, with high regioselectivity (85 ± 5%) at the primary hydroxyl groups, to give a water-soluble product with M_n 10 880 and M_w/M_n 1.6. Time-course studies of glycerol copolymerizations showed that, while the reaction is under kinetic control, linear chains (determined by NMR) were formed. Hence, the product formed at 18 h was linear. However, by extending the reaction to 42 h, pendant hydroxyl esterification transpired giving HBPs. In contrast using chemical methods, glycerol has been used to introduce branching into polyesters.[15]

Lipase regioselectivity can be varied by many parameters such as substrate structure, lipase structure, lipase immobilization, reaction medium, time and temperature. To better understand factors of reaction time and substrate structure (e.g. chain length, stereochemistry), a series of 4-, 5- and 6-carbon natural polyols were selected as monomers for N435 catalyzed polymerizations. The structures of alditol polyols used herein along with their stereochemical configurations are displayed in Scheme 1. Bulk-terpolymerizations of these substrates with 1,8-octanediol and adipic acid were performed. Adipic acid, 1,8-octandiol, erythritol, xylitol, ribitol, D-glucitol, D-mannitol and galactitol were purchased from the Aldrich Chemical Co. N435 was a gift from Novozymes (Bagsvaerd, Denmark). All reagents were purchased in the highest purity available and were used without further purification. Polymerizations were performed by the identical procedure published elswhere[14a-b] except for modifications in reactor hardware, vacuum regulation, and pressure used for water removal (see below). In summary, reactions were carried out in-bulk for up to 46 h at 90 °C using N435. Since N435 has 10%-by-wt CALB,[16] the weight ratio of CALB to monomer is 1%. Monomers were transferred to glass vessels and mixtures were heated to 130 °C for about 0.5 h. Solid reactants melted and/or dissolved forming a homogeneous liquid. The temperature was lowered to 90 °C and the reaction mixture remained as a homogeneous liquid but increased in viscosity. Polymerizations were performed in an Argonaut Advantage Series 2050 personal synthesizer. The unit has 5-parallel reactors each with independent temperature control and magnetic stirring. Each glass reactor (2.7 cm by 19.5 cm) was directly attached to one of five ports of a vacuum manifold. A digital vacuum regulator (J-KEM scientific, Model 200) was placed between the manifold and the vacuum pump. This system permitted reactions with different polyols, replicates and controls to be run side-by-side under nearly identical vacuum conditions. Polymerizations were performed with a molar ratio of adipic acid to 1,8-octanediol to polyol of 1.0:0.8:0.2 and 0.04 moles (about 6 g) of total reactants. Periodically, 50 mg aliquots were removed from reactions for analysis. Vacuum was applied in reactions according to the following schedule: 0 h $\xrightarrow{760mmHg}$ 2 h $\xrightarrow{100mmHg}$ 6 h $\xrightarrow{75mmHg}$ 18 h $\xrightarrow{50mmHg}$ 30 h $\xrightarrow{25mmHg}$ 46 h.

Although this series of reaction stages is non-optimized, it takes into account the following concerns that arose based on experimental observations. No vacuum during the first 2 h allows reactants to oligomerize without evaporation of volatile monomers. Slow decrease in the pressure from 2 to 30 h avoids turbulent bubbling and foaming. Furthermore, when low pressure (e.g. 25 mmHg) was applied throughout reactions of polyols with diacids there was an increased frequency of crosslinked product formation due to chemically-mediated reactions (unpublished results). To minimize these events, 25 mmHg was applied only during the last stage (30 - 46 h) to drive reactions towards formation of high molecular weight products.

Molecular weight increase as a function of time was used to assess differences in polyol reactivity. The relative weight average molecular weights were determined by size exclusion chromatography (SEC) in our lab[15b] using a PLgel HTS-D column. Figure 1 shows the increase in polyol-polyester weight average molecular weight (M_w) as a function of reaction time and polyol structure. Standard deviation, shown as error bars in Figure 1, was calculated

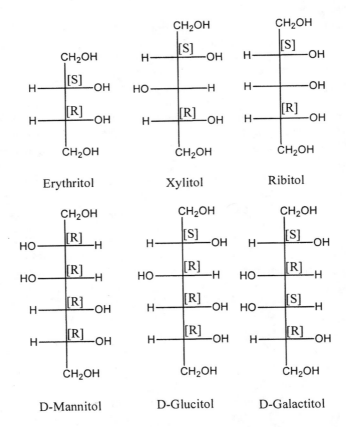

Scheme 1. Structure of alditols used in this study for N435-catalyzed polymerizations

from three experiments. Polyol-polyester M_w at 46 h decreased as follows: D-mannitol (73.0 ± 0.4 K) > erythritol (38.1 ± 4.4 K), xylitol (42.3 ± 2.2 K), ribitol (38.4 ± 2.9 K) > D-glucitol (27.7 ± 2.0 K) > galactitol (11.0 ± 0.9 K). This trend was identical at 18 and 30 h except for mannitol which, at these reaction times, did not give the highest polyol-polyester M_w. Instead, mannitol polyol-polyester M_w was similar to that for glucitol at 18 h and erythritol/xylitol/ribitol at 30 h. Thus, M_w for the mannitol copolymerization increased rapidly from 18 h relative to copolymerizations with other polyols.

Relative to higher temperature (>150°C) chemically-catalyzed polyesterification reactions, polymerizations between alcohol and acids catalyzed by CALB can be performed at 90 °C or lower. However, lowering the reaction temperature increases the reaction medium viscosity. This raised the question as to whether the applied vacuum, that largely determines water-removal efficiency, is a sensitive parameter controlling the polymerization rate. To probe this question, the standard deviation of triplicate reactions was determined by carrying out polymerizations using both a digital vacuum regulator and a conventional needle valve to control the applied vacuum. Values of %-error in Figure 1 ranged from 0.4 to 4.4 K. When these experiments were performed using a needle valve, the %-error values for the same series of reactions ranged from 13.7 to 39.6 K. Thus, without tight regulation of vacuum, experimental results were not reproducible. This result should be carefully considered by others that review related previous literature or begin new research on enzyme-catalyzed polyol-polyester condensation polymerizations.

As shown in Scheme 1, the alditols studied herein have different chain lengths and/or stereochemistry. Given the inherent selectivity of enzymes, it was anticipated that N435-catalyzed polyesterification of these substrates would occur at different rates. However, the relative order of their reactivity during polymerization reactions was unpredictable.

While this study does not include a systematic series of 4-, 5- and 6-carbon alditol polyols with all possible permutations of stereochemical configurations, this set of polyols does allow one to begin exploring general structural characteristics that may be responsible for different reactivity. First, the influence of substrate chain length was considered. erythritol, the only 4-carbon substrate, showed moderate reactivity for high polymer synthesis (see Figure 1). D-mannitol, D-glucitol, and galactitol, all 6-carbon substrates, showed high, intermediate, and low reactivity for high M_w polymer formation. From these results, no apparent correlation can be made between polyol chain length and its polymerization activity. Relative to terminal primary hydroxyl moieties, the nearest asymmetric and *pseudo*-asymmetric centers for chiral and achiral (*meso*) polyols, respectively, are at β-carbons. One explanation for polyol reactivity is it's determined by the stereochemical configuration of carbons closest to terminal primary hydroxyl groups (β-carbons). The stereochemical configuration of these carbons is [R]-[S] or [S]-[R] except for D-mannitol that is [R]-[R]

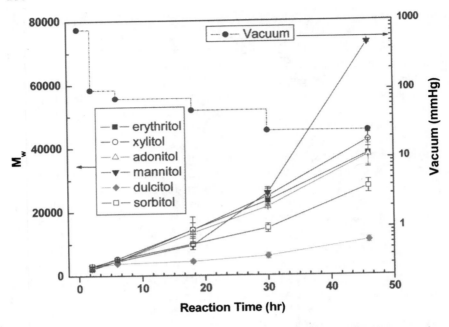

*Figure 1. Time-course study of molecular weight increase as a function of polyol structure. Polymerizations were performed at 90 °C, catalyzed by N435. The right Y-axis showed the schedule of applied vacuum during reactions. (Reproduced from Macromolecules **2006**, 39, 6789–6792. Copyright 2006 American Chemical Society.)*

(Scheme 1). None of the polyols have an [S]-[S] structure. Thus, it may be that the terminal [R]-[R] configuration of D-mannitol is a key factor that led to rapid formation of D-mannitol polyol-polyesters of high Mw. This coincides with Hult's research on the selective for R alcohols in the transesterification reactions between different alcohols and vinyl acetate catalyzed by CALB17, 18. When comparing the stereochemical configuration of these alditols, it is found that D-mannitol is the only polyol that has a C2 symmetry axis. No chain regioisomerisim is another possible reason of the rapid formation of D-mannitol polyol-polyesters of high Mw.

Certainly, the reactivity of secondary hydroxyl groups of alditols will also affect chain growth. Indeed, it is the combined reactivity of primary and secondary hydroxyl moieties that will ultimately determine chain growth and branching that will dictate polymer properties (e.g. viscosity). In order to evaluate differences in branching among the polyol-polyesters, their dilute solution properties were compared by using exponent *a* of the [η]-M-relationship. The [η]-M-relationship, also known as the Kuhn-Mark-Houwink-Sakurada-relationship, was determined by conducting size exclusion chromatography with online concentration (Wyatt Optilab DSP interferometer, Santa Barbara, CA), multi-angle light scattering (MALS) (Wyatt HELEOS) and viscosity (Wyatt Viscostar) detectors. Absolute molar mass and intrinsic viscosity of each elution moment was determined from RI, MALS and viscosity response by the corresponding detector. Even though the chemical composition may be the same, branched polymers have higher densities in solution than their linear structural isomers and, therefore, lower intrinsic viscosities[19]. Figure 2 displays plots of log[η] vs. logM_w ([η]-M-relationship) determined by SEC-MALS-Vis analyses of poly(octamethylene adipate) (POA, linear) and terpolymers with 80 mol-% OA and 20 mol% galactitol-adipate (DA), D-glucitol-adipate (SA) and D-mannitol-adipate (MA) (P[OA-20%DA], P[OA-20%SA] and P[OA-20%MA], respectively). Galactitol, D-glucitol and D-mannitol are each 6-carbon sugars that give copolymers of relatively low, intermediate, and high M_w, respectively (see above). Slopes of lines in Figure 2 gave exponent *a* values of 0.688, 0.547, 0.476 and 0.410 for POA, P(OA-20%DA), P(OA-20%SA) and P(OA-20%MA), respectively. Differences in *a* observed for different copolyesters could be due in part to configurational effects since conformation may be affected by the polyol stereochemistry. Meanwhile, the lower *a* values may indicate increased branching along chains. Therefore, plots in Figure 2 suggest reactivity at secondary hydroxyl groups increases in the following order: galactitol<D-glucitol<D-mannitol. These 6-carbon alditols have 2, 3 and 4 [R]-secondary hydroxyl groups, respectively. It maybe because that [R]-secondary hydroxyl groups are more reactive[17,18] to lead to higher molecular weight chains through branching.

In summary, an expanded set of naturally derived polyols was assessed for their potential to form high molecular weight polyol-polyesters by N435 catalysis. All substrates were polymerized forming polyol-polyesters with M_w that ranged from 11K (galactitol) to 73K (D-mannitol). No correlation was found between sugar reactivity and its chain length. By using a parallel reactor system and by performing replicate experiments with both a digital vacuum controller and a needle valve, the importance of precise vacuum control on reproducibility of experiments was established. Plots of log[η] vs. logM_w were prepared from SEC-MALS-Vis analyses of poly(octamethylene adipate) and adipate-1,8-octanediol-alditol (galactitol, D-glucitol, D-mannitol) terpolymers. Comparison of exponent *a* values from slopes of these plots showed copolymers from D-mannitol had the largest degree of branching and, therefore, greatest propensity for combined reactivity at both primary and secondary hydroxyl groups. An explanation proposed is that the higher reactivity of D-mannitol during copolymerizations was due to: i) the [R]-[R] stereochemical configuration of both carbons closest to terminal primary hydroxyl groups (β-carbons) ii) there is no chain regioisomerism in D-mannitol and iii) that all secondary hydroxyl group carbons are in the [R]-configuration. This study provides a foundation for future work with this expanded set of polyols and reproducible reaction

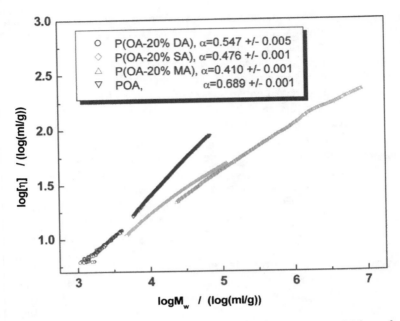

Figure 2. Intrinsic viscosity [η] as a function of the molar mass M for polyol containing polyesters and linear poly(octamethylene adipate) in THF solution at T=25 °C. (Reproduced from Macromolecules *2006, 39, 6789–6792. Copyright 2006 American Chemical Society.)*

conditions to illucidate details of regioselectivity and branching during enzyme-catalyzed polyol-polyester polymerizations.

Acknowledgement

We are grateful to the NSF-I/UCRC for Biocatalysis and Bioprocessing of Macromolecules and its industrial members (BASF, DNA 2.0, Johnson & Johnson, Rohm and Haas, Genencor, Estee Lauder, Novozymes, W.R.Grace) for financial support and intellectual input.

References

1. Wang, Q.; Dordick, J. S.; Linhardt, J. R. *Chem. Mater.* **2002**, *14*, 3232-3244.
2. (a)Kumar, R.; Gao, W.; Gross, R. A. *Macromolecules* **2002**, *35*, 6835-6844. (b) Shen, Y.; Chen, X.; Gross, R. A. *Macromolecules* **1999**, *32*, 2799-2802.

(c) Tian, D.; Dubois, P.; Grandfils, C.; Jerome, R. *Macromolecules* **1997**, *30*, 406-409. (d) Haines, A. H. *Adv. Carbohydr. Chem. Biochem.* **1981**, *39*, 13-70.

3. (a) Pavlov, D. J.; Gospodinova, N. N.; Glavchev, I. K. *Industrial Lubrication and Tribology* **2004**, *56*, 19-22. (b) Stumbé, J. F.; Bruchmann, B. *Macromol. Rapid Commun.* **2004**, *25*, 921-924. (c) Magnusson, H.; Malmström, E.; Hult, A. *Macromol. Rapid Commun.* **1999**, *20*, 453-457.

4. (a) Gao, C.; Yan, D. *Prog. Polym. Sci.* **2004**, *29*, 183-275. (b) Flory, P. J. *J. Am. Chem. Soc.* **1952**, *74*, 2718- 2723. (c) Flory, P. J. *Principles of Polymer Chemistry* Cornell University Press: Ithaca, NY, 1953.

5. Kim, Y. H. *J. Polym. Sci., Polym. Chem.* **1998**, *36*, 1685-1698.

6. Voit, B. *J. Polym. Sci., Polym. Chem.* **2000**, *38*, 2505-2525.

7. Kim, Y. H.; Webster, O. W. *Macromolecules* **1992**, *25*, 5561-5572.

8. Hult, A.; Johansson, M.; Malmströ, E. *Adv. Polym Sci* **1999**, *143*, 1-34.

9. Bohme, F.; Clausnitzer, C.; Gruber, F.; Grutke, S.; Huber, T.; Ootschke, P.; Voit, B. *High Perform. Polym.* **2001**, *13*, S21-S31.

10. Hong, Y.; Cooper-White, J. J.; Mackay, M. E.; Hawker, C. J.; Malmstrom, E.; Rehnberg, N. *Polymer* **2000**, *41*, 7705-7713.

11. Malmstrom, E.; Johansson, M.; Hult, A. *Macromolecules* **1995**, *28*, 1698-1703.

12. (a)Therisod, M.; Klibanov, A. M. *J. Am. Chem. Soc.* **1986**, *108*, 5638-5640. (b) Patil, D. R.; Dordick, J. S.; Rethwisch, D. G. *Macromolecules* **1991**, *24*, 3462- 3463.

13. (a) Kline, B. J.; Beckman, E. J.; Russell, A. *J. Am. Chem. Soc.* **1998**, *120*, 9475-9480. (b) Tsujimoto, T.; Uyama, H.; Kobayashi, S. *Biomacromolecules* **2001**, *2*, 29-31. (c) Uyama, H.; Inada, K.; Kobayashi, S. *Macromol. Biosci.* **2001**, *1*, 40-44. (d) Uyama, H.; Inada, K.; Kobayashi, S. *Macromol. Rapid Commun.* **1999**, *20*, 171-174. (e) Chaudhary, A. K.; Lopez, J.; Beckmann, E. J.; Russell, A. *J. Biotechnol. Prog.* **1997**, *13*, 318-325. (f) Kim, D. Y.; Dordick, J. S. *Biotechnol. Bioeng.* **2001**, *76*, 200-206. (g) Park, O. J.; Kim, D. Y.; Dordick, J. S. *Biotechnol. Bioeng.* **2000**, *70*, 208-216. (h) Uyama, H.; Klegraf, E.; Wada, S.; Kobayashi, S. *Chem. Lett.* **2000**, 800-801. (i) Morimoto, T.; Murakami, N.; Nagatsu, A.; Sakakibara, *J. Chem. Pharm. Bull.* **1994**, *42*, 751-753. (j) Patil, D. R.; Rethwisch, D. G.; Dordick, J. S. *Biotechnol. Bioeng.* **1991**, *37*, 639-646.

14. (a) Kumar, A.; Kulshrestha, A.; Gao, W.; Gross, R. A. *Macromolecules* **2003**, *36*, 8219-8221. (b) Kulshrestha, A. S.; Gao, W.; Gross, R. A. *Macromolecules* **2005**, *38*, 3193-3204.

15. Landry, C. J. T.; Massa, D. J.; Teegarden, D. M.; Landry, M. R.; Henrichs, P. M.; Colby, R. H.; Long, T. E. *Macromolecules* **1993**, *26*, 6294-6307.

16. Mei, Y.; Miller, L.; Gao, W.; Gross, R. A. *Biomacromolecules* **2003**, *4*, 70-74.

17. Vallikivi, I. et al. *Journal of Molecular Catalysis B: Enzymatic* **2005**, *35*, 62-69.
18. Ottosson, J.; Hult K. *Journal of Molecular Catalysis B: Enzymatic* **2001**, *11*, 1025-1028.
19. Kulicke, W. M.; Clasen, C. *Viscosimetry of Polymer and Polyelectrolytes*, Springer laboratory: New York, NY, 2004, pp 57-58.

Chapter 19

Candida antarctica Lipase B Catalyzed Synthesis of Poly(butylene succinate): Shorter Chain Building Blocks Also Work

Himanshu Azim, Alex Dekhterman, Zhaozhong Jiang, and Richard A. Gross

NSF I/UCRC for Biocatalysis and Bioprocessing of Macromolecules, Department of Chemical and Biological Sciences, Polytechnic University, Six Metrotech Center, Brooklyn, NY 11201

Lipase has been extensively used to convert aliphatic diacid and diol to biodegradable polyesters. In general, lipase catalysts show higher polymerization activity towards long chain substrates. Thus far, efforts to copolymerize short chain (≤ four carbons) diacid and diol monomers have only led to the formation of low molecular weight products. Herein we report the synthesis of high molecular weight ($M_w > 35,000$) poly(butylenes succinate) by using diethyl succinate, instead of succinic acid, as comonomer to avoid substrate phase separation, and by carrying out the copolymerization reaction at ≥ 95°C to prevent the product from precipitating out of the reaction solution.

Introduction

Poly(butylene succinate) (PBS) is an important member of biodegradable aliphatic polyester family. PBS and related copolymers have shown considerable promise for uses as environmentally biodegradable thermoplastics, as well as bioabsorbable/biocompatible medical materials (*1*). In both cases, practical applications require that the polymer possess a high molecular weight (M_n >20,000) so that it can have useful mechanical properties.

High molecular weight PBS materials, including commercial products sold by Showa Highpower Co., Ltd., have been synthesized by using organometallic catalysts at high reaction temperatures ($\geq$190°C) (*2,3*). Residual metals in these materials are difficult to remove due to strong metal-ester interactions. These metals can cause undesirable effects on the environment upon material disposal, limit polymer use in applications such as electronics, and may be unacceptable components of medical materials. Furthermore, the combination of high reaction temperatures and organometallic catalysts can lead to monomer/polymer decomposition reactions and, hence, discoloration and decreased product molecular weight. Moreover, many monomer building blocks that contain sensitive chemical entities (silicone, epoxy, vinyl, and more) are unstable and, therefore, can not be incorporated into copolymers under harsh polymerization conditions.

Lipase-catalyzed synthesis of aliphatic polyesters via diacid/diol polycondensation reactions have been explored by us and others at temperatures of 90°C and below (*4,5*). The substrates used in previous studies are α,ω-linear aliphatic diacid and/or diol monomers with six or more carbons. Herein we report a new method that allowed the first lipase-catalyzed synthesis of high molecular weight PBS from practical monomer precursors.

Experimental

Materials

Diethyl succinate, 1,4-butanediol, diphenyl ether, dodecane, and diglyme were purchased from Aldrich Chemical Co. in the highest available purity and were used as received. N435 (specific activity 10000 PLU/g) was a gift from Novozymes (Bagsvaerd, Denmark) and consists of *Candida antarctica* Lipase B (CAL-B) physically adsorbed within the macroporous resin Lewatit VPOC 1600 (poly[methylmethacrylate-co-butylmethacrylate], supplied by Bayer). Lewatit VPOC 1600 has a surface area of 110-150 m^2/g and an average pore diameter of 100 nm. N435 contains 10 wt% CAL-B that is located on the outer 100 μm of 600 μm average diameter Lewatit beads.

Instrumentation

^{1}H NMR and ^{13}C NMR spectra were recorded on a Bruker DPX 300 NMR spectrometer. The chemical shifts reported were referenced to internal tetramethylsilane (TMS, 0.00 ppm) or to the solvent resonance at the appropriate frequency. The number and weight average molecular weights (M_n and M_w, respectively) of polymers were measured by gel permeation chromatography (GPC) using a Waters HPLC system equipped with a model 510 pump, a Waters model 717 autosampler, and a Wyatt Optilab DSP interferometeric refractometer with 500, 10^3, 10^4, and 10^5 Å Ultrastyragel columns in series. Chloroform was used as the eluent at a flow rate of 1.0 mL/min. Calculation of molecular weights was based on a conventional calibration curve generated by narrow polydispersity polystyrene standards from Aldrich Chemical Company.

General procedure for N435-catalyzed condensation polymerization of diethyl succinate with 1,4-butanediol

N435 (10 wt% vs. total monomer), dried under 0.1 mmHg vacuum at 25°C for 24 h, was transferred into a 50 mL round-bottom flask containing diethyl succinate (4.00 g, 22.96 mmol), 1,4-butanediol (2.07 g, 22.96 mmol), with or without an organic solvent (200 wt% vs. total monomer). The reaction mixture was magnetically stirred and heated at a temperature between 60 and 90°C for a predetermined time. The reaction was carried out under atmospheric pressure for the first two hours to convert the monomers to oligomers. The reaction pressure was then reduced to 40 mmHg. A J-KEM vacuum regulator was employed to control the reaction pressure. At pre-selected time intervals during polymerization reactions, aliquots of about 20 mg were withdrawn, added to HPLC grade chloroform to dissolve products, and then filtered to remove the lipase catalyst. Catalyst particles were washed with chloroform twice and the combined filtrates (polymer solutions in chloroform) were directly analyzed by gel permeation chromatography (GPC).

The product mixture at the last time point was dissolved in chloroform and then filtered to remove the catalyst. The resulting chloroform solution was slowly added with stirring to methanol to precipitate white crystalline polymeric product. The precipitated polymer was washed with methanol three times and then dried *in vacuo* at 50°C for 16 h. The results of ^{1}H- and ^{13}C-NMR spectral analysis were as follows - ^{1}H-NMR (CDCl$_3$) (ppm): 1.71 (4H, m, -CH_2-), 2.62 (4H, s, -COCH_2-), 4.12 (4H, m, -CH_2O-), low intensity signals due to end groups were observed at 1.26 (t, CH_3CH$_2$O-), 3.67 (t, HOCH_2CH$_2$CH$_2$CH$_2$O-); ^{13}C-NMR (CDCl$_3$) (ppm): 25.2 (-CH$_2$-), 29.0 (-CH$_2$CO-), 64.2 (-OCH$_2$-), 172.3 (-CO-), low intensity signals due to end-groups at 14.2, 60.7 assigned to

CH_3CH_2O-CO- end groups, and resonances at 25.1, 29.1, 62.2, 64.6 that correspond to $HOCH_2CH_2CH_2CH_2O$- end groups.

Two stage enzymatic reaction to prepare high molecular weight poly(butylene succinate) (PBS)

N435 (0.61 g, dried under 0.1 mmHg vacuum at 25°C for 24 h) was added to a solution of diethyl succinate (4.00 g, 22.96 mmol) and 1,4-butanediol (2.07 g, 22.96 mmol) in diphenyl ether (12.14 g). Mixing in reactions was by magnetic stirring. The following sequence of reaction conditions was performed: *i*) 80°C for 2 h under atmospheric pressure, *ii*) reduce the pressure to 2 mmHg using a J-KEM vacuum regulator while maintaining the reaction at 80°C for 5 to 24 h, *iii*) increase the reaction temperature to 95°C for 20 to 30 h. During polymerization reactions, aliquots of about 20 mg were withdrawn and treated as was described in the previous section. The filtrates from these aliquots were analyzed by gel permeation chromatography (GPC).

Results and Discussion

Substrate Selection

The most convenient substrates for PBS synthesis are succinic acid and 1,4-butanediol. However, succinic acid was found to have low solubility in 1,4-butanediol under the reaction conditions, thus leading to two separate liquid phases in the reaction mixture. As a result, the copolymerization between the two monomers was slow and only low molecular weight oligomers were formed during the reaction. This problematic phase separation was avoided by replacing succinic acid with diethyl succinate as the acyl donor. The reaction mixture consisting of diethyl succinate and 1,4-butanediol, either in the presence or absence of an organic solvent, formed a single liquid phase which facilitated efficient mixing of reactants during the copolymerization.

Effects of reaction conditions on lipase-catalyzed copolymerization of diethyl succinate with 1,4-butanediol

Scheme 1 illustrates the general synthetic strategy for N435-catalyzed PBS synthesis.

The copolymerization reactions were performed at mild temperatures (< 100°C), under vacuum, either in solution or in bulk. The molar ratio of diethyl succinate to 1,4-butanediol was 1:1 and the catalyst was physically

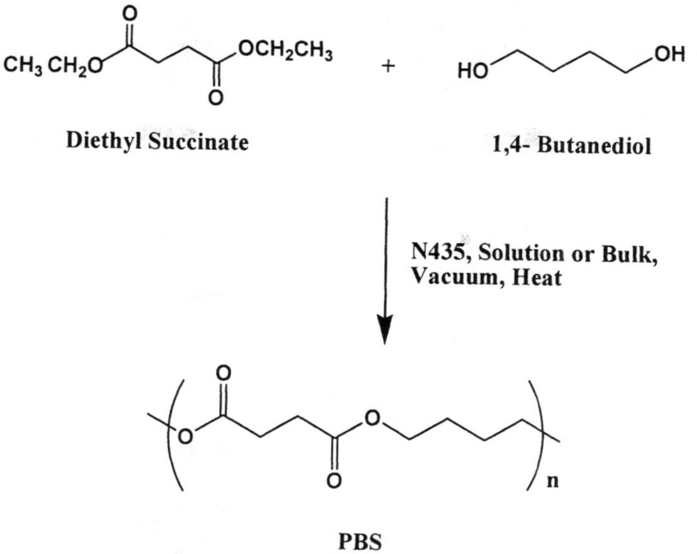

Scheme 1. Condensation copolymerization of diethyl
succinate with 1,4-butanediol to form PBS.

immobilized Lipase B from *Candida antarctica* (N435, 10 wt% catalyst or 1 wt% protein vs. total monomer). Prior to use, N435 was dried at 25°C under 0.1 mmHg vacuum for 24 h.

Diethyl succinate/1,4-butanediol copolymerization reactions were studied in bulk as well as in solution (200 wt% solvent vs. total monomer). The organic solvents evaluated include diphenyl ether, dodecane, and diglyme. They were selected since they have high boiling points (i.e. low volatility under vacuum), are chemically inert under the selected reaction conditions, they include a diverse range of structures (alkane, aliphatic and aromatic ethers), and have log P values ≥ 1.9. It's generally accepted that lipase activity is improved at higher log P values where the water structure around enzymes remains intact (*4c*). Polymerizations in diphenyl ether, dodecane, and diglyme, as well as a solvent-free reaction, were performed at 80°C under 40 mmHg for 72 h. The polymer of highest molecular weight was obtained in diphenyl ether. Values of product M_n (M_w/M_n) obtained from the reactions in dodecane, diglyme, diphenyl ether, and in-bulk are 2500 (1.4), 4400 (1.3), 10000 (1.6), and 3300 (1.2), respectively.

Effects of reaction temperature on PBS molecular weight averages were studied for polymerizations in diphenyl ether. Figure 1 delineates product M_n as a function of reaction time at different temperatures. In 24 h, the polymerization reaction at 60, 70, 80, and 90°C yielded PBS with M_n values of 2000, 4000, 8000, and 7000, respectively. Further increase in reaction time resulted in little change in the product molecular weight. This led us to focus on the course of reactions over shorter time intervals as is discussed below.

Since M_n was nearly unchanged with increase in reaction time from 24 to 72 h, a study was performed to characterize chain growth from 0 to 24 h at 80°C in diphenyl ether. Figure 2 shows that, during the first 10 h of the reaction, a rapid increase in M_n up to 5500 was observed. Thereafter, M_n increased slowly to 8000 and 10000 by 24 and 72 h, respectively. Close inspection of physical changes in the reaction mixture showed that, at about 8 h, precipitation of PBS occurred, and after 20 h, the reaction mixture became a gel. Phase separation of the product during the reaction causes infrequent collisions between substrate and catalyst molecules. This explains slowed chain growth beyond 10 and 24 h that was observed in Figures 2 and 1, respectively.

Temperature-varied two stage N435-catalyzed polymerizations between diethyl succinate and 1,4-butanediol

Based on observations discussed above, a temperature-varied two stage process was designed and studied to circumvent phase-separation of PBS during the reaction. The copolymerization of diethyl succinate with 1,4-butanediol in diphenyl ether solution was carried out first at 80°C for 5-24 h at 1.8-2.2 mm Hg to form PBS with M_n between 5000 and 10000. Subsequently, the reaction temperature was raised to 95°C at 1.8-2.2 mm Hg to re-form a monophasic reaction mixture. Figure 3 shows that, by this strategy, PBS with $M_w > 35000$ was synthesized. During the polymerization at 95°C, precipitation of PBS was not observed. Instead of nearly invariant molecular weight values with increasing reaction time above 21 h (see Figures 1 and 2), Figure 3 shows an abrupt increase in PBS molecular weight after 21 h. In other words, by increasing the reaction temperature to 95°C at 21 h, a monophasic reaction mixture was formed and PBS molecular weight increases up through 50 h. Furthermore, in contrast to M_w/M_n values of ≥ 1.8 observed using chemical catalysts for aliphatic polyester synthesis by step-condensation reactions (2a), PBS at 50 h has M_w/M_n of 1.39.

A control experiment without using N435 catalyst was performed in parallel under identical reaction conditions as described above. GPC analysis of the reaction products indicates that essentially no polymer ($M_n \geq 500$) was formed during the 50 hour period of the un-catalyzed reaction.

Conclusions

We report, for the first time, an efficient metal-free enzymatic route to synthesize high molecular weight PBS. Slow diffusivity of the substrates and

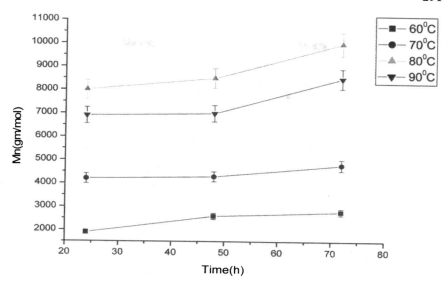

Figure 1. Effects of reaction temperature and time on PBS M_n for N435-catalyzed copolymerization of diethyl succinate with 1,4-butanediol

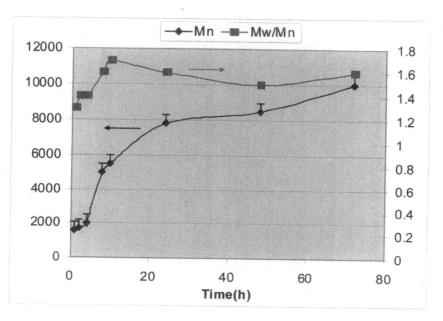

Figure 2. Effects of reaction time on PBS M_n and M_w/M_n for N435-catalyzed copolymerization of diethyl succinate with 1,4-butanediol at 80 °C in diphenyl ether.

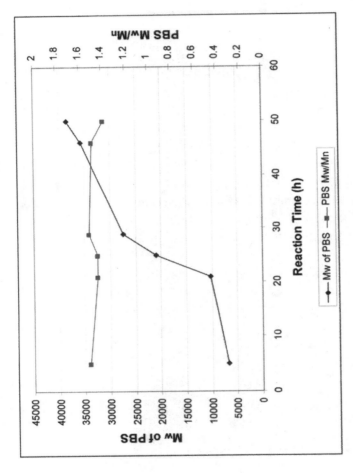

Figure 3. Temperature-varied two stage N435-catalyzed polymerization: Effects of reaction time on PBS M_w and M_w/M_n. Reaction was performed in diphenyl ether at 80°C during 0-21 h and at 95°C during 21-50 h.

growing polymer chains resulting from polymer precipitation was found to be a major factor limiting further molecular weight increase after 5 to 10 h reaction times. This problem was circumvented by carrying out polymerization reactions in solution using a temperature-varied, two stage process. In other words, by increasing the polymerization temperature to 95°C during the second-stage, the polymer remained soluble in the reaction medium so that the reaction proceeded in a single liquid phase that facilitated further molecular weight increases. Furthermore, by using the temperature-varied two stage process, the catalyst aging rate was minimized.

Acknowledgements

We are thankful to the members of NSF Industrial-University Cooperative Research Center for Biocatalysis and Bioprocessing of Macromolecules at the Polytechnic University for their financial support, encouragement, and advice during the performance of this research.

References

1. (a) Gross, R. A.; Kalra, B. *Science* **2002**, *297*, 803-806. (b) Gross, R. A.; Kumar, A.; Kalra, B. *Chem. Rev.* **2001**, *101*, 2097. (c) Kobayashi, S.; Uyama, H.; Kimura, S. *Chem. Rev.* **2001**, *101*, 3793. (d) Gross, R. A.; Kalra, B.; Kumar, A. *Appl. Microbiol. Biotechnol.* **2001**, *55*, 655.
2. (a) Takahashi, H.; Hayakawa, T.; Ueda, M. *Chem. Lett.* **2000**, 684. (b) Ishii, M.; Okazaki, M; Shibasaki, Y.; Ueda, M.; Teranishi, T. *Biomacromolecules* **2001**, *2*, 1267-1270 and the references therein.
3. Mochizuki, M.; Mukai, K.; Yamada, K.; Ichise, N.; Murase, S.; Iwaya, Y. *Macromolecules* **1997**, *30*, 7403.
4. (a) Mahapatro, A.; Kumar, A.; Kalra., B.; Gross, R. A. *Biomacromolecules* **2003**, *4*, 544-551. (b) Kumar, A.; Gross, R. A. *Biomacromolecules*, **2000**, *1*, 133. (c) Mahapatro, A.; Kumar, A.; Kalra, B.; Gross, R. A. *Macromolecules* **2004**, *37*(1), 35. (d) *Polymer Biocatalysis and Bio-materials*; H. N. Cheng and R. A. Gross, Eds; Oxford University Press: **2005**, Chapters 1, p 22-28.
5. (a) Linko, Y. Y.; Wang, Z. L.; Seppala, J. *J. Biotechnol.* **1995**, *40*, 133. (b) Uyama, H.; Inada, K.; Kobayashi, S. *Polym. J.* **2000**, *32*, 440. (c) Binns, F.; Harffey, P.; Roberts, S. M.; Taylor, A. *J. Polym. Sci., Part A: Polym. Chem.* **1998**, *36*, 2069.

Chapter 20

Enzyme-Catalyzed Oligopeptide Synthesis: Rapid Regioselective Oligomerization of L-Glutamic Acid Diethyl Ester Catalyzed by Papain

Geng Li, Alankar Vaidya, Wenchun Xie, Wei Gao, and Richard A. Gross[*]

NSF I/UCRC for Biocatalysis and Bioprocessing of Macromolecules, Polytechnic University, Six Metrotech Center, Brooklyn, NY 11201

Papain-catalyzed oligomerization of diethyl L-glutamate hydrochloride was conducted in phosphate buffer(0.9M, pH 7) at 40 °C. MALDI-TOF spectra of precipitated products showed two series of ion peaks separated by 157 m/z units, the mass of oligo(γ-ethyl-L-glutamate) repeat units. The most abundant signals were at DP 8 and 9, in excellent agreement with DP_{avg} values determined by ¹H-NMR. Oligo(γ-ethyl-L-glutamate) synthesis occurred rapidly and high product yields were observed over a broad range of pH values. Ionic strength had no significant effect on oligopeptide yield. The dominant role of phosphate buffer in reactions was its control of pH.

Poly(amino acid)s are normally biodegradable in the environment. Furthermore, they may be biocompatible and provide biological activities that are useful in therapeutics. A commonly practiced route to synthesize poly(α-amino acid)s of high molecular weight is by ring-opening polymerization of α-amino acid N-carboxylic anhydrides (NCAs).[1,2] However, this involves toxic phosgene or their equivalents for monomer synthesis, and the reaction conditions require strict removal of water and high monomer purity. An alternative route is by thermal or acid-catalyzed polymerizations. Thermal condensation polymerization of aspartic acid followed by alkaline hydrolysis was used to prepare poly(aspartic acid). This product is water-soluble, biodegradable, and is used as a metal chelator[3]. However, the harsh polymerization conditions results in racemization of aspartic acid and formation of chains with both α- and β-linked units. When poly(amino acid)s of defined stereochemistry and repeat unit composition are required, milder selective polymerization methods must be used. One such method is by protease-catalyzed oligopeptide synthesis.

Basic to protease catalyzed oligopeptide synthesis is equilibrium- or thermodynamic control to direct reversal of proteolysis [4-6]. Difficulties encountered include low reaction rates, high stoichiometric amounts of enzyme, and the need to apply direct approaches to shift the reactions towards formation of desired products. Reaction conditions that lead to product precipitation or extraction increase efficiency of the reverse reaction. Kinetically controlled syntheses has proved useful for serine and cysteine proteases that form activated acyl enzyme intermediates during catalysis[5]. This approach generally involves use of activated acyl moieties, such as esters, as donor components which significantly accelerate the reaction rate. This study makes use of principles from both kinetic and thermodynamically controlled reactions in that, reactants are activated by formation of esters and products precipitate from reactions.

Advantages of protease-mediated oligopeptide synthesis include: *i*) avoids racemization, *ii*) decreased requirement for protection-deprotection steps, *iii*) utilization of readily renewable and potentially inexpensive amino acid monomers, and *iv*) mild non-hazardous operating conditions[7, 8]. Aso et al [9] reported oligomerization of dialkyl L-glutamate hydrochloride by protease-catalysis. By increasing the organo-solubility of papain by its modification with poly(ethylene glycol), Uemura et al[10] showed oligomerizations of dialkyl L-aspartate and dialkyl L-glutamate can be performed in benzene to yield a mixture of oligomers with chain lengths from heptamer to decamer. Ester hydrochlorides of methionine, phenylalanine, threonine, and tyrosine were polymerized by papain-catalysis in buffer giving poly(α-amino acid)s with degree of polymerization (DP) less than 10[11-13]. Uyama et al[14] reported protease-catalyzed regioselective polymerization and copolymerization of diethyl L-glutamate hydrochloride. By ^1H-^1H COSY NMR they showed that oligomers formed from diethyl L-glutamate hydrochloride using papain, bromelain and α-chymotrypsin were exclusively α-linked. Matsumura et al[15] reported that alkalophilic protease from *Streptomyces* sp. catalyzed the oligomerization of diethyl L-aspartate in bulk forming a product with 88 % α-linkages. Soeda et al [16] found that diethyl L-aspartate was oligomerized by microbial protease BS in solutions containing small volumes of water. For example, oligomerizations of diethyl L-aspartate using protease BS were performed at 40 °C, for 2-days, in MeCN containing 4.5% by-volume water. The resulting product was α-linked poly(β-ethyl L-aspartate) with M_w up to 3 700 in 85%-yield.

The above cited publications demonstrate excellent progress has been made towards the synthesis of oligopeptides by protease-catalysis. However, many of these studies didn't take into account decreased pH during oligopeptide synthesis[14,15]. Reaction pH effects both catalyst activity and stability. Therefore, pH changes should be followed or eliminated by external addition of base during reactions. In addition, many of the above studies did not report catalyst activity on a standard substrate making it impossible for others to reproduce their work[9,14]. Given the importance of developing general, facile and selective routes to oligopeptides, our laboratory has begun studies to build on the above literature. However, given uncertainties from previous publications, this paper reports a reinvestigation of papain-catalyzed oligo(γ-ethyl-L-glutamate) synthesis under defined conditions. Activity of the papain-catalyst was determined using a standard substrate (Ac-Phe-Gly-*p*-nitroanaline). Oligo(γ-ethyl-L-glutamate) synthesis was studied as a function of buffer pH and normality. Changes in solution pH during reactions were recorded. Papain activity as a function of pH was determined using pH control by automated or manual addition of sodium hydroxide. By performing reactions with controlled pH, we explored whether the buffer causes 'salting out' of products or it is involved in non-specific protein-ion interactions. The extraordinarily rapid formation of γ-ethyl-L-glutamate under controlled reaction conditions is reported.

Experimental Section

General procedure for papain-catalyzed oligo(γ-ethyl-L-glutamate) synthesis

L-glutamic acid diethyl ester hydrochloride (600 mg, 2.5 mmol), papain (100 mg of water-soluble powder containing 40 mg protein), and 5 mL phosphate buffer solution set at a predetermined pH were transferred to a 15 mL Erlenmeyer flask. The flask was gently stirred in a water bath at 40 °C for a predetermined reaction time. Then, the reaction mixture was cooled to room temperature and deionized water (20 mL) was added. The insoluble product was separated by centrifugation (6000 rpm), washed once with dilute HCl (2 % v/v) and then twice with deionized water. The resulting product was lyophilized giving a beige powder.

Controlling system pH

Since the pH decreases during oligomerizations, manual and automatic pH control was employed. Manual control was performed by using a VWR SympHony SB301 pH meter and manual addition of 10M NaOH in response to observed pH changes. By manual control the reaction pH was maintained within ± 0.1 pH units of the set value. Automatic controlled pH was performed using a Tiamo titration control system and Metrohm CH9101 dosing unit. The dosing solution (10M NaOH) was added at 0.05 to 0.1 µL/min and the frequency at which the probe checked the pH was set to 1.0 second. By automatic controlled pH with this hardware and software the pH was controlled within ± 0.05 units of the set value.

Activity assay

A stock solution was prepared consisting of a 1:1 (v/v) mixture of Ac-Phe-Gly-p-nitroaniline (Ac-Phe-Gly-PNA, 0.3 mg/mL) in DMF and 50 mM phosphate buffer pH 6.8. To 1 mL of substrate stock solution, 50 µL of papain (Stock: 160mg/mL protein) was added and the absorbance was measured spectrophotometrically using a Shimadzu UV-1601 at 405 nm for 100 second. Enzyme activity per minute was calculated from the initial linear portion of the absorbance vs. time curve. Activity units are the amount of enzyme activity which catalyzes the transformation of 1 micromole of the p-nitroaniline per minute in phosphate buffer (pH 6.8) at 25 °C using 8 mg/mL protein catalyst.

Mass Assisted Laser Desorption/Ionization-Time-of-Flight (MALDI-TOF)

MALDI-TOF spectra were obtained on an OmniFlex MALDI-TOF mass spectrometer (Bruker Daltonics, Inc.). The instrument was operated in a positive ion linear mode with an accelerating potential of +20kv. The TOF mass analyzer had pulsed ion extraction. The linear flight path was 120 cm. OmniFLEX TOF control software was used for hardware control and calibration. X-massOminFLEX 6.0.0 was used for data processing. Spectra were acquired by averaging at least 200 laser shots. The pulsed ion extraction delay time was set at 200 ns. The spectrometer was externally calibrated using angiotensin II as a standard (1046.54 amu). To generate the matrix solution, a saturated solution of α-cyano-4-hydroxycinnamic acid (CCA) was prepared in trifluoroacetic acid/acetonitrile (TA) 1 to 10 v/v. Oligopeptide samples dissolved in dimethyl sulfoxide (DMSO, 5 μL) were diluted with TA solution to 1~5 pmol/μL and mixed with 5 μL saturated matrix solution. Then, 1 μl of this mixture was applied onto the clean target. The sample target was dried in a stream of cold air from a dryer. For details, see reference 22.

Results and Discussion

For cysteine-protease catalyzed peptide synthesis, the kinetic model mechanism (see Scheme 1)[5,17-19] reveals that, prior to peptide bond formation, activated amino acid and enzyme interact forming a Michaelis-Menten complex (i.e acyl enzyme intermediate). This complex is then competitively deacylated by water and a nucleophile. If the nucleophile is an amino acid or oligopeptide, a new peptide bond is formed. Precipitation of oligo(γ-ethyl-L-glutamate) from the reaction solution during synthesis further shifts the equilibrium towards peptide formation. Concurrent with peptide bond formation is release of HCl which can lead to changes in medium pH during reactions. Papain prefers α-position over γ of L-glutamate because of nature of its active site pocket. Uyama et al[14] also reported the regioselective production of the oligoglutamate having an exclusive α-peptide linkage.

Effect of Substrate Concentration

Figure 1 shows the effect of substrate concentration on %-yield precipitated product for 20 min reactions under non-controlled pH conditions (phosphate buffer 0.9 M). At initial pH 7, the %-yield increased from 56±3 to 71±4 % by increasing the substrate concentration from 0.03 to 0.3 M. The yield remained unchanged as the substrate concentration was increased from 0.3 to 0.5 M and then decreased to 58±3 % by increasing the substrate concentration to 0.7 M. A similar trend was observed when reactions were performed at initial pH 8, except, product yields were higher than those at pH 7 for substrate concentrations 0.03 to 0.5 M. The apparent maximum %-yields (71±4% and 81±4%) occurred at 0.3 M substrate concentration for reactions at pH 7 and 8, respectively.

Scheme 1 shows that chain growth is accompanied by liberation of HCl from L-glutamic acid diethyl ester hydrochloride. Hence, by using initial pH 8 instead of 7, the reaction pH may remain within an optimum range of values where papain has higher activity for longer periods of time. This might explain relatively higher yields at pH 8 for substrate concentrations up to 0.4. However, as the substrate concentration increases at constant product yield, more HCl is

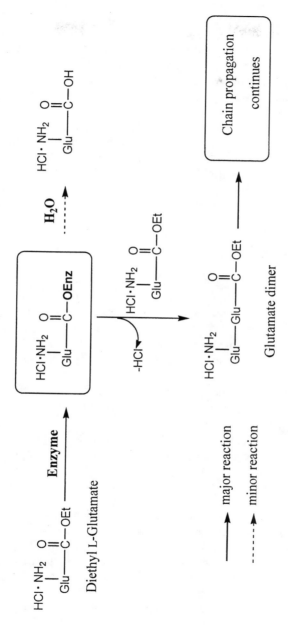

Scheme 1 . Mechanism for peptide bond formation (see references 5, 17-19)

liberated creating the potential for decreased solution pH and lower yields. This may explain identical values of %-yield at pH 7 and 8 at high substrate concentrations (0.5 to 0.7). However, without knowledge of pH values during the course of reactions, it is difficult to provide definitive explanations. Substrate concentration is not an independent variable since higher substrate concentrations can lead to larger changes in solution pH that effects enzyme activity. These results, where explanations required knowledge of pH change, motivated the studies below where pH was either monitored or controlled during reactions.

Figure 1 showed that, regardless of pH change, oligomerizations at initial pH 7 and 8 occur over a broad range of substrate concentrations with yields greater than 50%. Thus, kinetic constants for chain propagation forming oligomers that precipitate from solution must be large. Rapid oligomerization kinetics allows formation of product without relying on high substrate concentration to shift the equilibrium.

Structural analysis

The ^{1}H-NMR spectrum of the precipitated product synthesized by papain-catalysis (8 mg/mL catalyst, 0.9 M phosphate buffer, at 40 °C, for 20 min under non-controlled pH conditions) is obtained. Peaks positions and assignments are identical to those described in reference 14 for oligo(γ-ethyl-L-glutamate). Relative integration of methane proton signals F and D (4.3 and 3.8 ppm, respectively) was used to calculate DP_{avg} for oligo(γ-ethyl-L-glutamate)s. The DP_{avg}, 8.7 ± 0.3, varied little for oligo(γ-ethyl-L-glutamate) synthesized herein over a wide range of reaction conditions. The MALDI-TOF spectrum displayed in Figure 2 is of the identical product characterized above by ^{1}H-NMR. Two series of ion peaks, separated by 157 m/z units, equal to the mass of oligo(γ-ethyl-L-glutamate) repeat units, are observed and the peaks are isotopically resolved. Mass peaks corresponding to DP values 7-11 are seen in Figure 2. The most abundant signals were at DP 8 and 9, in excellent agreement with DP_{avg} values determined by ^{1}H-NMR. The major series of signals with m/z 1169, 1326, 1483, 1640 and 1797 was accompanied by a series of lower intensity peaks with m/z less by 28. This corresponds to hydrolysis of one ester group either at a chain-end or a pendant group along oligo(γ-ethyl-L-glutamate). Thus, results of MALDI-TOF suggest that the product is a mixture of completely esterified and mono-carboxylic acid oligomers. A route by which amide formation is the major pathway accompanied by less frequent hydrolysis is depicted in Scheme 1.

Aso et al [9] and Uyama et al [14] found that crude papain extracts catalyzed oligomerizations of L-glutamic acid diethyl ester (conditions: pH 7~8.5, 0.5M, ionic strength 2.0 M phosphate buffer, 25°C and pH 7.0, 40°C, 3 hours) giving oligomers with DP 5 to 9 and DP_{avg} 9.5 from FAB-Mass spectroscopy and ^{1}H-

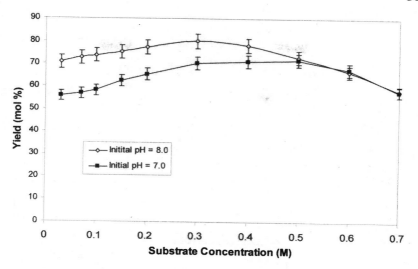

Figure 1. Effect of substrate concentration on %-yield of oligo(γ-ethyl-L-glutamate). Reactions were conducted with 8 mg/mL catalyst, 0.9 M phosphate buffer, at 40 °C, for 20 min, under non-controlled pH conditions. Values reported are the mean from at least duplicate experiments and error bars define the maximum and minimum values obtained.

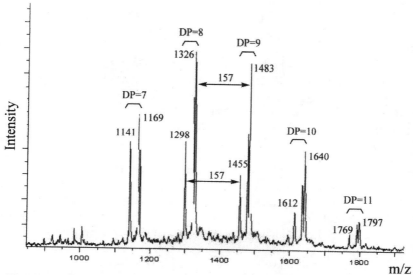

Figure 2. MALDI-TOF spectrum of oligo(γ-ethyl-L-glutamate) synthesized using 0.5M L-glutamic acid diethyl ester hydrochloride, 8 mg/mL catalyst, at 40 °C for 20 min, pH 7.0 under non-controlled pH conditions.

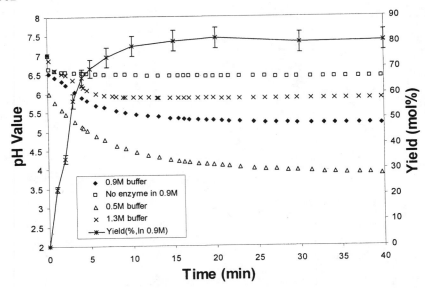

Figure 3. Time course of oligo(γ-ethyl-L-glutamate) synthesis and change in pH during oligo(γ-ethyl-L-glutamate) synthesis for the control (0.9M) and reactions in 0.5, 0.9 and 1.3M phosphate buffer. Synthesis was performed at 40 °C in 0.9M phosphate buffer at pH 7 under non-controlled pH conditions. Values reported are the mean from at least duplicate experiments and error bars define the maximum and minimum values obtained. (Reproduced from reference 22. Copyright 2006 American Chemical Society.)

NMR analysis, respectively. Hence, the molecular weights of products formed in this and related published work are in good agreement. This is largely due to that, molecular weight of products are determined by their solubility in the reaction medium. In other words, when chains grow to a length at which they are poorly soluble, they precipitate from solution.

Time Course of oligo L-(glutamic acid ester) synthesis

To investigate the time course of oligo(γ-ethyl-L-glutamate) synthesis, 8 mg/mL papain catalyst and 0.5 M substrate were incubated at 40 °C in phosphate buffer (0.9M, pH 7). No induction period was required and, by 5, 10 and 20 min, the yield reached 70, 78 and 81±5%, respectively (Figure 3). Further increase in the reaction time beyond 20 min did not significantly increase product yield. Hence, under these conditions, the oligomerization is rapid making this reaction attractive from both fundamental and commercial viewpoints. Investigation of reaction progress at short time intervals (≤20 min) is crucial to accurately assess effects of reaction parameters on product yield. However, previous work used longer reaction times for such studies. For example, Uyama et al[14] used higher concentrations of papain (40 mg/mL) then this study and 3 h reaction times to assess effects of monomer concentration, enzyme concentration and ionic strength on oligo(γ-ethyl glutamate) yield. Similarly, even though Aso et al[9] reported high oligomer yield after 1 h, they assessed effects of substrate concentration, reaction time and buffering agent for 3 or 24 h reactions. In these cases, it may be that differences in polymer yields due to changes in reaction parameters are blurred since time is given so that slow reactions can '*catch up*' to those under more optimal conditions.

The possibility that products may be hydrolyzed by extended time exposure to protease was studied. Products obtained at 20 min and 3 h were both analyzed by MALDI-TOF and ^{1}H NMR. However, no observable difference in %-hydrolysis of ester side chains was found for these products. This is explained by precipitation from solution of oligo(γ-ethyl glutamate) limiting its accessibility to protease.

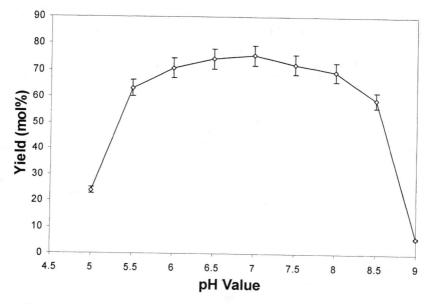

Figure 4. Relationship between reaction pH and yield of oligo(γ-ethyl-L-oligoglutamate.. Rea ctions were conducted using an automated pH-stat, 0.9M phosphate buffer, 0.5M L-glutamic acid diethyl ester hydrochloride, 8 mg/mL catalyst, at 40 °C for 20 min. Values reported are the mean of at least duplicate experiments; error bars define the maximum and minimum values obtained.

Interplay between pH, buffer concentration and ionic strength

Biotransformations should be performed within the optimum activity range of the enzyme-catalyst. In the present system, pH decreases as the reaction progresses due to HCl liberation. One method to control pH change is by adjusting the buffer strength. However, increasing the buffer concentration also increases salt concentration which might alter enzyme activity. Furthermore, after attaining a certain molecular weight, the product precipitates from solution. The salt concentration in reactions may change product yield and molecular weight by lowering the products solubility in the reaction medium (i.e. 'salting out' effect). This complex set of variables was systematically studied as described below.

Oligomerizations were performed as a function of reaction time and phosphate buffer concentration. Figure 3 shows that, at 1.3 M phosphate buffer, the reaction pH decreased from 7.0 to 5.9 in just 20 min. By further reducing the phosphate buffer concentration, pH values in reactions decreased to larger extents. For example, at 0.5 M phosphate buffer, the reaction pH decreased from 7.0 to 4.1 in 20 min. Even using a moderately high phosphate buffer concentration (0.9 M), decrease in pH by 20 minutes is substantial (from 7 to 5.3). A no-enzyme control, performed at 0.9 M phosphate buffer, showed no change in reaction pH to greater than 40 min. Excellent correlation was found between oligomer formation and pH decrease over time. For example, at 0.9M phosphate buffer (initial pH 7.0), the pH decreases as oligomer formation increases to 15 min, after which, neither showed any further change.

The relationship between reaction pH and product yield was studied to establish a suitable pH range at which papain-catalyzed γ-ethyl-oligo-glutamate synthesis can be performed (Figure 4). The reaction pH was maintained at different pH values (5.0 to 9.0) using an automated pH-stat (see experimental section). Reactions were performed as follows: 0.5M L-glutamic acid diethyl ester hydrochloride monomer, 8 mg/mL catalyst, 40 °C, 20 min. High product yields were observed over a broad pH range. As long as the pH was maintained at from 5.5 to 8.5, the product yield was $\geq$ 60%. Maximum product yields between 70 and 75% were reached by maintaining the pH between 6 and 8. Product yields decreased to 63±3% and 24±1% by performing reactions at pH 5.5 and 5, respectively. Similarly, product yields decreased to 59±3% and 6.8±0.5% by performing reactions at pH 8.5 and 9, respectively. Therefore, care must be taken to ensure that the reaction pH does not fall below 5.5 or above 8.5.

To study the effect of ionic strength (I) on product yield (Figure 5), phosphate buffer (pH 7.0) was prepared in which the phosphate ion concentration of buffer was maintained at 0.1 and 0.3 M respectively. The ionic strength of each buffer was then adjusted to 1.9 M by addition of NaCl. Reactions were performed as above with 0.5 M L-glutamic acid diethyl ester hydrochloride, 8 mg/mL catalyst, at 40 °C for 20 min, however, pH of reactions was not controlled. Results in Figure 5 show that the ionic strength had no significant effect on yields of precipitated oligopeptide. Furthermore, DP_{avg},

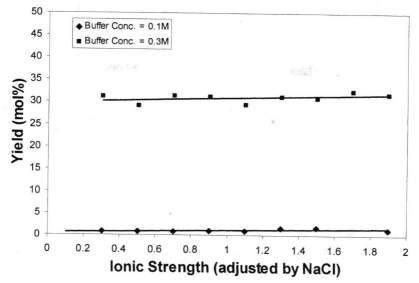

Figure 5. Effect of ionic strength (I) on oligo(γ-ethyl-L-oligoglutamate) yield. The ionic strength of each buffer was adjusted to 1.9 M by addition of NaCl. Reactions were performed with 0.5 M L-glutamic acid diethyl ester hydrochloride, 8 mg/mL catalyst, at 40 °C for 20 min, without external pH control.

calculated from ^{1}H-NMR, was found to be independent of ion strength. At 0.3 M phosphate buffer, product yield was unchanged irrespective of the NaCl concentration. Furthermore, the rapid decrease of pH for reactions run at 0.1M resulted in little product yield for ionic strengths ranging from 0.1 to 1.9M. Thus, it's not ionic strength (adjusted by NaCl) that determines oligomer yield but, instead, it's the capacity of regulating system pH that is crucial. In contrast, Uyama et al[14] reported that buffer ionic strength (pH 7.0) plays a decisive role in determining the yield of the same oligomerization reaction. In addition, they did not consider the role of pH change on the oligomerization.

To further clarify relationships between phosphate concentration, pH change and oligomers yield as well as to investigate phosphate buffer concentration as an independent variable, experiments were performed at varying phosphate buffer concentrations with and without external pH control. Reactions were performed with 0.5M L-glutamic acid diethyl ester hydrochloride, 8 mg/mL protein catalyst, 40 °C, pH 7.0 for 20 min. Without external pH control, the %-yield oligomers increased from about 4 to 82% as phosphate buffer concentration was increased from 0.1 to 1.1M (Figure 6). Results above showed that, by maintaining the pH at between 5.5 and 8.5, the product yield was ≥ 60% (see Figure 4).

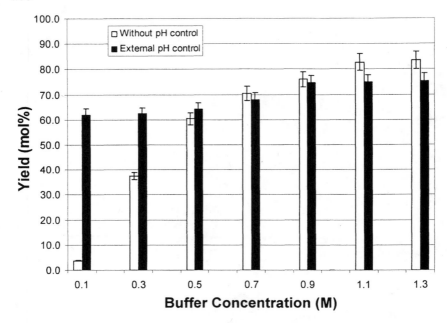

Figur e 6. Effect of phosphate buffer concentration on oligo(γ-ethyl-oligoglutamate) yield. Reactions were performed with 0.5M L-glutamic acid diethyl ester hydrochloride, 8 mg/mL catalyst, 40 °C, pH 7.0 for 20 min with and without external pH control. Values reported are the mean from at least duplicate experiments and error bars define the maximum and minimum values obtained. (Reproduced from reference 22. Copyright 2006 American Chemical Society.)

However, papain activity decreased when the pH dropped to 5.0. Without pH control, using 0.9M phosphate buffer, the pH remained at ≥5.5 to 10 min and gave product yields at 10 min of 79% (see Figure 3). It follows that nearly optimal yields were attained without pH control by using 0.9M phosphate buffer. At phosphate buffer concentrations >0.9M (i.e. 1.3M, see Figure 3), reaction pH remains above 5.5, papain activity remains high, and product yields are optimal. In contrast, by decreasing phosphate buffer concentration from 0.9 to 0.5M, oligomer yield decreased from 76 to 60%. This decrease in oligomer yield is due to the relatively poor buffering capacity of 0.5 M phosphate buffer (Figure 3). Hence, by performing reactions with 0.3 and 0.1M phosphate buffer, oligo(γ-ethyl-L-glutamate) yield decreased to 37.5 ± 3.0 % and 3.8 ± 2.0, respectively.

Increasing the buffer concentration also increases the phosphate content in reactions which might alter enzyme activity. Also, after attaining a certain molecular weight (see below), oligo(γ-ethyl-L-glutamate) precipitates from solution. By increasing the phosphate salt concentration oligo(γ-ethyl-L-glutamate), the yield might increase due to 'salting out' of product. To investigate these possibilities, reactions were carried out with controlled pH at different buffer concentrations (Figure 6). Without pH control, at 0.1 and 0.3M phosphate buffer concentrations, product yields were quite low: 3.8 ± 2.0 and 37.5 ± 3.0 % respectively. However, by externally controlling the pH at 7.0, product yield at these low buffer concentrations increased to about 62 %. This proves that high product yields can also be achieved at low phosphate buffer concentrations as long as the pH of the reaction is controlled. At phosphate buffer concentrations ≥ 0.5 M, differences between %-product yield with and without pH control are small. Therefore, the dominant role of phosphate buffer in reactions was its control of pH. Other influences of phosphate ions on papain, such as non-specific salt interactions, or a 'salting out' of product, appear to be unimportant.

The high sensitivity of papain to pH changes in reactions is consistent with what is known about enzyme structure and mechanism of action. The catalytic activity of papain is based on a nucleophilic attack of the cystein residue, and the activity requires two ionizable residues: Cys-25 and His-159. Only near neutral pH do these key amino acids have the appropriate ionization state that favors synthesis[20]. At low or acidic pH values, His-159 which must be in the deprotonated state for acyl-enzyme complex formation, remains protonated leading to low enzyme activity. In contrast, at high or alkaline pH values, a proton is abstracted from Cys-25 destabilizing the intermediate transition state of the acyl-enzyme complex.

Conclusions

Papain-catalyzed oligomerization of diethyl L-glutamate hydrochloride was conducted in 0.9M phosphate buffer at 40 °C to give oligo(γ-ethyl-L-glutamate) in 80% yield by 10 min. Concurrent with peptide bond formation is release of HCl. Changes in medium pH must be managed as this can affect enzyme activity. Nearly optimal yields were obtained without pH control at 0.9M phosphate buffer. Decrease in phosphate buffer concentration from 0.9 to 0.5M resulted in a decrease in oligomer yield from 76 to 60%. This is explained by decrease in reaction pH below 5.0 within 5 min. Values of pH below 5.5 result in decreased papain activity for oligo(γ-ethyl-L-glutamate) synthesis. The dominant role of phosphate buffer in reactions was its control of pH. Other influences of phosphate on papain such as non-specific salt interactions, or a 'salting out' of product, appear to be of little or no importance. The predominant species in the molecular weight distribution by MALDI-TOF analysis are DP 8 and 9. This agrees with DP_{avg} calculations from ^{1}H-NMR.

Acknowledgements

We are grateful to the NSF-I/UCRC for Biocatalysis and Bioprocessing of Macromolecules and its industrial members (BASF, DNA 2.0, Johnson & Johnson, Rohm and Haas, Genencor, Estee Lauder, Novozymes, W.R.Grace) for financial support and intellectual input.

References

1. H.-A. Klok, S. L. *Advanced Materials* **2001**, *13*, 1217.
2. Gallot, B. *Progress in Polymer Science* **1996**, *21*, 1035.
3. Swift, G. Polymer Degradation and Stability **1998**, *59*, 19.
4. Rozzell, J. D., Wagner, F., Eds., Biocatalytic Production of Amino Acids and Derivatives. *Hanser Publishers: Muenchen* **1992**.
5. Bordusa, F., Proteases in Organic Synthesis. *Chem. Rev.* **2002**, *102*, 4817.
6. Hans-Dieter Jakubke, P. K., Andreas Konnecke. *Angew. Chem. Int. Ed. Engl* **1985**, *24*, 85.
7. Iqbal Gill, R. L. p.-F., Xavier Jorba, and Evgeny N. Vulfson. *Enzyme and Microbial Technology* **1996**, *18*, 162.
8. Tepanov, V. M. S. *Pure & Appl. Chem.* **1996**, *68*, 1335.
9. Aso, K. U., T.; Shiokawa, Y. *Agric. Biol. Chem.* **1988**, *52*, 2443.
10. Uemura, T. F., M.; Lee, H.-H.; Ikeda, S.; Aso, K. *Agric. Biol. Chem.* **1990**, *54*, 2277.
11. Sluyterman, L. A. A. W., J. *Biochimica et Biophysica Acta, Enzymology* **1972**, *289*, 194.
12. Gregory Anderson, P. L. L. *Helvetica Chimica Acta* **1979**, *62*, 488.
13. Rolf Jost, E. B. J. C. M. P. L. L.*Helvetica Chimica Acta* **1980**, *63*, 375.
14. Uyama, H.; Fukuoka, T.; Komatsu, I.; Watanabe, T.; Kobayashi, S. *Biomacromolecules* **2002**, *3*, 318.
15. Shuichi, M.; Yasuhiro T.; Nobuyuki O.; *Macromolecular Rapid Communications* **1999**, *20*, 7.
16. Soeda, Y.; Toshima, K.; Matsumura, S. *Biomacromolecules* **2003**, *4*, 196.
17. M.L.Bender, G. E. C.; C.R.Gunter, G.J.Kezdy *J.Am. Chem.Soc.* **1964**, *86*, (1964), 3697.
18. J.Fastrez, A. R. F. *Biochemistry* **1973**, *12*, 2025.
19. Jakubke, H.-D. *Angew. Chem. Int. Ed. Engl* **1995**, *34*, 175.
20. Lewis, S.D.; J. F.; Shafer, J.A. *Biochemistry* **1976**, *5*, 5009.
21. Wartchow, C. A., Wang, P., Bednarshki, Mark D, Callstrom, Matthew R., *J. Org. Chem* **1995**, *60*, 2216.
22. Geng L., A. Vaidya, K. Viswanathan, J. Cui, W. Xie,W. Gao, and Richard A. Gross. *Macromolecules* **2006**, *39*, 7915-7921

Chapter 21

Enzyme-Catalyzed Polyamides and Their Derivatives

Qu-Ming Gu, W. W. Maslanka, and H. N. Cheng[*]

Hercules Incorporated Research Center, 500 Hercules Road,
Wilmington, DE 19808–1599

In this work lipase was found to be an effective catalyst for the polycondensation of a diester with a diamine to form a polyamide. The reaction is relatively mild and can be achieved at a high yield at relatively low temperatures. The resulting polyamide is amenable to derivatization reactions, thereby generating structures that can impart additional properties to the polymer. This enzyme-catalyzed polymerization reaction is applicable to a wide range of water-soluble polyamides, including several that cannot be previously prepared via chemical methods.

Introduction

Polyamides are well-known industrial products having applications in many areas (*1*). For instance, the Nylon polymers (water-insoluble polyamides) are widely used in fibers. A water-soluble poly(aminoamide), derived from adipic acid and diethylene triamine, is the precursor to a well-known industrial resin (*2*). This poly(aminoamide) is currently produced by a chemical reaction at elevated temperatures which is accompanied by the formation of some branched structures. Subsequent derivatization of this polyamide produces a water-soluble resin, known for its ability to impart wet strength to paper and paper products (*2a,2b*) and shrink proofing to wools and other textiles (*2c*).

In nature, hydrolases are used to degrade or hydrolyze particular functionalities, e.g., lipases to hydrolyze esters and proteases to hydrolyze peptide (amide) linkages (3). It is well known that under suitable reaction conditions lipase can catalyze reverse reactions to form polyesters (4,5). Some of these methods have relied upon the use of organic solvents, whereas others are solvent-free processes. However, it is not previously known that proteases can catalyze reverse reactions to form amide polymers.

In this work we attempted to make polyamides using both lipases and proteases. To our knowledge no one previous to our work (6,7) has made a pure polyamide from a diacid/diester and a diamine via these enzymes.

Results and Discussion

Synthesis of Polyamides

In the chemical approach, a poly(aminoamide) is formed through the polycondensation of adipic acid and diethylene triamine (DETA).

The reaction was carried out at 180°C, and a polyamide with molecular weight of about 9000 was obtained (2). In our work, we sought to use enzymes to catalyze the polymerization. To facilitate the reaction, we started with dimethyl adipate and DETA. Typically, lower temperatures (50-100°C) were used. In our initial work, we thought a protease might be suitable for amide formation and thereby screened several proteases, e.g., chymotrypsin, trypsin, subtilisin, and papaya latex papain. Interestingly, only lower-molecular-weight polymers were obtained under the experimental conditions used. The breakthrough came when we used a lipase instead and obtained a high-molecular-weight polymer. This was unexpected because no polyamide had been previously reported that was catalyzed via a lipase.

Typically, the process entailed reacting the diester with the diamine in the absence of extraneous solvents and in the presence of a lipase. The molar ratio of the ester group to the primary amine group in the reactants was 1:1 or thereabouts. The efficiency of the reaction depended on the lipase used. We screened a large number of commercial lipases, and found the lipases from the following sources to be active: *Pseudomonas fluorescens, Candida antarctica, Pseudomonas sp., Candida rogasa, Mucor javanicus, Aspergillus niger,* and *Mucor miehei*. Highest activities were found for immobilized lipases from *Candida antarctica* (e.g., Novozym® 435) and *Mucor miehei* (e.g., Amano Lipase M) The reaction conditions have been optimized (7). Some examples of the reaction conditions are given in the Experimental Section.

The enzymatic polycondensation reaction has several advantages over the chemical process (Table 1). It is carried out at lower temperatures (50-110°C

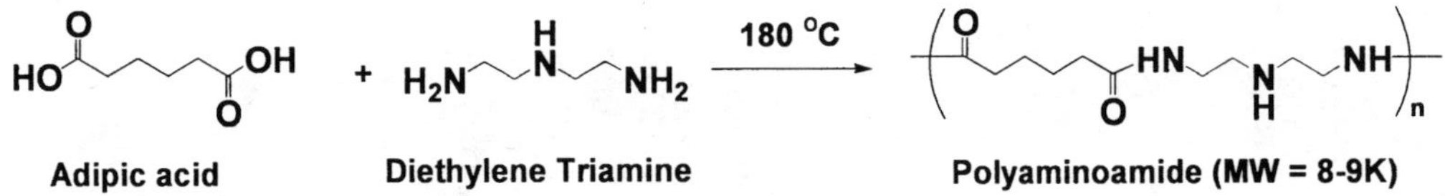

Figure 1. Polycondensation reaction of adipic acid and DETA to form the poly(aminoamide)

Figure 2. Enzyme-catalyzed polymerization of dimethyl adipate and DETA to form the poly(aminoamide)

versus 180°C). The enzymatic product has a narrower molecular weight distribution as compared to the chemical product. If we use an immobilized enzyme (e.g., Novozym 435), the enzyme can be optionally retrieved by dissolving the polymer product in water and fitration after the reaction and recycled in order to decrease cost.

Table 1. Comparison of chemical versus enzymatic polycondensation

	Chemical	Enzymatic
temperature	~ 180°C	50 – 110°C
molecular weights		
M_n	~ 3000	~3800
M_w	8000 – 9000	8000 – 9000
M_w/M_n	2.6 – 3.0	2.2 – 2.7
polymer structure	more branching	less branching

Apparently this enzyme-catalyzed polycondensation reaction is applicable to the synthesis of a wide range of polyamides. Some examples are given in Figure 3. Thus, changes can be made in both the diamine structure and the diester structure.   It may be noted that malonic-based polyamide and fumaric-based polyamide cannot be made at high molecular weights via the conventional high temperature chemical process.   In the former case, cyclization can occur at chain terminus during polymerization, and in the latter case the amine can add to the double bond of the fumarate due to Michael reaction. However, both can be made via the current enzymatic process. Other examples of polymers that can be made via the enzymatic process and not via the chemical process have been reported earlier (7).

Another possibility is to apply the enzymatic reaction to dimethyl adipate and triethylene glycol diamine (TEGDA) to produce a polyamide that is water-soluble, but less amenable to further derivatization.

Further variations can be made by using mixtures of diesters (e.g., adipate/malonate or malonate/oxalate), or mixtures of diamines (e.g., ethylene diamine/DETA, DETA/TETA, or DETA/piperazine). These combinations (with various stoichiometric ratios) permit different polymer structures (and properties) to be obtained, particularly after the derivatization reaction (below).

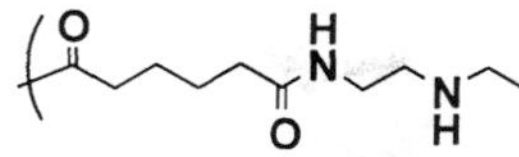

DETA-based

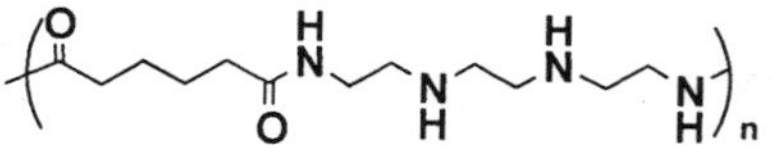

TETA-based

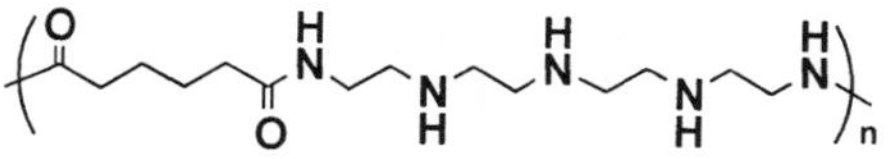

TEPA-based

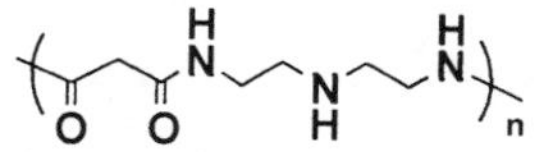

Malonic acid-based

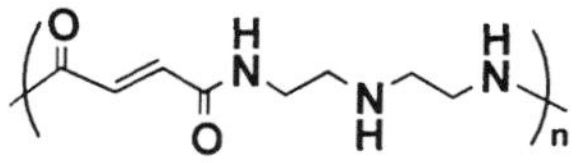

Fumaric acid-based

Figure 3. Other polyamides that can be synthesized via enzyme-catalyzed polycondensation. TETA = triethylene tetraamine, TEPA = tetraethylene pentaamine.

Synthesis of Polyamide Derivatives

The aforementioned poly(aminoamides) can be subjected to additional derivatization reactions. Some potential reactions include reaction with epichlorohydrin (to generate reactive functionalities), degradation of the polyamide chain (for repulping purposes), reaction with Quat 188 (to add pendant quaternary ammonium groups to the polymer chain), and reaction with 1,3-dichloropropane (to increase the molecular weight and to crosslink the polymer).

The specific case of epichlorohydrin reaction will be examined herein (Figure 4). The resulting products are water-soluble resins which can impart wet strength to paper and shrinkproof property to wool and other textiles (*2*). The detailed preparative procedure is given in the Experimental Section.

This enzyme-catalyzed resin has been evaluated for its wet strength relative to the resin made via the chemical means. Two resins (A and B) are prepared according to the procedures given in the Experimental Section. Table 2 shows the comparison between Resin A and its chemical counterpart.

The tensile strength data show comparable values for the two samples. Similar dry and wet tensile strength data (lbs/in) are observed at the 0.25%, 0.50% and 0.75% levels of the two resins. In particular, the wet tensile strength of both resins increases after curing. Thus, enzymatic synthesis produces fully acceptable polymers with commercially viable properties.

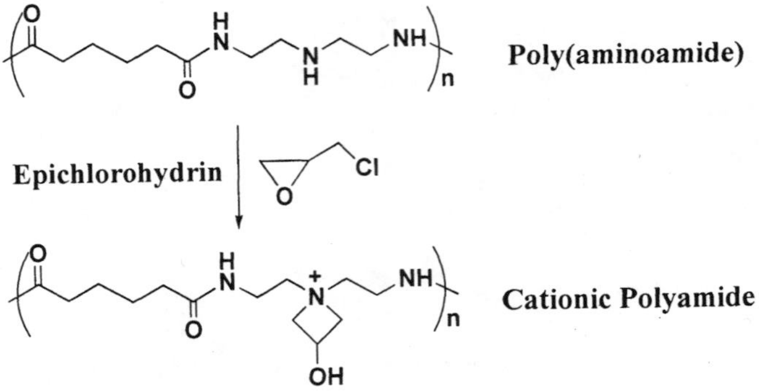

314

Figure 4. Formation of the poly(aminoamide)-epichlorohydrin product

Table 2. Wet and dry strengths of enzymatic and chemical polyamides

	% Dosage[a]	Basis Weight[b]	Dry Tensile [c]		Wet Tensile [c]	
			Not Cured [d]	Cured [e]	Not Cured [d]	Cured [e]
Resin A (enzyme-catalyzed)	0.25	40.0	16.2	15.1	2.70	3.02
	0.50	40.0	17.0	17.0	3.21	3.58
	0.75	40.0	16.8	17.0	3.39	3.78
Resin via chemical route	0.25	40.0	14.9	15.9	2.82	3.20
	0.50	40.0	16.9	16.8	3.49	3.79
	0.75	40.0	16.8	17.3	3.57	3.87

[a] Based on dry pulp. [b] lbs/3000 sq. ft. [c] lbs/in. [d] 47 days natural aging. [e] 80°C/30 min

Table 3 gives the comparison between Resin B (enzyme-catalyzed product) and the chemical counterpart. The tensile strength test also shows comparable values between the enzymatic and the chemical products. The wet tensile strength of both resins also increases after curing.

Table 3. Wet and dry strengths of enzymatic and chemical polyamide

	% Dosage[a]	Basis Weight[b]	Dry Tensile [c]		Wet Tensile [c]	
			Not Cured [d]	Cured [e]	Not Cured [d]	Cured [e]
Resin B (enzyme-catalyzed)	0.50	40.0	16.4	17.6	3.07	3.40
Resin via chemical route	0.50	40.0	14.5	16.7	3.15	3.48

[a] Based on dry pulp. [b] lbs/3000 sq. ft. [c] lbs/in. [d] 9 days natural aging [e] 80°C/30 min

In addition, evaluation of these resins as creping adhesives for paper towel applications has also been done. Both the resins from the chemical and enzymatic syntheses give comparable results, again confirming the viability of enzyme-catalyzed products.

The epichlorohydrin reaction has also been carried out on other poly(aminoamides) made via the enzymatic route. For example, the malonate-DETA resin can react readily with epichlorohydrin (Figure 5).

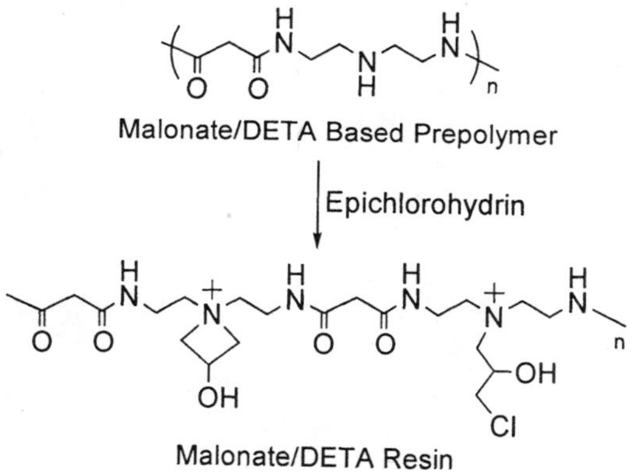

Figure 5. *Reaction of epichlorohydrin with the malonate/DETA resin*

The reactions for other poly(aminoamides) in Figure 3 are similar. Each of the polymers has somewhat different properties and may be of interest in different contexts.

Experimental

Materials

All chemical compounds were obtained from commercial sources (e.g., from Sigma-Aldrich). The enzymes came from Novozymes A/S, Amano Enzyme USA Co. Ltd., and Sigma-Aldrich.

Structure Characterization

The chemical structure of each polyamide was characterized by infrared and ^{1}H and ^{13}C NMR spectroscopy. The formation of a secondary amide was shown via IR by a strong absorption at 3300 cm^{-1} (N-H stretch), 1650 cm^{-1} (O=CNHR stretch band I) and 1560 cm^{-1} (O=CNHR stretch band II). ^{1}H NMR (in D_2O) showed only four multiplets at 1.35-1.55 ppm, 2.05-2.20 ppm, 2.55-2.70 ppm and 3.12-3.30 ppm, corresponding to central methylene in the adipic moiety, methylene adjacent to carbonyl, methylene adjacent to the central amine of diethylene triamine, and methylene adjacent to amide nitrogen, respectively. ^{13}C NMR showed five peaks at 25.4, 35.8, 39.1, 47.5 and 177.0 ppm, corresponding to central carbons in the adipic moiety, carbons adjacent to carbonyl, carbons adjacent to the central amine of diethylene triamine, carbons adjacent to amide nitrogen, and amide carbons, respectively.

Physical Testing

For wet and dry strengths, the resins are evaluated in handsheets prepared from a 50:50 softwood Kraft/hardwood blend, which is beaten to 450 ml Canadian standard freeness at pH 7.5. The basis weight is 40 lbs/3000 sq. ft.. The handsheets are then subjected to stress-strain tests on an Instron.

For evaluation of a resin as a creping adhesive, a 1% solids aqueous solution of the formulated resin is sprayed onto the surface of a Yankee Dryer along with the appropriate amount of release agent to provide a good balance of adhesion and release. This application optimizes both the productivity of the paper machine and the quality of the product produced on the machine. The solid consists of about 50% polyvinyl alcohol (PVOH) and 10-50% (polyamide-epihalohydrin resin). A cellulosic fibrous web is pressed against the drying surface to adhere the web to the surface. The dry web is then dislodged from the drying surface with a doctor blade and is wound on a reel. After the tissue material is converted to the final product (e.g., facial tissue, bath tissue, or kitchen towel), it is then subjected to a sensory panel to obtain a softness rating.

Synthesis of DETA-Adipate Polyamide

Dimethyl adipate (43.55 g, 0.25 mol), diethylene triamine (28.33 g, 0.275 mol) and 2.5 g of Novozym® 435 (immobilized *Candida antarctica* lipase) were mixed in a 250-ml flask and heated in an oil bath to 90°C. The viscous mixture was stirred at 90°C for 16 hrs in an open vessel with a steam of nitrogen. The product turned into a yellowish solid at the end of the reaction. 150 ml of methanol was added to dissolve the polyamide product. The immobilized enzyme was insoluble in the methanol solution and was removed by filtration. Methanol was removed by evaporating using a rotary evaporator under a reduced pressure to give the product as a yellowish solid. The yield was 48 g: M_w 8,400; M_w/M_n 2.73.

Synthesis of TETA-Adipate Polyamide

Dimethyl adipate (39.40 g, 0.30 mol), triethylene tetraamine (43.87 g, 0.30 mol) and 2.0 g of Novozym 435 (immobilized *Candida antarctica* lipase) were mixed in a 250 ml flask and heated in an oil bath to 80°C. The mixture was stirred at 80°C and the color of the reaction mixture turned to light brown in one hour. After being stirred at this temperature for 16 hrs in an open vessel, the reaction mixture solidified. 150 ml of methanol was added to dissolve the polyamide product. The immobilized enzyme was insoluble in the methanol solution and was removed by filtration. Methanol was removed by evaporation using a rotary evaporator under reduced pressures to give the product as a brown solid. The yield was 50 g; M_w, 8000; M_w/M_n, 2.10.

Synthesis of TEGDA-Adipate Polyamide

The dimethyl adipate (17.42 g, 0.10 mol), triethylene glycol diamine (15.60 g, 0.105 mol) and 1.0 g of Novozym 435 (immobilized *Candida antarctica* lipase) were mixed in a 250-ml open vessel and heated in a stream of nitrogen in an oil bath to 70°C for 24 hours with stirring. After cooling down, the mixture became very viscous. 100 ml of methanol was added. The immobilized enzyme (being insoluble in methanol) was removed by filtration. The methanol in the reaction mixture was removed by using a rotary evaporator under reduced pressures. The product was obtained as a semi-solid. The yield was 28 g: M_w, 4540; M_w/M_n, 2.71.

Synthesis of Poly(DETA-Adipamide)-Epihalohydrin

The polyamides in this example are prepared as described above. The polyamide-epihalohydrin resins are prepared in a one-step process in which a polyamide and an epihalohydrin are reacted in an aqueous medium until the desired molecular weight is achieved. The reaction is then diluted and/or stabilized at an acidic pH with an added acid. Two examples are given below.

Resin A. Poly(DETA adipamide) (42.6g, 0.20 mole) is added to a reaction vessel with 171 ml of water (pH 10.5). The solution is warmed to about 36 - 37°C , and epichlorohydrin (23.3g, 0.252 mole) is added. The reaction mixture is heated to 61 -64°C and the viscosity is monitored until a Gardner-Holdt viscosity of J is achieved. Water (250 ml) is then added and the pH is adjusted to 4.4 with concentrated sulfuric acid. The reaction is diluted with water for a final product solids concentration of 12.5% solids, and Brookfield viscosity of 49 cps.

Resin B. Poly(DETA adipamide) (29.0g, 0.169 mole) is added to a reaction vessel with 126 ml of water (pH 10.0). The solution is warmed to about 36°C, and epichlorohydrin (19.7g, 0.213 mole) is added. The reaction mixture is heated to 70°C and the viscosity is monitored until a Gardner-Holdt viscosity of J is achieved. Water (210 ml) is then added and the pH is adjusted to 4.5 with concentrated sulfuric acid. The final product is diluted with water for a final product solids concentration of 10.5% solids, and Brookfield viscosity of 55 cps.

Synthesis of Poly(DETA-Malonamide)-Epihalohydrin

The monomers, DETA (30.90 g, 0.30 mol) and methyl malonate (39.40 g, 0.30 mol) are added together in the presence of 0.7 g of Amano Lipase M (*Mucor miehei* lipase). The mixture is heated to 55-70°C for 2 hours. Water is then added with stirring to form a 50% solution and the pH is adjusted to 4.5 with concentrated sulfuric acid. Epichlorohydrin (34.69 g, 0.375 mol) is added to the solution slowly to form the desired adduct.

Conclusions

In this work we have shown that it is possible to use an ezyme to catalyze the polycondensation reaction to form polyamides. The large number of polyamides that have been made indicate that lipases from *Candida antarctica* and *Mucor miehei* are rather nonspecific and can be used generally for polyamide synthesis. In a particular example, a water-soluble polyamide has been produced from dimethyl adipate and ethylene triamine via this enzyme-catalyzed reaction at 50-110°C. The enzymatic polymerization is easy to do and permits the synthesis of several new high-molecular-weight polyamides that cannot be made through conventional chemical methods. Derivation of the polymer to form epichlorohydrin derivatives is readily accomplished. The enzymatic product is shown to be equivalent in physical properties to the chemical product when added to paper.

Acknowledgements

The authors would like to thank Robert G. Nickol and Ronald R. Staib for helpful discussions and Sadhana Mital and Gordon Tozer for technical assistance.

References

1. For example, a) Page, I. B. "Polyamides as Engineering Thermoplastic Materials", *Rapra Review Report*, Vol. 11, No. 1, Report 121, 2000. b) Steinbüchel, A.; Rhee, S. K. (Eds.), "Polysaccharides and Polyamides in the Food Industry. Properties, Production, and Patents." Wiley-VCH, Weinheim, 2005,

2. For example, a) Espy, H. H. *TAPPI J.* **1995**, *78*, 90. b) Riehle, R. J. *ACS Symp. Ser.* **2005**, *900*, 302. c) De la Maza, A.; Parra, J. L.; Leal, J.S.; Comelles, F. *Textile Res. J.* **1989**, *59*, 687.

3. For example, a) Roberts, S. M.; Turner, N. J.; Willetts, A. J.; Turner, M. K. "Introduction to Biocatalysis." Cambridge Univ. Press, Cambridge, 1995. b) Malcata, F.X. "Engineering of/with Lipases". Kluwer,Dordrecht, 1996.

4. Earlier articles include: a) Chaudhary, A.K.; Lopes, J.; Beckman; E. J. and Russell, A. J. *Biotechnol. Prog.*, **1997**, *13*, 318. b) Brazwell, E. M.; Filos D. Y.; and Morrow, C. J., *J. Polym. Sci., Part A: Polym. Chem.* **1995**, *33*, 89. c) Linko, Y. Y;. Wang, Z. L. and Seppala, J., *Enzyme Microb. Technol.* **1995**, *17*, 506. d) Binns, F.; Roberts, S. M.; Taylor, A.; Williams, C. F., *J. Chem. Soc., Perkin Trans.* **1993**, 1. e) Geresh, S. and Gilboa,,Y. *Biotechnol. Bioeng.*, **1991**, *37*, 883. f) Taylor, A.; Binnis, F., WO 94/12652.

5. For example, a) Cheng, H. N.; Gross, R. A. "Polymer Biocatalysis and Biomaterials". ACS Symp. Ser. 900, Am. Chem. Soc., Washington, 2005. b) Gross, R. A.; Cheng, H. N.; "Biocatalysis in Polymer Science". ACS Symp. Ser. 840, Am. Chem. Soc., Washington, 2002.

6. Cheng, H. N.; Gu, Q.-M.; Maslanka, W. W. *US Patent 6,677,427.* January 13, 2004.

7. Gu, Q.-M.; Maslanka, W. W.; Cheng, H. N. *ACS Polym. Prepr.* **2006**, *47*(2), 234.

Polysaccharides, Glycopolymers, and Sugars

Chapter 22

Synthesis of Unnatural Hybrid Polysaccharides via Enzymatic Polymerization

Akira Makino[1] and Shiro Kobayashi[1,2,*]

[1]Department of Material Chemistry, Graduate School of Engineering,
Kyoto University, Nishikyo-ku, Kyoto 615–8510, Japan
[2]Current address: R & D Center for Bio-based Materials, Kyoto Institute
of Technology, Sakyo-ku, Kyoto 606–8585, Japan

Various natural polysaccharides and their derivatives have been synthesized via in vitro enzymatic polymerization. The results indicate that the glycosidic hydrolases can catalyze the reaction of unnatural substrates, in spite of the high substrate specificity of the enzyme in vivo. This paper describes the enzymatic polymerization to produce unnatural hybrid polysaccharides having a disaccharide repeating unit consisted from components of different two natural homo-polysaccharides. Newly synthesized monomers designed on the basis of the transition-state analogue substrate (TSAS) concept were successfully polymerized, giving rise to unnatural hybrid polysaccharides in a regioselective and stereocontrolled manner. Such unnatural polysaccharides having a well-defined structure are hardly synthesized via a conventional organic method. Thus, they are expected as new functional materials to exhibit various biological activities originated from both natures.

Introduction

Polysaccharides play important roles as energy storage and structural supports in the living system. Recent developments of carbohydrate chemistry revealed that they are also associated with cell recognition, immuno-adjuvant of stimulation in the immune system including anti-cancer or allergic responses (*1,2*). For the detailed investigation of their functions at molecular level, synthetic polysaccharides with well-defined structure are essential. Therefore, facile and efficient methods to synthesize such polysaccharides are urgently required. However, studies on chemistry and biochemistry of polysaccharides have been slowly progressed because of their synthesis difficulty due to complicated structures with high molecular weight.

During the decades, glycobiology has made rapid progress; many kinds of carbohydrate-associated proteins such as glycosyltransferases, glycoside hydrolases, and other glycoproteins have been cloned and characterized (*3*), leading to the success of biochemical and chemo-enzymatic synthesis of various carbohydrate compounds (*4–6*). Among them, we focused on glycoside hydrolases, which are preferentially utilized to the in vitro reaction because of its easy availability and handling character (*7,8*). We developed a new glycoside hydrolase-catalyzed synthetic method of polysaccharides via non-biosynthetic pathway. That is called as "enzymatic polymerization" (*9–14*). Carbohydrate chains produced in the living system are formed by the catalysis of glycosyltransferases (*15,16*). Glycoside hydrolases are enzymes which catalyze in vivo hydrolysis reaction of glycosidic linkages. Therefore, the key point of the polymerization is how to cause a reverse reaction by the enzyme; the in vivo bond-cleavage catalysis function of the hydrolase is to induce the in vitro bond-formation catalysis for the polymerization. To accomplish the reverse reaction, a monomer is to be designed as having a structure close to a transition-state during the hydrolysis based on the transition-state analogue substrate (TSAS) concept (*9–14*). Also, appropriate conditions of the polymerization are essential so as not to hydrolyze the monomer and resulting polysaccharide (addition of organic solvents, regulation of pH value, reaction temperature and so forth). Utilizing enzymatic polymerization, cellulose was synthesized from β-cellobiosyl fluoride catalyzed by cellulase (*17*). This is the first example of the in vitro polysaccharide synthesis. Various natural polysaccharides like xylan (*18*), chitin (*19*), hyaluronic acid (*20–22*) and chondroitin (*23*) were also synthesized for the first time via xylanase, chitinase and hyaluronidase-catalyzed polymerizations, respectively.

Glycoside hydrolases have two types, one is an *endo*-type and the other is an *exo*-type enzyme (*24*). For the enzymatic polymerization, *endo*-type enzymes are normally utilized, whose shape at the catalytic domain looks like cleft (*25*). At the catalytic domain of the enzyme, both sites (donor and acceptor sites) recognize substrates precisely, and a new glycosidic linkage is formed in a regioselective and stereocontrolled manner. Enzymes show high substrate specificity as well as the high reaction specificity, which is the important mechanistic function in order not to catalyze undesirable reaction in vivo as pointed out by the key and lock theory (9,12). But, it encounters the limitation of substrates adaptable to the enzymatic reaction and becomes an obstacle for the development of this method to the general polysaccharide synthesis. However, compared to the other kinds of carbohydrate-related enzymes, glycoside hydrolases show dynamic nature in the substrate specificity and often recognize unnatural substrates. Thus, the synthesis of polysaccharide derivatives such as 6-O-methylated cellulose and 6-fluorinated chitin via cellulase- and chitinase-catalyzed polymerizations were achieved (26,27). Such polysaccharide derivatives with a well-defined structure are hardly possible to synthesize via conventional methods and this is one of the most remarkable advantages for the enzymatic polymerization.

The above background motivated us to synthesize hybrid polysaccharides via enzymatic polymerization. Hybrid polysaccharides are defined as unnatural polysaccharides consisting from components of two different natural homo-polysaccharides aligned alternatingly. Therefore, it is a hybrid of original two natural homo-polysaccharide at a molecular level, and is expected to have functions originated from the both natures. Furthermore, these unnatural polysaccharides with well-defined structure will become potent tools for the investigation of the relationships between biological activities of polysaccharides and their chemical structures. In this paper, we describe the syntheses of unnatural hybrid polysaccharides via enzymatic polymerization from newly designed TSAS monomers (Figure 1).

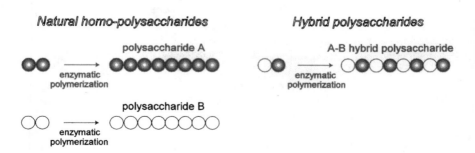

Figure 1. Synthesis of hybrid polysaccharides via enzymatic polymerization.

Reaction Mechanism of Glycoside Hydrolases

Endo-type glycoside hydrolases, which have been utilized for the enzymatic polymerization, are classified into two groups from the type of structure in the transition state of during the hydrolysis reaction; a glycoside-enzyme intermediate type and an oxazoline intermediate type (*28,29*).

Cellulase, xylanase, and amylase are classified into the first group. For example, cellulase (EC 3.2.1.4) from *Trichoderma viride* belonging to the glycoside hydrolase family 5 normally catalyzes cleavage of $(1\rightarrow4)$-β-glucoside linkage in cellulose (*30,31*). Hydrolysis mechanism of cellulose is considered to proceed through a double displacement mechanism as illustrated in Figure 2.

In the polymerization, how to form a glycoside-enzyme transition-state (or intermediate) smoothly at the donor site of the enzyme. The structure of the glycoside must be close to the transition-state of the hydrolysis, i.e., stage a to stage b, based on the TSAS concept. Therefore, a fluoride type monomer having a leaving group at the anomeric carbon is designed as a TSAS monomer. After the recognition of the monomer at the donor site of cellulase, the reducing end C1 is nucleophilically attacked by the carboxylate (catalytic nucleophile) from α-side (stage a′), resulting in formation of a glycosyl-enzyme intermediate (or transition state) (stage b′) as well as in leaving the fluorine atom as hydrogen fluoride. Then, the C4 hydroxy group of Glc in the growing chain end or in another monomer molecule placed at the acceptor site attacks the anomeric carbon of the intermediate nucleophilically from β-side (stage b′), leading to the formation of a $\beta(1\rightarrow4)$-glycosidic linkage (stage c′). Repetition of this regio- and stereoselective glycosylation process is a polycondensation, giving rise to synthetic cellulose. To design a fluoride type monomer as a TSAS monomer is a key-point for the glycoside-enzyme intermediate type polymerization.

Enzymes classified into oxazoline intermediate type are typically chitinase and hyaluronidase. Chitinase from *Bacillus* sp. (EC 3.2.1.14) belonging to the glycoside hydrolase family 18 primarily catalyzes hydrolysis of $(1\rightarrow4)$-β-*N*-acetylglucosaminide linkage in chitin (*30,32,33*). The hydrolysis mechanism is postulated to proceed through the substrate-assisted mechanism as follows (Figure 3).

In the polymerization, a monomer having already an oxazoline moiety is used. Immediately after the recognition of monomer at the donor site of the enzyme, nitrogen atom on the oxazoline ring is protonated and forms the corresponding oxazolinium ion (stage b′). The anomeric carbon of the oxazolinium is attacked from β-side by the C4 hydroxy group of *N*-acetyl-D-glucosamine (GlcNAc) in another monomer or in the growing chain end placed at the acceptor site, resulting in formation of a $\beta(1\rightarrow4)$-glycosidic linkage (stage c′). Repetition of this glycosylation is a ring-opening polyaddition, providing a synthetic chitin. Designing oxazoline type compounds as TSAS monomers are

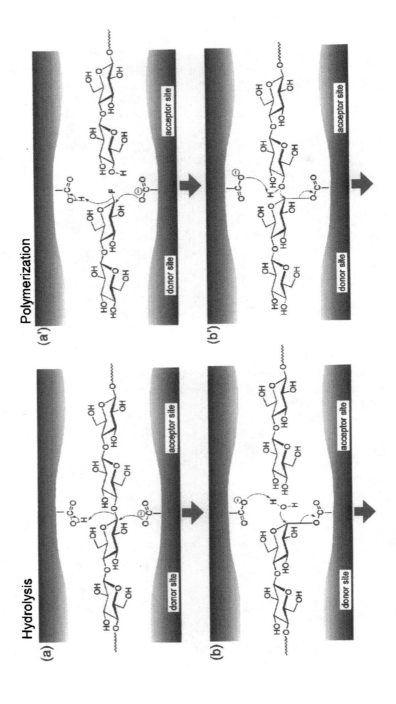

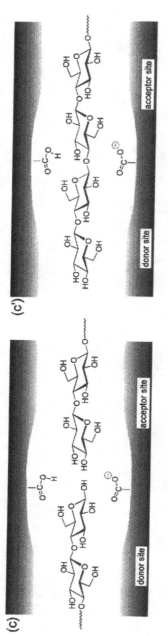

Figure 2. Postulated reaction mechanism of cellulase catalysis.

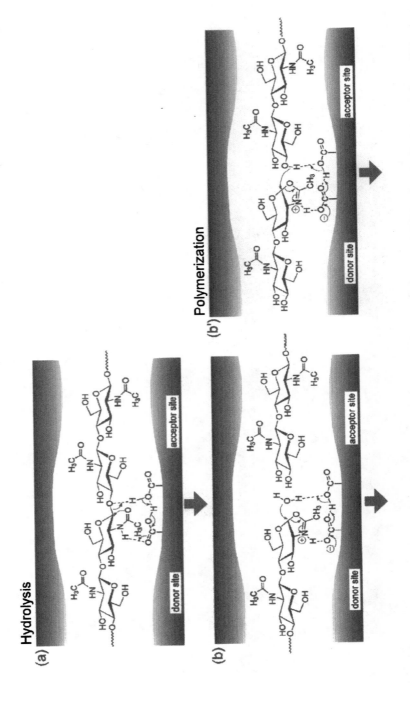

329

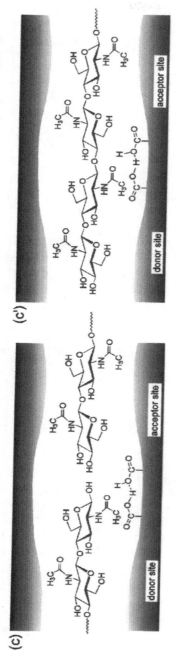

Figure 3. Postulated reaction mechanism of chitinase catalysis.

necessary for chitinase- and hyaluronidase-catalyzed polymerization. The important point is that an oxazoline species is involved in both hydrolysis (stage b) and polymerization (stage b') as a common intermediate.

Enzymatic Polymerization to Several Hybrid Polysaccharides

Cellulose–Chitin Hybrid Polysaccharide (*34*)

Cellulose is a linear polysaccharide having a $\beta(1\rightarrow4)$-linked D-glucose (Glc) as a repeating unit, which is the most abundant organic substance on earth (*35*). Chitin is a $\beta(1\rightarrow4)$-linked *N*-acetyl-D-glucosamine (GlcNAc) polysaccharide and widely distributed in animal world as a supporting material (*36*). The structural difference between these two polysaccharides is the C2 substituent of a pyranoside unit; a hydroxy group on the cellulose and an acetamido group on the chitin. This acetamido group provides chitin with various biological activities. However, the acetamido group participates in formation of inter- and intramolecular hydrogen bonding with the C6 hydroxy groups (*37*). As the result, chitin is hardly soluble in almost all solvents and its utilization is partly restricted.

Recently, the polymer blend of cellulose and chitin was prepared by spinning an aqueous alkaline solution of their sodium xanthates (*38,39*). The blend showed good mechanical properties as materials for wound-healing and improving quality of soil and water. Therefore, a hybrid polysaccharide of cellulose and chitin, having a disaccharide repeating unit composed of $\beta(1\rightarrow4)$-linked Glc and GlcNAc in a molecule, will be expected to bring about a new functional unnatural polysaccharide.

Monomer Design

The target cellulose–chitin hybrid polysaccharide here has two kinds of glycosidic linkages; $(1\rightarrow4)$-β-D-glucoside linkage and $(1\rightarrow4)$-β-*N*-acetyl-D-glucosaminide linkage (Figure 4). For the $(1\rightarrow4)$-β-D-glucosidic bond formation, cellulase is chosen and GlcNAc$\beta(1\rightarrow4)$Glc-β-fluoride (**1**) is designed as a TSAS monomer for the enzyme. On the other hand, $(1\rightarrow4)$-β-*N*-acetyl-D-glucosaminide linkage will be produced by chitinase. In this case, Glc$\beta(1\rightarrow4)$GlcNAc oxazoline (**2**) is designed as a TSAS monomer. Monomers **1** and **2** were synthesized via conventional organic chemistry.

Enzymatic Polymerization of 1 Catalyzed by Cellulase from Trichoderma Viride

Monomer **1** was subjected to cellulase-catalyzed polymerization. Without the enzyme, because of the instability of anomeric fluorine atom, monomer **1**

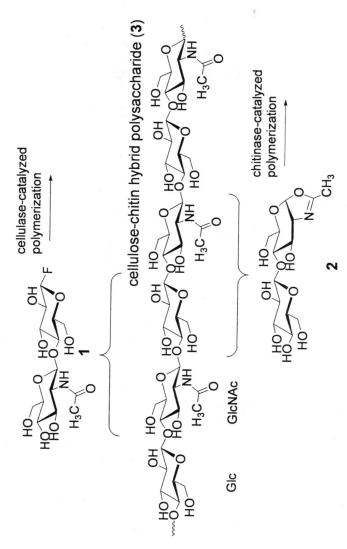

Figure 4. Possible monomer design for the synthesis of a cellulose–chitin hybrid polysaccharide 3.

331

was gradually hydrolyzed in aqueous media to afford hydrolysate **4** (Figure 5a, route 1). With the enzyme catalysis, enzymatic polymerization of **1** provides corresponding polymer **3** (Figure 5a, route 2). At the same time, nonenzymatic hydrolysis of **1** (Figure 5a, route 1) and enzymatic hydrolysis of **1** or **3** are conceivable to occur (Figure 5a, route 3).

Figure 6a illustrates the concentration change of **1** during the polymerization. With the enzyme catalysis, consumption rate of **1** was significantly accelerated, and it disappeared within 9 h with formation of a white precipitate. In contrast, monomer **1** was gradually decomposed without enzyme and any precipitates were not formed.

Enzymatic Polymerization of 2 Catalyzed by Chitinase from Bacillus sp.

Monomer **2** was also subjected to *Bacillus* sp. derived chitinase-catalyzed polymerization (Figure 5b). Figure 6b illustrates the concentration change of **2** during the polymerization. Under the weakly alkaline conditions, an oxazoline ring is stable and **2** was hardly hydrolyzed by aqueous media without enzyme. In contrast, the consumption rate of **2** was greatly accelerated by the addition of chitinase and it disappeared within 3 h. As the reaction progressed, a white precipitate was deposited.

Characterization of Cellulose–Chitin Hybrid Polysaccharide 3

After completion of the enzymatic reaction, a white precipitate was separated by centrifugation, and the supernatant was analyzed by MALDI-TOF/MS, ^{1}H and ^{13}C NMR measurements. From the MALDI-TOF/MS spectrum, the peaks appeared at every *m/z* 365, corresponded to the molecular mass of the repeating disaccharide unit of **3**. ^{1}H and ^{13}C NMR spectra supported that the resulting product is a cellulose–chitin hybrid polysaccharide. Furthermore, it has been proven by CP/MAS ^{13}C NMR spectrum that the white precipitate is also hybrid polysaccharide **3**.

Characterization data of the resulting cellulose–chitin hybrid polysaccharide **3** are summarized in Table 1. The average molecular weight value (M_n) of the resulting product **3** via cellulase- and chitinase catalyzed polymerization reached to 2840 and 4030, which correspond to 15–16 and 22 saccharide units, respectively.

Chitin–Chitosan Hybrid Polysaccharide (*40*)

Chitin, a $\beta(1{\rightarrow}4)$-linked *N*-acetyl-D-glucosamine (GlcNAc) polysaccharide, is well-known for the excellent characters such as biodegradability and biocompatibility (*36*). Chitosan is a $\beta(1{\rightarrow}4)$-linked D-glucosamine (GlcN) polysaccharide, which is *N*-deacetylated polysaccharide of chitin (*41*). Chitosan

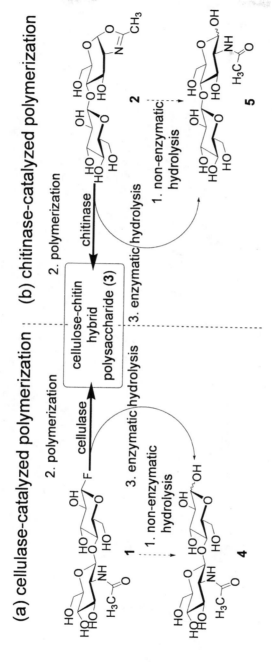

Figure 5. Two kinds of reactions are possible to occur during (a) cellulase-catalyzed reaction and (b) chitinase-catalyzed reaction.

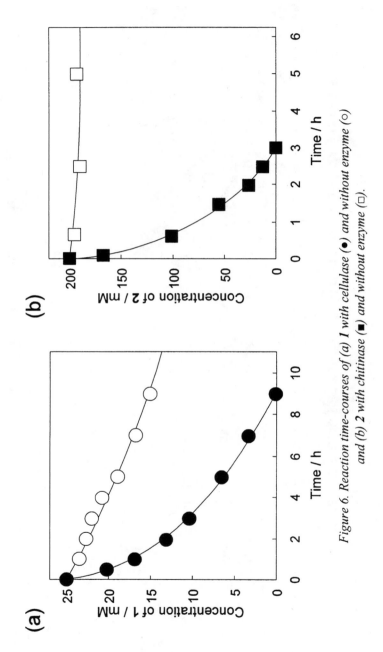

Figure 6. Reaction time-courses of (a) 1 with cellulase (●) and without enzyme (○) and (b) 2 with chitinase (■) and without enzyme (□).

Table 1. Enzymatic Polymerization of 1 or 2 to 3

polymerization[a]				product (3)			
enzyme	pH	initial conc. /mM	time[b] /h	Insoluble polymer			soluble polymer
				Yield[c] /%	$M_n{}^d$	$M_w{}^d$	yield[e]/%
cellulase	5.0	25	9	55	2820	3070	28
	5.0	50	8	63	2840	3220	25
chitinase	10.5	200	3	34	3750	4170	45
	11.0	200	10	46	4030	4660	33

[a] In an CH_3CN-acetate buffer (50 mM) mixture (5:1, v/v), (cellulase-catalyzed polymerization) or in a carbonate buffer (10 mM) (chitinase-catalyzed polymerization). Amount of enzyme: 5.0 wt% for 1 or 2. Reaction temperature: 30 °C. [b] Indicating the time for disappearance of corresponding monomer. [c] Isolated yields. [d] Determined by SEC using pullulan, cellooligosaccharides and chitooligosaccharides standards. [e] Determined by HPLC (more than tetrasaccharides, RI detector, calibrated by cello- and chito-oligosaccharides).

is produced from chitin in vivo by the catalysis of chitin deacetylase (*42*). Due to difference of chemical structures between C2 acetamido and amino groups, chitosan shows various attractive bioactivities. However, chitosan contains GlcNAc unit in small amounts due to incomplete action of the deacetylase, causing structural diversities of these polysaccharides (*43*). Furthermore, both polysaccharides are insoluble in almost all solvents. The poor solubility of these polysaccharides makes their chemical modification difficult. It is known that 45–55% deacetylated chitin is soluble to water (*44*). Therefore, the hybrid polysaccharide of chitin and chitosan, chitin–chitosan hybrid polysaccharide (**8**), is expected to soluble in water.

Monomer Design

The target glycosidic linkages for the synthesis of a chitin–chitosan hybrid polysaccharide are (1→4)-β-D-glucosaminide and (1→4)-β-N-acetyl-D-glucosaminide linkages (Figure 7). For the (1→4)-β-D-glucosaminide linkage formation, chitosanase with GlcNAcβ(1→4)GlcN-β-fluoride monomer (**6**) will be appropriate for the catalysis. On the other hand, the (1→4)-β-N-acetyl-D-glucosaminide linkage will be formed by chitinase catalysis. In this case, GlcNβ(1→4)GlcNAc oxazoline (**7**) is designed as a TSAS monomer. However, the hydrolysis mechanism of commercially available chitosanase does not involve a double displacement mechanism, which is suitable for the enzymatic polymerization. Therefore, chitinase-catalyzed polymerization is designable for the synthesis of chitin-chitosan hybrid polysaccharide (**8**).

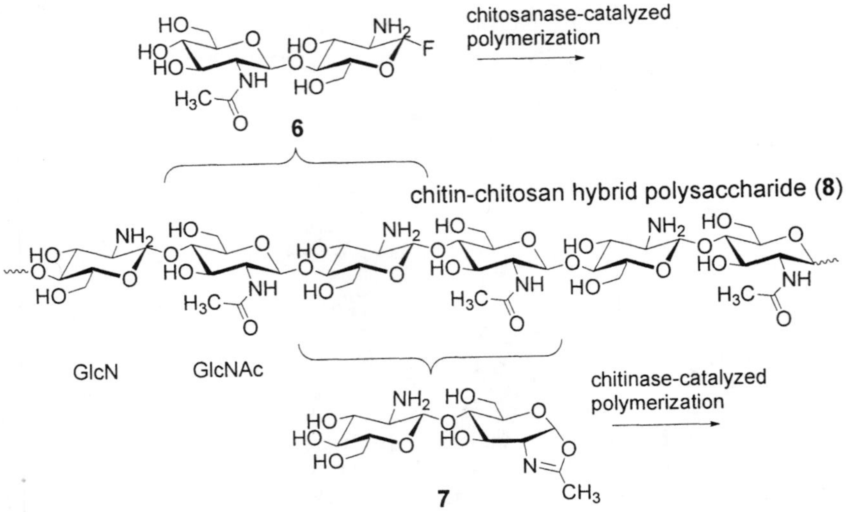

Figure 7. *Possible monomer design for the synthesis of chitin–chitosan hybrid polysaccharide 8.*

Enzymatic Polymerization of 7 Catalyzed by Chitinase from Bacillus sp.

Monomer **7** synthesized via conventional organic technique, was subjected to the *Bacillus* sp. derived chitinase-catalyzed polymerization. Without enzyme, the monomer was gradually decomposed by non-enzymatic hydrolysis from aqueous media. In contrast, the ring opening reaction of oxazoline monomer was drastically activated by chitinase, giving rise to polymer **8** within 0.7 h. During the polymerization, the reaction mixture was kept homogeneous.

After the complete consumption of **7**, the reaction mixture was analyzed by MALDI-TOF/MS, ^{1}H and ^{13}C NMR measurements. The MALDI-TOF/MS spectrum showed peaks at every *m/z* 364, which is corresponding to the molecular mass of the repeating disaccharide unit of **8**. ^{1}H and ^{13}C NMR spectra also supported that the resulting product is a chitin–chitosan hybrid polysaccharide **8**.

Enzymatic Polymerization of 7 Catalyzed by Chitinase from Other Origins

Besides chitinase from *Bacillus* sp., there are some kinds of commercially available family 18 chitinases from other origins. Among the chitinases, it has been reported that chitinase from *Serratia marcescens* hydrolyzes chitosan having *N*-acetyl group in 35–65% (*45*) and its hydrolysis rate was accelerated with decreasing the amount of *N*-acetyl group. Then, *Serratia marcescens* derived chitinase-catalyzed polymerization of **7** was examined. The ring-opening reaction of oxazoline monomer was accelerated by the addition of the enzyme, giving rise to a chitin–chitosan hybrid polysaccharide **8**.

Characterization of Chitin–Chitosan Hybrid Polysaccharide 8

The yields and M_n value of the resulting chitin–chitosan hybrid polysaccharides **8** under various reaction conditions were summarized in Table 2. The yields of **8** were around 60–75% and there were little significant differences in catalysis between these two enzymes except for the reaction time. However, M_n value of **8** reached 1600 and 2020 with using chitinases from *Bacillus* sp. and *Serratia marcescens*, respectively, which correspond to 8–10 (n = 4–5) and 10–12 (n = 5–6) saccharide units, respectively. These results indicated that M_n values of polysaccharides are controllable to some extent by selecting enzymes from different origins.

Table 2. Enzymatic Polymerization of 7 to 8

enzyme origin	pH	polymerization[a]			polymer (8)		
		enzyme amount / wt% for 7	temp. / °C	time[b] / h	yield[c] / %	$M_n{}^d$	$M_w{}^d$
Bacillus	8.5	10	30	0.7	64	1450	1550
	8.5	5	30	1.0	70	1450	1580
	8.5	5	10	3.0	58	1600	1730
Serratia	8.5	10	30	16	62	1490	1870
	9.0	10	30	18	64	1580	2610
	9.0	5	30	24	75	2020	4280

[a] In a phosphate buffer (50 mM), initial concentration of **7**: 100 mM. [b] Indicating the time for disappearance of corresponding monomer. [c] Isolated yields. [d] Determined by SEC using pullulan, chitooligosaccharide and chitosanoligosaccharide standards.

Cellulose–Xylan Hybrid Polysaccharide (*46*) and Chitin–Xylan Hybrid Polysaccharide (*47*)

Xylan is a D-xylose (Xyl) polysaccharide connecting through $\beta(1{\rightarrow}4)$ glycosidic linkage. It is one of the most important components of hemicellulose in plant cell walls (*48*). The structural difference between Xyl and Glc is C6 methylol group. The substitution position in disaccharide TSAS monomers is

338

only C-2 in the non-reducing end (*34,40*). Therefore, the synthesis of cellulose–xylan (**11**) and chitin–xylan hybrid polysaccharides (**14**) is an important challenge for the widely application of enzymatic polymerization and will provide important information of enzymes.

Monomer Design

Cellulose–xylan hybrid polysaccharide has alternatingly aligned Glc and Xyl structure connecting through $\beta(1\rightarrow4)$ glycosidic linkage. On the other hand, chitin–xylan hybrid Polysaccharide has alternatingly aligned GlcNAc and Xyl structure connecting through $\beta(1\rightarrow4)$ glycosidic linkage. On the basis of a TSAS monomer concept, fluoride-type monomers, Glc$\beta(1\rightarrow4)$Xyl-β-fluoride (**9**), Xyl$\beta(1\rightarrow4)$Glc-β-fluoride (**10**), and GlcNAc$\beta(1\rightarrow4)$Xyl-β-fluoride (**12**) were designed to form $(1\rightarrow4)$-β-D-glucoside or $(1\rightarrow4)$-β-D-xyloside linkages. Furthermore, an oxazoline type monomer, a Xyl$\beta(1\rightarrow4)$GlcNAc-oxazoline (**13**) is conceivable for the $(1\rightarrow4)$-β-N-acetyl-D-glucosaminide linkage formation.

Here, compounds **10** and **13** were chemically synthesized as monomers for the synthesis of cellulose–xylan hybrid polysaccharide **11**, and chitin–xylan hybrid polysaccharide **14**, respectively (Figure 8).

Enzymatic Polymerization of 10 Catalyzed by Xylanase from Trichoderma Viride

In order to polymerize a fluoride-type monomer **10**, cellulase and xylanase were chosen for the catalyst. In an acetonitrile–acetate buffer mixed solution, the consumption of **10** was accelerated by the addition of xylanase from *Trichoderma viride*, giving rise to a white precipitate as the progress of the reaction in a 58% yield. From the analysis of the resulting products, it is a cellulose–xylan hybrid polysaccharide **11** and polymers with the degree of polymerization up to 12 (24 saccharide units) were detected from the MALDI-TOF/MS.

Enzymatic Polymerization of 13 Catalyzed by Chitinase from Bacillus sp.

Chitinase from *Bacillus* sp. was added to a solution of Xyl$\beta(1\rightarrow4)$GlcNAc oxazoline **13** dissolved into a phosphate buffer. The consumption rate of **13** was accelerated by the enzyme addition and chitin–xylan hybrid polysaccharide **14** was produced. During the polymerization, the reaction proceeded homogenously throughout the reaction. The yield was 76% and its average molecular weight value of **14** was $M_n = 1500$ with $M_w/M_n = 1.76$, determined by GPC. Furthermore, MALDI-TOF mass spectrum of **14** indicated the peaks at every m/z 335, which corresponds to the molecular mass of the repeating disaccharide unit.

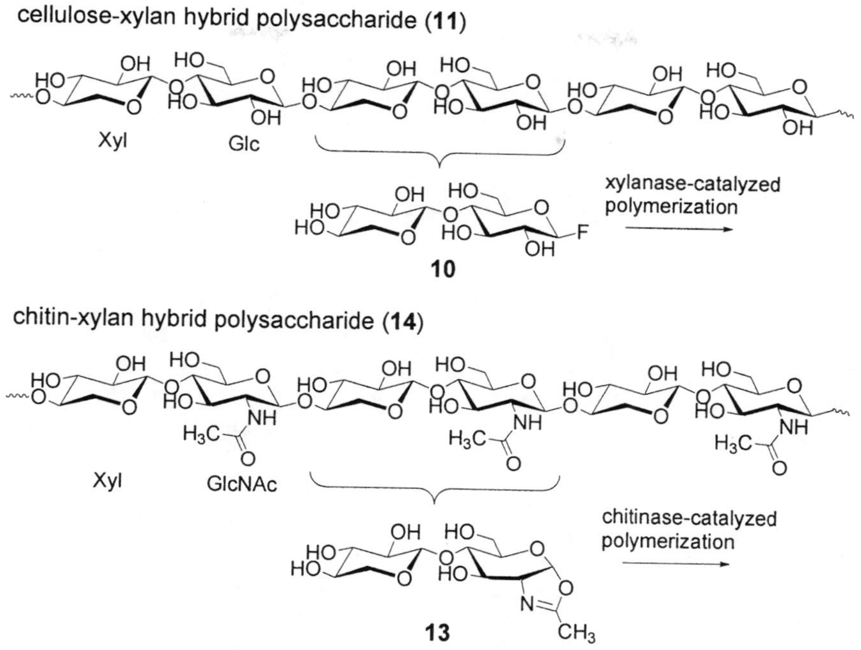

cellulose-xylan hybrid polysaccharide (11)

chitin-xylan hybrid polysaccharide (14)

Figure 8. Chemical structures of cellulose–xylan hybrid polysaccharide 11 and chitin–xylan hybrid polysaccharide 14 as well as designed monomers 10 and 13 for their synthesis.

Conclusion

The present paper described the synthesis of unnatural hybrid polysaccharides via cellulase, xylanase, and chitinase-catalyzed polymerizations. For the enzymatic polymerization, compounds **1, 2, 7, 10**, and **13** were newly designed and synthesized as TSAS monomers on the basis of the hydrolysis reaction mechanisms of corresponding enzymes. Under the appropriate reaction conditions, these monomers were successfully polymerized in the regioselecitve and stereocontrolled manner, giving rise to polysaccharides having alternatingly aligned two monosaccharide components from two different natural homo-polysaccharides. Therefore, these newly synthesized unnatural polysaccharides are hybrid polysaccharides of the natural two classes at a molecular level. To synthesize such polysaccharides are normally difficult by means of conventional synthetic methods. This type of enzymatic polymerization technique will broaden the scope of polysaccharide synthesis and give new functional polysaccharides.

340

References

1. Hurtley, S.; Service, R.; Szuromi, P. *Science* **2001**, *291*, 2337.
2. Dwek, R. A. *Chem. Rev.* **1996**, *96*, 683–720.
3. Narimatsu, H. *Glycoconjugate J.* **2004**, *21*, 17–24.
4. Seeberger, P. H.; Haase, W. C. *Chem. Rev.* **2000**, *100*, 4349–4394.
5. Hanson, S.; Best, M.; Bryan, M. C.; Wong, C. H. *Trends Biochem. Sci.* **2004**, *29*, 656–663.
6. Zhang, Z. Y.; Ollmann, I. R.; Ye, X. S.; Wischnat, R.; Baasov, T.; Wong, C. H. *J. Am. Chem. Soc.* **1999**, *121*, 734–753.
7. Ajisaka, K.; Yamamoto, Y. *Trends Glycosci. Glycotechnol.* **2002**, *14*, 1–11.
8. Trincone, A.; Giordano, A. *Curr. Org. Chem.* **2006**, *10*, 1163–1193.
9. Kobayashi, S. *J. Polym. Sci., Polym. Chem. Ed.* **1999**, *37*, 3041–3056.
10. Kobayashi, S.; Uyama, H.; Ohmae, M. *Bull. Chem. Soc. Jpn.* **2001**, *74*, 613–635.
11. Kobayashi, S.; Uyama, H.; Kimura, S. *Chem. Rev.* **2001**, *101*, 3793–3818.
12. Kobayashi, S.; Sakamoto, J.; Kimura, S. *Prog. Polym. Sci.* **2001**, *26*, 1525–1560.
13. Kobayashi, S.; Ohmae, M. *Adv. Polym. Sci.* **2006**, *194*, 159–210.
14. Kobayashi, S.; Shoda, S.; Uyama, H. *Adv. Polym. Sci.* **1995**, *121*, 1–30.
15. Scheible, W. R.; Pauly, M. *Curr. Opin. Plant Biol.* **2004**, *7*, 285-295.
16. Davies, G. J.; Charnock, S. J.; Henrissat, B. *Trends Glycosci. Glycotechnol.* **2001**, *13*, 105–120.
17. Kobayashi, S.; Kashiwa, K.; Kawasaki, T.; Shoda, S. *J. Am. Chem. Soc.* **1991**, *113*, 3079–3084.
18. Kobayashi, S.; Wen, X.; Shoda, S. *Macromolecules* **1996**, *29*, 2698–2700.
19. Kobayashi, S.; Kiyosada, T.; Shoda, S. *J. Am. Chem. Soc.* **1996**, *118*, 13113–13114.
20. Kobayashi, S.; Itoh, R.; Morii, H.; Fujikawa, S. I.; Kimura, S.; Ohmae, M. *J. Polym. Sci., Polym. Chem. Ed.* **2003**, *41*, 3541–3548.
21. Ochiai, H.; Ohmae, M.; Mori, T.; Kobayashi, S. *Biomacromolecules* **2005**, *6*, 1068–1084.
22. Kobayashi, S.; Morii, H.; Itoh, R.; Kimura, S.; Ohmae, M. *J. Am. Chem. Soc.* **2001**, *123*, 11825–11826.
23. Kobayashi, S.; Fujikawa, S.; Ohmae, M. *J. Am. Chem. Soc.* **2003**, *125*, 14357–14369.
24. Crout, D. H. G.; Vic, G. *Curr. Opin. Chem. Biol.* **1998**, *2*, 98–111.
25. Davies, G.; Henrissat, B. *Structure* **1995**, *3*, 853-859.
26. Okamoto, E.; Kiyosada, T.; Shoda, S. I.; Kobayashi, S. *Cellulose* **1997**, *4*, 161–172.
27. Makino, A.; Ohmae, M.; Kobayashi, S. *Macromol. Biosci.* **2006**, *6*, 862–872.

28. Vasella, A.; Davies, G. J.; Bohm, M. *Curr. Opin. Chem. Biol.* **2002**, *6*, 619–629.

29. Rye, C. S.; Withers, S. G. *Curr. Opin. Chem. Biol.* **2000**, *4*, 573–580.

30. Henrissat, B. *Biochem. J.* **1991**, *280*, 309–316.

31. Schulein, M. *Biochim. Biophys. Acta-Protein Struct. Molec. Enzym.* **2000**, *1543*, 239-252.

32. Tews, I.; vanScheltinga, A. C. T.; Perrakis, A.; Wilson, K. S.; Dijkstra, B. W. *J. Am. Chem. Soc.* **1997**, *119*, 7954–7959.

33. Sakamoto, J.; Watanabe, T.; Ariga, Y.; Kobayashi, S. *Chem. Lett.* **2001**, 1180–1181.

34. Kobayashi, S.; Makino, A.; Matsumoto, H.; Kunii, S.; Ohmae, M.; Kiyosada, T.; Makiguchi, K.; Matsumoto, A.; Horie, M.; Shoda, S. I. *Biomacromolecules* **2006**, *7*, 1644–1656.

35. Klemm, D.; Heublein, B.; Fink, H. P.; Bohn, A. *Angew. Chem.-Int. Ed.* **2005**, *44*, 3358–3393.

36. Muzzarelli, R. A. A. *Chitin*; Pergamon Press: Oxford, U.K., 1977.

37. Kameda, T.; Miyazawa, M.; Ono, H.; Yoshida, M. *Macromol. Biosci.* **2004**, *5*, 103–106.

38. Hirano, S.; Usutani, A.; Yoshikawa, M.; Midorikawa, T. *Carbohydr. Polym.* **1998**, *37*, 311–313.

39. Hirano, S.; Midorikawa, T. *Biomaterials* **1998**, *19*, 293–297.

40. Makino, A.; Kurosaki, K.; Ohmae, M.; Kobayashi, S. *Biomacromolecules* **2006**, *7*, 950–957.

41. Kumar, M.; Muzzarelli, R. A. A.; Muzzarelli, C.; Sashiwa, H.; Domb, A. J. *Chem. Rev.* **2004**, *104*, 6017–6084.

42. Amorim, R. V. S.; Ledingham, W. M.; Fukushima, K.; Campos-Takaki, G. M. *J. Ind. Microbiol. Biotechnol.* **2005**, *32*, 19–23.

43. Sato, H.; Mizutani, S.; Tsuge, S.; Ohtani, H.; Aoi, K.; Takasu, A.; Okada, M.; Kobayashi, S.; Kiyosada, T.; Shoda, S. *Anal. Chem.* **1998**, *70*, 7–12.

44. Sannan, T.; Kurita, K.; Iwakura, Y. *Polym. J.* **1977**, *9*, 649–651.

45. Sorbotten, A.; Horn, S. J.; Eijsink, V. G. H.; Varum, K. M. *Febs J.* **2005**, *272*, 538–549.

46. Fujita, M.; Shoda, S.; Kobayashi, S. *J. Am. Chem. Soc.* **1998**, *120*, 6411–6412.

47. Kobayashi, S.; Makino, A.; Tachibana, N.; Ohmae, M. *Macromol. Rapid Commun.* **2006**, *27*, 781–786.

48. Ebringerova, A.; Heinze, T. *Macromol. Rapid Commun.* **2000**, *21*, 542–556.

Chapter 23

Glycopolymers: The Future Antiadhesion Drugs

Yalong Zhang, Jinhua Wang, Chengfeng Xia, and Peng George Wang[*]

The Department of Chemistry and Biochemistry, The Ohio State University, Columbus, OH 43210

Glycopolymers refer to glycosylated   linear polymers or hydrogels. They can interact with biomacromolecules by the polyvalency effect.   Here we present a review of applying glycopolymers as anti-adhesion reagents towords human antibodies, bacteria as well as viruses in our lab. These include applying   Galα1-3Galβ1-4Glcβ (α-Gal) polymer to inhibit human antidoby IgG, IgM and IgA for protecting cells in xenotransplantaion; an α-Gal co Mannose (Man) glycol-polymer to target bacteria by mannose part and trigger antibody mediated immune defence against the bacteria by α-Gal part; and a sugar blood B antigen glycosylated hydrogel to trap norovirus inside the hydrogel pocket. Our works indicate that the glycopolymer can inhibit biomacromolecules from binding host cells. Thus, the glycopolymer can be applied as an anti-adhesion drug in the future.

Introduction

Carbohydrates in glcolipides and glycoproteins are exposed on cell surfaces like molecular antennae. There they form ligands, which can be recognize by a number of diffirent molecules such as lectins and selectins, enzymes, hormones, or antibodies which use oligosaccharides as antigenic markers to distinguish 'own' from 'foreign'. Carbohydrate-protein interactions not only have a function in physiological processes, but are also of importance in pathological processes, such as the adhesion of microbes, including viruses, bacteria, protozoa, or fungi to the host cells.

In physiological processes, carbohydrate-protein recognition is key to a variety of biological processes and the first step in numerous phenomena based on cell–cell interactions, such as fertilization, embryogenesis, cell migration, organ formation, immune defense(1). For example, Trisaccharides Galα1-3Galβ1-4Glcβ-R and Galα1-3Galβ1-4GlcNAcβ-R' and pentasaccharide Galα1-3Galβ1-4GlcNAcβ1-3Galβ1-4Glcβ-R'' have been identified as the major α-Gal epitopes (2-5), which are abundantly expressed on the cells of most mammals with the exception of humans, apes, and other old world primates. It was found that the initial step for severe immunological rejection in xenotransplantation is the recognition of human natural anti-Gal antibody IgG to α-Gal epitopes. This adhesion will trigger the antibody-dependent cell-mediated cytotoxicity by human blood monocytes and macrophages. The IgM isotype of anti-Gal is believed to be responsible for the complement activation that leads to complement-mediated lysis of the xenograft cells.

In pathological processes, the initial step for many bacterial pathogens infection is to attach or adhere to the oligosaccharides epitols on the host cells and tissues (1). Adhesion is required so that the organisms are not swept away by the natural cleansing mechanisms of the host, such as airflow in the respiratory tract or urine flow in the urinary tract. It also provides the pathogens with better access to sources of nutrition, facilitates the delivery of toxic agents into the host tissues and eventually the penetration of the bacteria into the tissues. The proteins involved, generally named lectins, are most frequently found on cell surfaces of the bacteria or virus. They have the ability to bind specifically and non-covalently to carbohydrates. In this review, we will discuss two examples: E. Coli, selectively binding with mannose (6-7) and norovirus, binding with human sugar blood B epitols (8).

It is easy to consider applying oligosaccharides to inhibit those antiobody, bacteria or virus from adhering host cells. However, most saccharides only weakly bind to their protein receptors, seldom showing association constants beyond 10^{-6} M^{-1} (9-10), which is not sufficient to effectively control the in vivo events mediated by protein–carbohydrate binding.

Indeed, in biological systems the interactions between cell surface carbohydrates and sugar-binding proteins are of polyvalent nature. Although low affinity and poor specificity are intrinsic in protein binding, carbohydrates

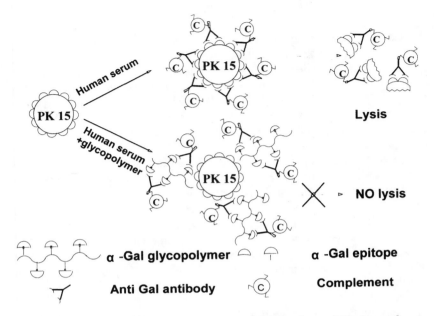

Figure 1. Protecting pig cells from human antibodies by α–Gal glycopolymers

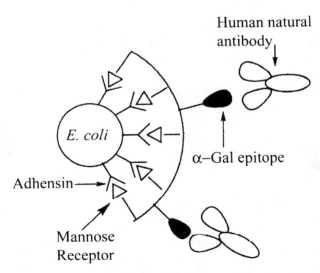

Figure 2. Trigerring human natural immune response for antibacterial by an α-Gal-co-Man glycopolymer

function as very important signaling molecules in a wide variety of biological recognition events. Their high avidity can be attributed, at least in part, to their multivalent presentation pattern on the cell surface. Carbohydrates are typically expressed on the cell surface in clusters; thus their overall binding capacity with protein receptors (commonly with multiple binding sites) is enhanced over the affinity of individual monovalent ligands through cooperative multiple interactions *(11-17)*. Therefore, it is of logical choice to use multivalent ligands as potent synthetic inhibitors to effectively block the recognition process. The multivalent inhibitors of carbohydrate-binding proteins, bacteria, and viruses have been well studied primarily by Lee *(18-19)*, Whitesides *(20, 21)*, Roy *(22, 23)*, and other *(24, 25)* in recent years. They have demonstrated that the multivalent forms of carbohydrate ligands, either polymers or dendrimers, often have amplified inhibitory effects over their monovalent counterparts, although the levels of enhancement vary.

We will discuss three examples about glycopolymer in this review: a) inhibition of human antibody by linear α-Gal glycopolymer *(26)*; (Figure 1) b) an α-Gal co Man glycopolymers was enzymatically synthesized for anti bacterial agaist *E. col (27)* (Figure 2); c) a three dimentional glycopolymer, glycosylated hydrogel, was applied to trap noroviruses *(28)*. (Figure 3)

Inhibition of human anti-Gal antibody by α-Gal glycopolymer

The syntheses of α-Gal-containing polymers were achieved by the reaction of preactivated poly [*N*-(acryloyloxy) succinimide] (PNAS **2**) with an α-Gal trisaccharide derivative (**3**), followed by capping the active esters with aqueous ammonia (Figure 4). All α-Gal polymers (**4A-4F**) used in this study were prepared from one single batch of PNAS, which was obtained by polymerization of *N*-(acryloyloxy) succinimide (**1**). The molecular weight of this "parent" polymer was determined by gel filtration chromatography after its complete hydrolysis to poly (acrylic acid) sodium salt. The average molecular weight of the hydrolyzed polymer was (M_w = 252 kD) with a relatively narrow molecular weight distribution (M_w/M_N = 1.5). The degree of polymerization is 1.8×10^3. By varying the ratio of α-Gal trisaccharide to active esters in PNAS, a series of polymers were obtained with different densities of α-Gal. The ratios of α-Gal unit to acrylamide unit calculated from the integration of trisaccharide signals and acrylamide signals in ^{1}H NMR spectra were 1:1.8 (**4A**), 1:2.5 (**4B**), 1:3.3 (**4C**), 1:6.4 (**4D**), 1:15 (**4E**), and 1:29 (**4F**), respectively. The degrees of functionalization of polymers withα-Gal trisaccharide were 36% (**4A**), 28% (**4B**), 23% (**4C**), 13% (**4D**), 6% (**4E**), and 3% (**4F**), respectively.

The chemical synthesis of α-Gal trisaccharides (Figure 5) derivative {*N*-[*O*-α-D-galactopyranosyl)-(1, 3)-*O*-(*β*-D-galactopyranosyl)-(1, 4)-1-*β*-D-glucopy-ranosyl]-5-aminopentamide} **3** started with a glycosylation reaction between

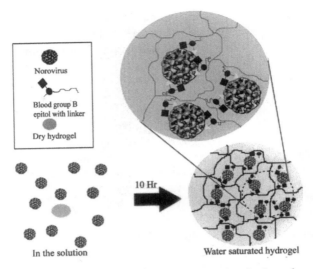

Figure 3. Trapping norovirus in the glycohydrogels

donor **5** and acceptor **6** promoted by NIS/TfOH at -30 °C. Trisaccharide derivative **7** was afforded with 20:1 (α/β) selectivity. A mild hydrogenation condition with PtO_2 reduced the azido group of **7** to primary amine, which was immediately reacted with 5-chlorovaleryl chloride to give compound **8**. Sequential debenzylation and acetylation of **8** produced peracetylated trisaccharide **9**. Nucleophilic substitution of **9** with sodium azide and fully deprotected acetyl groups gave compound **10**. α-Gal trisaccharides **3** was obtain by hydrogenation of **10**. Overall it needs 8 steps with about 50 % yield.

A shortcut to compound **3** using the chemo-enzymatic synthesis is illustrated in Figure 6. Compound **3** was achieved using this bifunctional fusion enzyme. This enzyme contained both uridine-5'-diphospho-galactose 4-epimerase (GalE) and $\alpha(1,3)$ galactosyltransferase (GalT) and was cloned in our laboratory by in-frame fusion of the *Escherichia coli* gene *galE* to the 3'-terminus of a truncated bovine *galT* gene within a high-expression plasmid. It has dual functions both as an epimerase and as a galactosyltransferase that allows the use of relatively inexpensive UDP-glucose instead of the high-cost UDP-galactose as the donor. Reaction of disaccharide derivative **11** with UDP-glucose gave compound **3** in 52% yield.

The evaluation of the binding affinities of synthesized α-Gal-containing polymers **7** to anti-Gal antibodies was accomplished by inhibition ELISA (enzyme-linked immunosorbent assay) with purified human (male, blood type AB) anti-Gal antibody as the primary antibody and mouse laminin as a natural source of α-Gal. The concentrations of α-Gal polymers at 50% inhibition (IC_{50}) of anti-Gal antibody binding to α-Gal epitopes on mouse laminin were measured and summarized in Table 1.

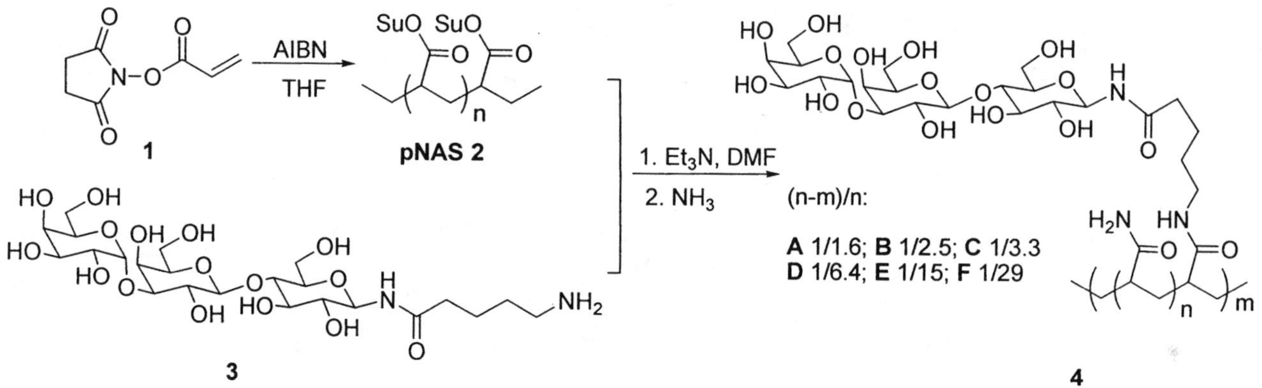

Figure 4. Synthesis of a-Gal glycopolymer from PNAS

348

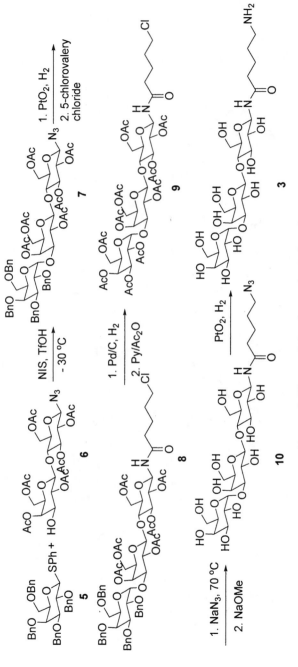

Figure 5. Chemical Synthesis of α-Gal analogs.

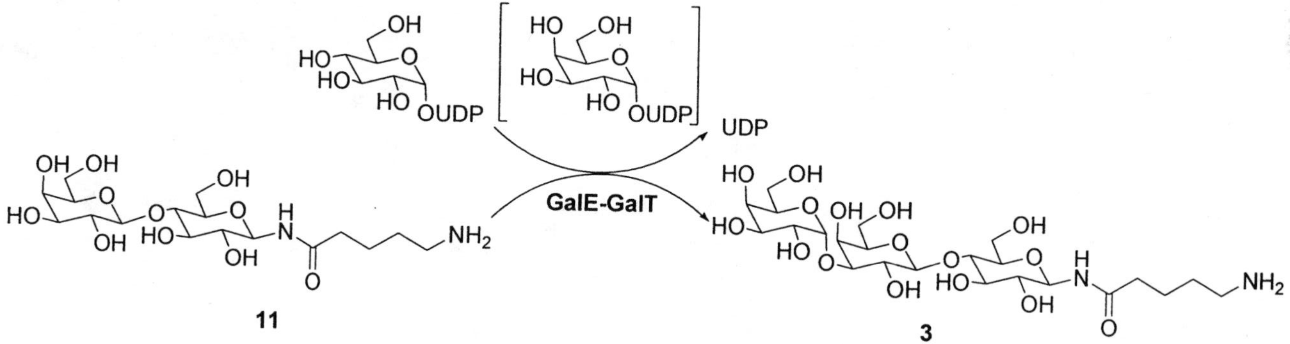

Figure 6. Chemo-enzymatic synthesis of a-Gal analogs

Table 1. Inhibition of α-Gal glycopolymer agaist human anti-Gal antibody (Male, Blood type: AB) to mouse laminin

Compound	Ratio[a]	IC$_{50}$ [µM][b]		
		IgG	IgA	IgM
α-Gal		69±5	55±16	277±24
polymer 4A	1/1.8	0.37±0.008	0.016±0.006	0.0068±0.0016
polymer 4B	1/2.5	0.28±0.005	0.0053±0.0023	0.0063±0.0015
polymer 4C	1/3.3	2.8±0.9	0.0070±0.0032	0.0056±0.0014
polymer 4D	1/6.4	7.2±2.7	0.043±0.017	0.027±0.018
polymer 4E	1/15	60±5	0.86±0.28	165±13
polymer 4F	1/29	93±16		
polylactose	1/2.5	N/I	N/I	N/I

[a] Ratio of α-Gal to acrylamide; [b] the concentration of 50% inhibition of binding between anti-Gal antibody (16 g/mL) and mouse laminin measured by ELISA; N/I: no inhibition.

All of the IC$_{50}$ data presented in the text are the net α-Gal trisaccharide concentrations (micromolar) calculated from the degree of functionalization and polymerization of each polymer compared with the corresponding α-Gal monomer explicitly. Results indicated that polymers **4A-4D** inhibit the antibody/antigen binding better than the α-Gal monomer, Galα1-3Galβ1-4Glcβ-NHAc, in all cases. The efficacy can be illustrated by the IC$_{50}$ result for polymer **4C** of 5.6 nM, which is 5.0 x 10^4-fold better than its monomeric analogue in inhibiting anti-Gal IgM. Polymer **4B** has an IC$_{50}$ of 5.3 nM with 10^4-fold enhancement over the α-Gal monomer in inhibiting anti-Gal IgA. The enhancement has been proven to be greater for anti-Gal IgM and IgA than for anti-Gal IgG. For example, the activity enhancement of polymer **4B** is 246-, 1.0 x 10^4-, and 4.4 x 10^4-fold to anti-Gal IgG, IgA, and IgM (activity of monomer α-Gal as 1). This observation is consistent with the increasing numbers of binding sites from IgG to IgA to IgM. In human serum, IgM exists as a pentamer with 10 equivalent binding sites in one molecule; 80% of IgA exists as a monomer with two binding sites and 20% as higher oligomers with a higher number of binding sites; IgG exists as a monomer with two binding sites. Therefore, the "multivalent" effect is certainly more pronounced for antibodies (IgM, IgA) with more protein carbohydrate interaction sites.

To assess the practical application of α-Gal polymers, intact human serum (male, blood type: AB) instead of purified human anti-Gal antibody using the inhibition ELISA (Table 2). The results indicated the activity enhancement of the α-Gal polymers as compared to the monomer. The same trends were observed in the interaction of α-Gal polymers with different isotypes of the antibody and with the varied densities of the α-Gal epitope conjugated to the polymer. Interestingly, the IC$_{50}$ observed with the purified antibodies were consistently lower than the IC_{50} with human sera. One explanation is that the purified antibodies were obtained from affinity column immobilized with α-Gal trisaccharide similar in structure to the epitope on the polymer. Therefore, subsets of antibodies selected during the purification would bind most tightly to the polymer. Since the purified antibodies from our affinity column contained 89% of IgM, 50% of IgA, and 42% of IgG of corresponding anti-Gal in human sera, the increased proportion of IgM in the purified anti-Gal would contribute to the greater inhibitory effect of the α-Gal polymers.

Table 2. Inhibition of α-Gal glycopolymer agaist binding of Human serum (Male, Blood type: AB) to pig kidney (PK 15) cells

Compound	Ratio[a]	IC_{50} [μM][b]		
		IgG	IgA	IgM
α-Gal		~ 1000	> 1000	> 1000
polymer 4A	1/1.8	595		5.2
polymer 4B	1/2.5	407	63	2.2
polymer 4C	1/3.3	774		> 1000
polylactose	1/2.5	N/I	N/I	N/I

[a] Ratio of α-Gal to acrylamide; [b] the concentration of 50% inhibition of binding between anti-Gal antibody (16 g/mL) and mouse laminin measured by ELISA; N/I: no inhibition.

In summary, the synthetic α-Gal-conjugated polymers significantly enhanced activities in the inhibition of human anti-Gal antibody binding to mouse laminin glycoproteins and pig PK15 cells. Such enhancement is greater for anti-Gal IgA and IgM than for IgG. The amplified binding differences among the three anti-Gal isotypes can be utilized to selectively inhibit and remove particular isotype antibodies. Moreover, it was demonstrated through flow cytometry analysis that certain α-Gal polymers are effective in inhibiting anti-Gal antibodies in human serum binding to pig kidney cells.

Apply α-Gal-co-mannose glycopolymer as targeted anti-bacteria agent

The α-Gal epitope glycopolymers were synthesized via two different routes. The first approach enzymatically synthesize Gal (α 1,3) Gal epitope 13 by α Galactosidase from green cofee beans, (Figure 7), then the nitro group of compound 13 was hydrogenated, followed by the treatment with acryloyl chloride in methanol to give an acryl glycomonomer. This monomer will be copolymerized with mannose monomer and acrylamide to give Gal (α 1,3) Gal co Man glycopolymer.

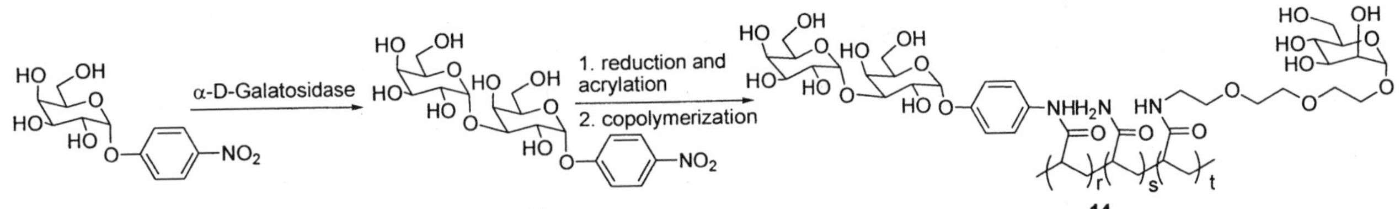

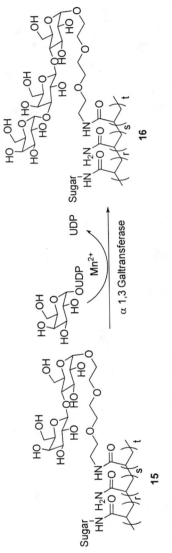

Figure 7. Chemoenzymatical synthesis of Gal(α 1,3)Gal co Man glycopolymer

Figure 8. Chemoenzymatical synthesis of α-Gal co Man glycopolymer

Another enzymatic method was used to synthesize α-Gal-co-Man glycopolymer from a lactose-co-Mannose glycopolymer **15** (Figure 8). Then the polymer was treated with uridine 5' diphosphate galactose (UDP-Gal) and α 1,3 galactotransferase to give α-Gal-co-Man glycopolymer **16** in quantitive yield. The enzymatic reactions exhibited excellent selectivity in modifying glycopolymers bearing mannose moieties. To the best of our knowledge, this is the first reported case of selective enzymatic transformations on a polymeric substrate. In order to enhance the transfer efficiency, the biocompatible PEG linker was employed to provide the lactosyl acceptor adequate space from the polymer chain.

Firon et al. *(30)* found that the mannose binding sites of *E. coli.* is a hydrophobic region. Therefore, mannose with hydrophobic linker, such as *p*-nitrophenyl-α-D-mannopyranoside, will exibit stronger binding. We designed two different linker for mannose: a PEG linker (hydrophilic, such as **16A**) and p-amidophenyl linker (hydrophobic, such as **16B**).

The α-Gal-co-Man glycopolymers were tested to bind *E. coli* K-12 HB101 bacterial cells. The *E. coli* strain contains mannose-binding sites located on the surface of the bacteria, which can bind yeast (*Saccharomyces cerevisiae*), resulting in visible agglutination. Thus, the inhibition ability of the synthetic mannose glycopolymer or analogs to the agglutination caused by yeast can give a criteria to evaluate binding strength between synthetical molecules with *E. coli.* (Table 3) The results indicate that *p*-Nitrophenyl-α-D-Gal almost show no inhibition to the agglutination. The *p*-nitrophenyl α-D-Man, glycopolymers **15B** and **16B** have lowest inhibition concentration, around 100 μM. While Methyl-α-D-Man and glycopolymer **14**, **15A** and **16A** have an inhibition concentration around 1000 μM. Here, two conclusion can be drawn from the result: 1) mannose-binding sites on *E. coli* are highly sugar specific; 2) hydrophobic binding pocket on the surface of the bacteria play a more important role than polyvalent effect of glycopolymer to the binding between mannose and bateria.

ELISA inhibition assay was applied to evaluate the activity of α-Gal-co-Man glycopolymer binding to the anti a-Gal antibodies. Mouse laminin having a-Gal epitopes were fixed on ELISA plate as solid-phase antigens. Glycopolymers were then incubated with human anti-Gal antibodies on the ELISA plate. The plate was washed and incubated with horseradish peroxidase (HRP) conjugated anti-human IgG antibody. After additional washing, the color was developed by HRP reaction using a chromogenic compound. The results are shown in Table 4. It was found that the glycopolymer shown highly selectivity of binding with human IgG antibody. Besides this, glycopolymer **16 A** and **B** demonstrate polyvalent effect of polymer contributing significantly to the binding.

We have used chemoenzymatic synthesis of α-Gal-co-Man glycopolymers. They were subsequently proved to bind both *E. coli.*, K-12 and human natural anti-Gal antibodies. Cross reactivities of mannose binding to anti-Gal antibodies and α-Gal residue binding to bacteria were not observed. The glycopolymer

354

Table 3. Inhibiting agglutination of *E. coli.* with yeast by mannose
containing glycopolymer

Compound	Ratio	Inhibition conc. (μM)
Methyl-α-D-Man		1500
p-Nitrophenyl-α-D-Man		90
p-Nitrophenyl-α-D-Gal		>20000
Glycopolymer **14**[b]	1/28/5	1500
Glycopolymer **15A**[a]	3/1	1000
Glycopolymer **15B**[a]	12/1	100
Glycopolymer **16A**[b]	1/12/2	1000
Glycopolymer 16B[b]	1/5.8/1	100

[a] Ratio of Man to acrylamide; [b] Ratio of α-Gal or Gal (α 1,3) Gal to acrylamide to Man.

Table 4. Inhibiting human natural anti-Gal antibody IgG by α-Gal
co mannose glycopolymer

Compound	Ratio	OD_{655}(% inhibition)
α-Gal (0.1 mM)		0.799 (0%)
α-Gal (1.0 mM)		0.427 (47%)
Glycopolymer **15A**[a] (0.1 mM)	3/1	0.784 (2%)
Glycopolymer **15B**[a] (1.0 mM)	12/1	0.729 (9%)
Glycopolymer **16A**[b] (0.1 mM)	1/12/2	0.241 (70%)
Glycopolymer 16B[b] **(1.0 mM)**	1/5.8/1	0.183 (77%)

[a] Ratio of Man to acrylamide; [b] Ratio of α-Gal or Gal (α 1,3) Gal to acrylamide to Man.

with dual binding capacity will be viable ligands to direct human natural immunity against bacterial pathogens.

Trapping norovirus by glycohydrogel: a potential oral antiviral drug

Acute gastroenteritis is a common disease in humans. Each year there are 700 million acute diarrhea cases; including 2.5~3.2 million deaths of children under the age of 5 *(31, 32)*. Recent studies indicated that the majority of acute viral gastroenteritis episodes were caused by noroviruses *(33)*. Noroviruses cause large scale outbreaks of acute diarrhea through water and food contamination, which results in a public panic. The outbreaks can also bring unnecessary stress situations, such as foreign military operations. The highly contagious nature and high resistance to disinfectants make norovirus associated diseases difficult to control. Thus, they have been categorized as B agents *(34, 35)* in the NIH/CDC biodefense program.

Norovirus, also called Norwalk-like virus, has been found to recognize human Histo-Blood Group Antigens (HBGA). HBGAs are complex carbohydrates linked to glycoproteins or glycolipids located on the surface of red blood cells and mucosal epithelial cells, or are present as free antigens in biological fluids such as blood, saliva, intestinal contents and milk. Huang and coworkers demonstrated that different noroviruses can recognize different HBGAs *(8)*. Lee*(18,19)*, Whitesides*(20,21)* demonstrated that the polyvalent form of carbohydrate ligands, either polymer or dendrimer based, could significantly increase carbohydrate binding to bacterial or viral lectins. Eklink and coworkers prepared Lewis B derivatives copolymerized with acrylamide to inhibit *Helicobacter pylor (36)*. However, there are two limitations for applying linear polymers or dendrimers as oral antiviral drugs: 1) a large amount of the polymers or dendrimers are required to efficiently bind noroviruses located in the intestinal track due to the high solubility of the polymer in aqueous solution; and 2) because of the polydispersity, the low molecular weight portion of the linear glycosylated polymer, especially the size of which are less than 200 nm , may enter the blood vessels and cause severe side effects.

In order to overcome these limitations, glycosylated hydrogels were applied to trap noroviruses. Hydrogels are the crosslinked polymers. They are insoluble in water but can swell from several to a few thousand times its original weight by absorbing water. Hydrogels are also stable in acidic or alkaline environments due to their covalent texture. All these characteristics make hydrogels resemble living tissues in their physical properties, biocompatibility and non-toxicity. As a result, hydrogels are widely used as drug delivery systems *(37-40)*. Herein we report a new method to "trap" norovirus in HBGAs containing hydrogel (Figure

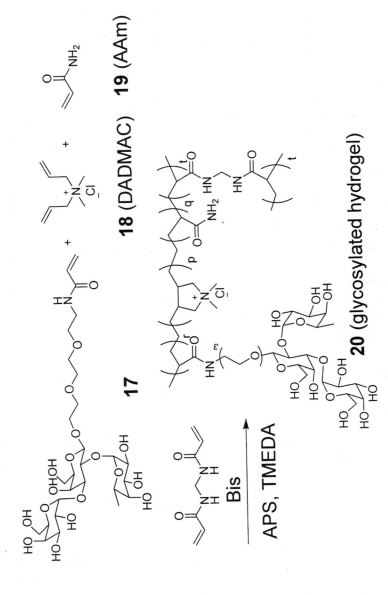

Figure 9. Synthesis of blood group B epitope glycohydrogel

3). Potentially, glycosylated hydrogels can be used as an oral prophylactic antiviral drug by entrapping the norovirus inside their mesh pocket through a "caged polyvalent effect". The noroviruses entrapped in the hydrogels can be excreted from the body through the normal GI track.

The acrylic group conjugated human HBGA type B **17** was chemically synthesized and used to prepare the glycosylated polydiallyldimethyl ammonium chloride (DADMAC) **18** -acrylamide (AAm) **19** hydrogel at different crosslinker densities (Figure 9). The hydrogel was synthesized through a free radical polymerization in deionized aqueous solution with ammonium peroxidisulfate (APS)-N,N,N',N'-tetramethyl ethylenediamine (TMEDA) as the initiators along with the presence of a crosslinker: N,N'-methylenebis acrylamide (Bis), at 37 °C for 24 h. Swelling ratio, the parameter in direct proportion to the mesh size of the hydrogel *(41)*, were determined by calculating the ratio between the mass of the deionized water saturated hydrogel and that of the dry hydrogel. All the hydrogels were immersed in deionized water for 48 h. The water was changed every 12 h, to wash out the residue of the initiators and unreacted monomers as well as the linear polymers before performing the trapping experiments.

The optimal synthetic conditions for the highest mesh size of the non-glycosylated hydrogels were obtained by varying the crosslinker density and the ratio of DADMAC-AAm with a constant total monomer concentration (3 M). The results indicated that the hydrogel synthesized with 7:3 DADMAC-AAm and 0.2 mol % Bis has the largest mesh size (highest swelling ratio of 300). The glycosylated hydrogel was obtained in a milligram scale by adopting this optimal condition (swelling ratio of 320).

To test the hydrogel's ability to trap noroviruses, recombinant virus-like particles (VLPs) from strain VA387, the protein shell of norovirus and a type B blood sugar binder was used to interact with the hydrogel in solution. The concentration of VLPs VA387 was determined by enzyme linked immunosorbent assays (ELISA) before and after trapping by the glycosylated hydrogels using the procedure described by Jones and coworkers *(42)*, with the non-glycosylated hydrogels being the control.

Two experiments were performed to determine the effects of entrapping VLPs in the glycosylated hydrogels. First, we planed to determine how the mesh size will affect the entrapment of VLPs (Figure 10). Since the diameter of the VLPs is the same as a norovirus (about 20 nm), it is important to know whether the mesh size of the hydrogel can accommodate the VLPs. Thus, glycosylated hydrogels of similar mass (about 1 mg) but with a different amount of crosslinker were immersed in a 1.0×10^{-4} μg μL^{-1} VLPs solution overnight. The VLPs concentration of the solution was calculated through the calibration curve obtained by ELISA. Only a minor entrapment difference was observed by the glycosylated hydrogels with different crosslinker densities, ranging from 0.75 % to 2 % (Figure 10). This result indicates that the mesh size of the hydrogels in this range is sufficient to accommodate the VLPs VA387.

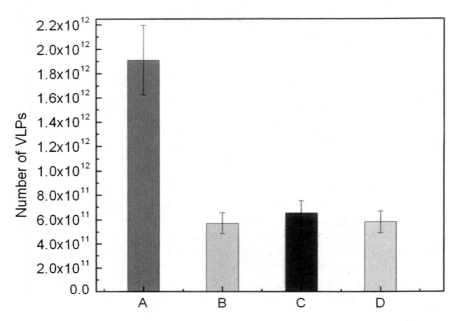

Figure 10. Absorption of the VLPs VA387 by hydrogels of different crosslinker density A): controlled sample; B): glycosylated hydrogel with 0.75% of crosslinker; C): with 1.0% of crosslinker; D): with 2.0% of crosslinker.

The second experiment evaluated the maximum amount of VLPs VA387 entrapped by measuring the entrapment of the VLPs by hydrogel with the same crosslinker density (1.0 mol % to the total monomers) in varying concentrations of the VLPs VA387 solution (Figure 11). It was found that the amount of VLPs left in the solution treated with the glycosylated hydrogels decreased significantly. The trapping of the VLPs increased proportionally to their concentration, which may indicate a diffusion controlled trapping instead of a surface absorption. The trapping capacity of the hydrogel is about 4×10^{12} VLPs mg^{-1} glycosylated hydrogel in a VLPs solution (8.0×10^{-4} µg µL^{-1}, 1:1000). Thus in practice, multiple doses of hydrogel can be applied to completely entrap the virus in practice.

Therefore, the entrapment of VLPs, the protein shell of norovirus, by the glycosylated hydrogels in aqueous solution indicates that the mesh size of the chemical crosslinked hydrogel may be more than 20 nm and can accommodate the norovirus inside the hydrogel. Furthermore, the insolubility of the hydrogel in water simplifies the purification procedure. This result as well as the

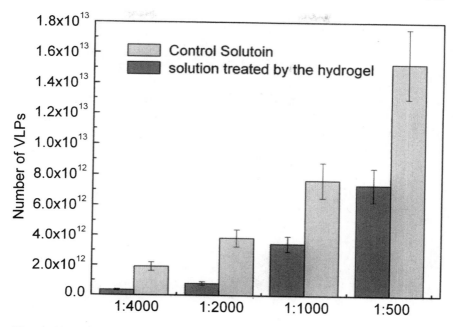

Figure 11. Immersing the glycohydrogel (5 mol% of sugar blood B) in different concentrations of VLPs solution. All solution was diluted from VLPs solution (0.8 μg/μL).

biocompatibility of poly DADMAC-AAm hydrogels suggest that the glycosylated hydrogels can be prophylactic antinorovirus drugs.

Conclusion

We have shown that glycopolymer can be applied to inhibit the adhesion of antibody or pathogenic organisms to host tissues. Since the adhesion is the prerequisite for the initiation for further lysis caused by immune response or infectious diseases, our glycopolymer has potential to be applied in xenotransplantation or anti-bacterial and anti-virus oral drugs. From a medical point of view, intervention of carbohydrate-protein interaction may be considered mild and gentle, and more sound ecologically, as well as safer, compared with present chemotherapy approaches. Moreover, since anti-adhesive agents do not act by killing or arresting the growth of the pathogens, it is very likely that strains resistant to such agents will emerge at a markedly lower rate than those that are resistant to antibiotics.

References

1. Sharon, N. *Biochimica et Biophysica Acta,* 2006, *1760*, 527–537.
2. Cooper, D. K. C.; Koren, E.; Oriol, R. *Immunol. Rev.* 1994, *141*, 31-58.
3. Galili, U. *Immunol. Today* 1993, *14*, 480-482.
4. Sandrin, M. S.; Vaughan, H. A.; McKenzie, I. F. C. *Transplant. Rev.* 1994, *8*, 134-149.
5. Samuelsson, B. E.; Rydberg, L.; Breimer, M. E.; Backer, A.; Gustavsson, M.; Holgersson, J.; Karlsson, E.; Uyterwaal, A.-C.; Cairns, T.; Welsh, K. *Immunol. Rev.* 1994, *141*, 151-168.
6. Sharon, N. *FEBS Lett.* 1987, *217*, 145-157.
7. Ofek, I.; *Doyle, R. J. In Bacterial Adhesion to Cells and Tissues*; Chapman and Hall: New York, 1994, 94-135.
8. Huang, P.; Farkas, T.; Marionneau, S.; Zhong, W.; Ruvoen-Clouet, N.; Ardythe, L. M.; Altaye, M.; Larry, K. P.; David, S. N.; LePendu, J.; Jiang, X. *J. Infect. Dis.* 2003, *1*, 19-31.
9. Preissner, K.; Chhatwal, G., *Cellular Microbiology*, ASM Press: Washington, D.C., 2000, 49-64.
10. Finlay, B.; Caparon, M., *Cellular Microbiology*, ASM Press: Washington, D.C., 2000, 67-80.
11. Sabesan, S.; Duus, J.; Neira, S.; Domaille, P.; Kelm, S.; Paulson, J.; Block, K. *J Am Chem Soc* 1992, *114*, 8363-8375.
12. Nagy, J. O.; Wang, P.; Gilbert, J. H.; Schaefer, M. E.; Hill, T. G.; Callstrom, M. R.; Bednarski, M. D. *J Med Chem* 1992, *35*, 4501-4502.
13. Lees, W. J.; Spaltenstein, A.; Kingery-Wood, J. E.; Whitesides, G. M., *J Med Chem* 1994, *37*, 3419-3433.
14. Mochalova, L. V.; Tuzikov, A. B.; Marinina, V. P.; Gambaryan, A. S.; Byramova, N. E.; Bovin, N. V.; Matrosovich, M. N. *Antiviral Res* 1994, *23*, 179-190.
15. Mammen, M.; Dahmann, G.; Whitesides, G. M. *J Med Chem* 1995, *38*, 4179-4190.
16. Tuzikov, A. B.; Byramova, N. E.; Bovin, N. V.; Gambaryan, A. S.; Matrosovich, M. N. *Antiviral Res* 1997, *33*, 129-134.
17. Reuter, J. D.; Myc, A.; Hayes, M. M.; Gan, Z.; Roy, R.; Qin, D.; Yin, R.; Piehler, L. T.; Esfand, R.; Tomalia, D. A.; Baker, J. R., Jr., *Bioconjug Chem.* 1999, *10*, 271-278.
18. Lee, Y. C.; Townsend, R. R.; Hardy, M. R.; Lonngren, J.; Arnarp, J.; Haraldson, M.; Lonn, H. *J. Biol. Chem.* 1983, *258*, 199-202.
19. Lee, Y. C.; Lee, R. T. *Acc. Chem. Res.* 1995, *28*, 321-327.
20. Pollak, A.; Blumenfeld, H.; Wax, M.; baughn, R. L.; Whitesides, G. M. *J. Am. Chem. Soc.* 1980, *102*, 6324-6336.

21. Mammen, M.; Dahmann, G.; Whitesides, G. M. *J. Med. Chem.* **1995**, *38*, 4179-4190.

22. Zanini, D.; Roy, R. *Carbohydrate mimics: Concepts and Methods*; Chapleur, Y., Ed.; Verlag Chemie: Weinheim, Germany, **1998**; 385-415.

23. Roy, R. *Top. Curr. Chem.* **1997**, *187*, 241-274.

24. Kiessling, L. L.; Pohl, N. L. *Chem. Biol.* **1996**, *3*, 71-77.

25. Bovin, N. V. *Glycoconjugate J.* **1998**, *15*, 431-446.

26. Wang, J. Q.; Chen, X.; Zhang, W.; Zacharek, S.; Chen, Y. S.; Wang, P. G. *J. Am. Chem. Soc.* **1999**, *121*, 8174-8181.

27. Li, J.; Zacharek, S.; Chen, X.; Wang, J. Q.; Zhang, W.; Janczuk, A.; Wang, P. G. *Bioorgan. & Medic. Chem.* **1999**, *7*, 1549-1558.

28. Zhang, Y. L.; Yao, Q. J.; Xia, C. F.; Jiang, X.; Wang, P. G. *ChemMedChem.* **2006**, *1*, 1361-1366.

29. Win, K. Y.; Feng, S. S.; *Biomaterials*, **2005**, *26*, 2713-2722.

30. Firon, N.; Ofek, I.; Sharon, N. *Infect. Immun.* **1984**, *43*, 1088-1090.

31. Parashar, U. D.; Monroe, S. S. *Rev. Med. Virol.* **2001**, *4*, 243-252

32. Wilhelmi, I.; Roman, E.; Sanchez-Fauquier, A. *Clin. Microbiol. Infect.* **2003**, *4*, 247-262.

33. Matson, D. O.; Szucs. G. *Curr. opin. in infect. Dis.* **2003**, *16*, 241-6.

34. Hutson, A. M.; Atmar, R. L.; Estes, M. K. *Trends in Microbiol.* **2004**, *6*, 279-287.

35. King, L.; David, T.; Loren, K.; Chris, C.; Martin, P.; Mary, W.; Mike, B. *Clin. Infect. Dis.* **2005**, *40*, 1471-1480.

36. Eklind, K.; Gustafsson, R.; Tiden, A. K.; Norberg, T.; Aaberg, P. M.; Astra Arcus, A. B.; Soedertaelje, S. *J. Carb. Chem.* **1996**, *15*, 1161-1178.

37. Blanchette, J.; Kavimandan, N.; Peppas, N. A.; *Biomedicine & Pharmacotherapy* **2004**, *3*, 142-151.

38. Delgado, M.; Lee, K. J.; Altobell, L. III; Spanka, C.; Jr. Wentworth, P.; Janda, K. D.; *J. Am. Chem. Soc.* **2002**, *18*, 4946-4947.

39. Aaron, C. F.; Goto, T.; Morishita, M.; Peppas, N. A. *Eur. J. Pharm. Biopharm.* **2004**, *2*, 163-169.

40. Gupta, P.; Vermani, K.; Garg, S. *Drug Discovery Today* **2002**, *10*, 569-579.

41. Patil, N. S.; Dordick, J. S; Rethwisch. D. G. *Biomaterials* **1996**, *17*, 2343-2350.

42. Jones, G.; Wortberg, M.; Kreissig, S. B.; Hammock, B. D.; Rocke, D. M.; *Anal. Chim. Acta* **1995**, *3*, 197-207.

Chapter 24

Novel Materials Based on Enzymatically Synthesized Amylose and Amylopectin

Jeroen van der Vlist and Katja Loos

Department of Polymer Chemistry, Zernike Institute for Advanced Materials, Faculty of Mathematics and Natural Sciences, University of Groningen, Nijenborgh 4, 9747AG Groningen, The Netherlands

Oligo- and polysaccharides are important macromolecules in living systems, showing their multifunctional characteristics in the construction of cell walls, energy storage, cell recognition and their immune response. Starch, the most abundant storage reserve carbohydrate in plants, is composed of two types of alpha-glucan, amylose and amylopectin. The branched structure, high amount of functional groups, biocompatibility of starch architectures makes them suitable for applications in the biomedical field and in the food industry. The exact way of starch biosynthesis in plants is still not known today. Here we are using a tandem reaction of two enzymes to synthesize "artificial" starch or rather (hyper)branched amylose in vitro. One enzyme is responsible for building the linear (amylose) part while the other enzyme introduces the branches. With the described tandem reaction of two enzymes, we are currently synthesizing hybrid copolymeric materials bearing (hyper)branched polysaccharide structures.

Introduction

Oligo- and polysaccharides are important macromolecules in living systems, showing their multifunctional characteristics in the construction of cell walls, energy storage, cell recognition and their immune response.

Saccharides as organic raw materials can open new perspectives on the way to new biocompatible and biodegradable products which could help overcome the problems resulting from the upcoming restrictions of petrochemical resources. Large amounts of carbohydrates are commercially available and a large part of the surplus of their agricultural production could be used for hybrid structures with synthetic materials.

Starch is the most abundant storage reserve carbohydrate in plants. Carbohydrates such as starch function as a reservoir of energy for later metabolic use. It is found in many different plant organs, including seeds, fruits, tubers and roots, where it is used as a source of energy during periods of dormancy and regrowth.

Starch granules are composed of two types of α-glucan, amylose and amylopectin, which represent approximately 98–99% of the dry weight. The ratio of the two polysaccharides varies according to the botanical origin of the starch.

Amylose is a linear molecule in which the glucose units are joined via α-(1→4) glucosyl linkages. Amylopectin is a branched molecule in which about 5% of the glucose units are joined by α-(1→6) glucosyl linkages (see figure 1).

In animals, a constant supply of glucose is essential for tissues such as the brain and red blood cells, which depend almost entirely on glucose as an energy source.

The mobilization of glucose from carbohydrate storage provides a constant supply of glucose to all tissues. For this glucose units are mobilized by their sequential removal from the non-reducing ends of starch. For this process three enzymes are required *in vivo*:

1. Glycogen phosphorylase catalyzes glycogen phosphorolysis (bond cleavage of the α-(1→4) bonds by the substitution of a phosphate group) to yield glucose-1-phosphate.

    glycogen $(Gluc)_n + P_i \leftrightarrows$ glycogen $(Gluc)_{n-1} +$ glucose-1-phosphate

    Phosphorylase is just able to release glucose if the unit is at least five units away from a branching point.

2. Glycogen debranching enzyme removes α-(1→6) glycogen branches, thereby making additional glucose residues accessible to glycogen phosphorylase.

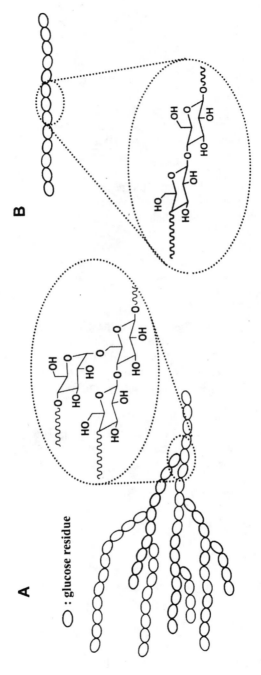

Figure 1. Structure of A) amylopectin and B) amylose

A

○ : glucose residue

B

3. Phosphoglucomutase converts glucose-1-phosphate into glucose-6-phosphate which has several metabolic fates.

The glycogen phosphorolysis of phosphorylase can be reverted, which makes it possible to enzymatically polymerize amylose as well as hybrid structures with amylose as outlined in the following.

Results and Discussions

Enzymatic polymerization of amylose with glycogen phosphorylase

The existence of a phosphorylating enzyme (phosphorylase) in a higher plant was first reported by Iwanoff who observed that an enzyme, he found in the germinating vetches Vicia sativa, liberates inorganic phosphate from organic phosphorous compounds.[1] Shortly after, the same enzyme was found in other vetches and wheat,[2,3] rice and coleseed,[4] barley and malt etc. Bodňar was the first to report a progressive disappearance of inorganic phosphate (thus the reverse reaction) while incubating suspended flour from ground peas in a phosphate buffer.[5] Cori and Cori demonstrated that animal tissues contain an enzyme which acts upon glycogen as well.[6-9] Cori, Colowick and Cori suggested that the product of this reaction is α-glucopyranose-1-phosphoric acid (also called Cori-ester), which was confirmed later by Kiessling[10] and Wolfrom and Pletcher[11].

Glycogen phosphorylases belong to the group of vitamin B_6 enzymes bearing a catalytic mechanism that involves the participation of the phosphate group of pyridoxal-5'-phosphate (PLP). The proposed mechanism is a concerted one with front-side attack as can be seen in figure 2.[12] In the forward direction, e.g., phosphorolysis of α-1,4-glycosidic bonds in oligo- or polysaccharides, the reaction is started by protonation of the glycosidic oxygen by orthophosphate, followed by stabilization of the incipient oxocarbonium ion by the phosphate anion and subsequent covalent binding of the phosphate to form glucose-1-phosphate. The product, glucose-1-phosphate dissociates and is replaced by a new incoming phosphate.

In the reverse direction, protonation of the phosphate of glucose-1-phosphate destabilizes the glycosidic bond and promotes formation of a glucosyl oxocarbonium ion-phosphate anion pair. In the subsequent step the phosphate anion becomes essential for promotion of the nucleophilic attack of a terminal glucosyl residue on the carbonium ion. This sequence of reactions brings about α-1,4-glycosidic bond formation and primer elongation.

This mechanism accounts for retention of configuration in both directions without requiring sequential double inversion of configuration. It also provides for a plausible explanation of the essential role of pyridoxal-5'-phosphate in glycogen phosphorylase catalysis, as the phosphate of the cofactor, pyridoxal-5'-

phosphate, and the substrate phosphates approach each other within a hydrogen-bond distance allowing proton transfer and making the phosphate of pyridoxal-5'-phosphate into a proton shuttle which recharges the substrate phosphate anion.

The fact that glycogen phosphorylase can be used to polymerize amylose was first demonstrated by Schäffner and Specht[13] in 1938 using yeast phosphorylase. Shortly after, the same behavior was also observed for other phosphorylases from yeast by Kiessling[14, 15], muscles by Cori, Schmidt and Cori[16], pea seeds[17], potatoes by Hanes[18] and preparations from liver by Ostern and Holmes[19], Cori, Cori and Schmidt[20] and Ostern, Herbert and Holmes[21]. These results opened up the field of enzymatic polymerizations of amylose using glucose-1-phosphate as monomer and can be considered the first experiments ever to synthesize biological macromolecules *in vitro*.

One of the remarkable properties of phosphorylase is that it is unable to synthesize amylose unless a primer is added (poly- or oligomaltosaccharide).

$$n \text{ (glucose-1-phospate)} + \text{primer} \leftrightarrows \text{amylose} + n \text{ (orthophosphate)}$$

The kinetic behavior of the polymerization of amylose with potato phosphorylase with various saccharides as primers was first studied by Hanes.[18] Green and Stumpf[22] failed to detect priming action with maltose but were able to confirm all other results by Hanes. Weibull and Tiselius[23] found that the maltooligosaccharide of lowest molecular weight to exhibit priming activity was maltotriose which was confirmed by Whelan and Bailey[24], who also showed that maltotriose is the lowest member of the series of oligosaccharides to exhibit priming activity (see figure 3)

Whelan and Bailey were also able to clarify the polymerization mechanism of the enzymatic polymerization with phosphorylase.[24] Their results showed that the polymerization follows a "multi chain" scheme in contrast to a "single chain" scheme that was also proposed by some authors. In the "multi chain" polymerization scheme the enzyme-substrate complex dissociates after every addition step, whereas in the "single chain" scheme each enzyme continuously increases the length of a single primer chain without dissociation. By studying the polydispersities of amyloses obtained by enzymatic polymerization with potato phosphorylase from maltooligosaccharides of various length Pfannemüller and Burchard were able to show that the reaction mechanism of the polymerization with maltotriose as primer varies from its higher homologes.[25] While the amyloses build by polymerization from maltotetraose or higher showed a Poisson distribution[26] that can be expected from a polymerization following a "multi chain" scheme (random synthesis occurs and all the primer chains grow at approximately equal rates) a bimodal broad distribution was observed when maltotriose was used as primer. The authors found that in the case of maltotriose as a primer the reaction can be divided into

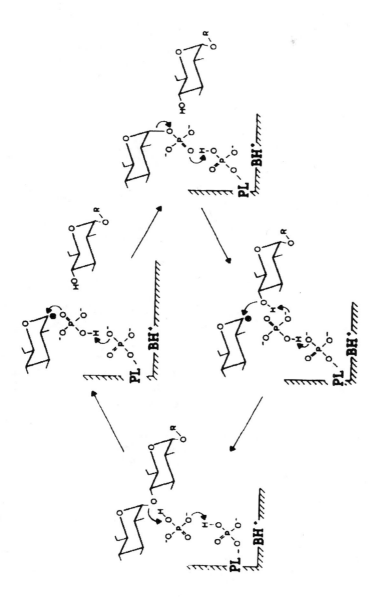

Figure 2. Catalytic mechanism of glycogen phosphorylases. The reaction scheme accounts for the reversibility of phosphorolysis of oligosaccharides (R) in the presence of orthophosphate (upper half) and primer-dependent synthesis in the presence of glucose-1-phosphate (lower half). PL = enzyme-bound pyridoxal; BH⁺ = a general base contributed by the enzyme protein. (Reproduced with permission from 12. Copyright 1990 American Chemical Society.)

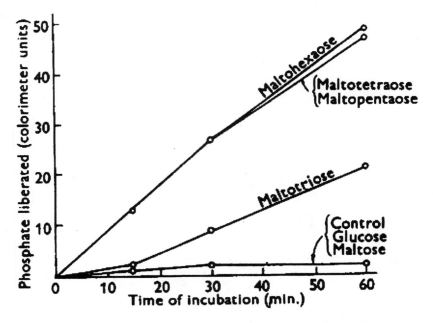

Figure 3. Priming activity of glucose and maltooligosaccharides in the enzymatic polymerization with potato phosphorylase and glucose-1-phosphate as monomer. (Reproduced with permission from reference 24. Copyright 1954.)

a start reaction and the following propagation, the rate of the first reaction being 400 times slower then the rate of the propagation. Due to this start reaction not all chains start to grow at the same time which results in a broader distribution. The propagation follows again a "multi chain" reaction scheme. Suganuma et al.[27] were able to determine the exact kinetic parameters of the synthetic as well as the phosphorolytic reaction using maltotriose and higher maltooligo-saccharides as primer and were able to confirm the results of Whelan & Bailey and Pfannemüller & Burchard.

Hybrid structures with amylose blocks

The strict primer dependence of the glycogen phosphorylases makes them ideal candidates for the synthesis of hybrid structures of amylose with non natural materials (e.g. inorganic particles and surfaces, synthetic polymers). For this, a primer functionality (maltooligosaccharide) can be coupled to a synthetic structure and subsequently elongated by enzymatic polymerization resulting in amylose blocks.

Various examples on these types of hybrid materials are reported which are outlined in the following.

Amylose hybrids with short alkyl chains

Pfannemüller et al. showed that it is possible to obtain carbohydrate containing amphiphiles with various alkyl chains via amide bond formation. For this maltooligosaccharides were oxidized to the according aldonic acid lactones which could subsequently be coupled to alkylamines.[28-36] Such sugar based surfactants are important industrial products finding their applications in cosmetics, medical applications etc.[37-39] The authors were also able to extend the attached maltooligosaccharides with enzymatic polymerization with potato phosphorylase which resulted in products with very interesting solution properties.[40, 41]

Amylose brushes on inorganic surfaces

Amylose brushes (a layer consisting of polymer chains dangling in a solvent with one end attached to a surface is frequently referred to as a polymer brush) on spherical and planar surfaces can have several advantages, such as detoxification of surfaces etc. The modification of surfaces with thin polymer films is widely used to tailor surface properties such as wettability, biocompatibility, corrosion resistance and friction.[42-44] The advantage of polymer brushes over other surface modification methods like self-assembled monolayers is their mechanical and chemical robustness, coupled with a high degree of synthetic flexibility towards the introduction of a variety of functional groups.

Commonly, brushes are prepared by grafting polymers to surfaces by for instance chemical bonding of reactive groups on the surface and reactive end groups of the attached polymers. This 'grafting to' approach has several disadvantages as it is very difficult to achieve high grafting densities and/or thicker films due to steric crowding of reactive surface sites by already adsorbed polymers.

The so-called 'grafting from' approach (polymers are grown from initiators bound to surfaces) is a superior alternative as the functionality, density and thickness of the polymer brushes can be controlled with almost molecular precision.

The first surface initiated enzymatic polymerization reported was the synthesis of amylose brushes on planar and spherical surfaces.[45] For this silica or silicone surfaces were modified with self assembled monolayers of (3-aminopropyl)trimethoxysilane or chlorodimethylsilane respectively. To these functionalities oligosaccharides were added via (a) reductive amidation of the oligosaccharides to surface bound amines, (b) conversion of the oligosaccharide to the according aldonic acid lactone and reaction with surface bound amines and (c) incorporation of a double bond to the oligosaccharide and subsequent hydrosililation to surface bound Si-H functions. The surface bound oligosaccharides could be enzymatically elongated with potato phosphorylase and glucose-1-phosphate as monomer to amylose chains of any desired length. The degree of polymerization could be determined by spectrometric measurement of the liberated amount of inorganic phosphate[46] which was confirmed by cleavage of the amylose brushes (either enzymatically or by prior incorporation of light sensitive spacers) and subsequent characterization of the free amylose chains. The obtained amylose modified surfaces showed good chiral discrimination when employed as column materials in chiral affinity chromatography. Modification of the OH-Groups of the amylose brushes even enhanced the separation strength of the developed column materials.[47] The results were recently confirmed by Breitinger who attached maltooligosaccharides to surfaces via acid labile hydrazide linkers and enzymatically extended the chains with potato phosphorylase. [48]

Copolymers with amylase

The combination of oligo- or polysaccharides with non natural polymeric structures opens up a novel class of materials. By varying the chain topology of the individual blocks as well as of the whole copolymer, the type of blocks, the composition etc. a complete set with tailor made properties can be designed.

Amylose is a rod-like helical polymer consisting of α-(1→4) glycosidic units. A measurement of the stiffness of a polymer is afforded by the so-called persistence length, which gives an estimate of the length scale over which the tangent vectors along the contour of the chains backbone are correlated. Typical values for persistence lengths in synthetic and biological systems can be several orders of magnitude larger than for flexible, coil-like polymers. Rod-like polymers have been found to exhibit lyotropic liquid crystalline ordered phases such as nematic and/or layered smectic structures with the molecules arranged with their long axes nearly parallel to each other. Supramolecular assemblies of rod-like molecules are also capable of forming liquid crystalline phases. The main factor governing the geometry of supramolecular structures in the liquid crystalline phase is the anisotropic aggregation of the molecules.

Copolymeric systems with amylose are therefore systems in which at least one component is based on a conformationally rigid segment, which are generally referred to as rod-coil systems.[49-52] By combining rod-like and coil-like polymers a novel class of self-assembling materials can be produced since the molecules share certain general characteristics typical of diblock molecules and thermotropic calamitic molecules. The difference in chain rigidity of rod-like and coil-like blocks is expected to greatly affect the details of molecular packing in the condensed phases and thus the nature of thermodynamically stable morphologies in these materials. The thermodynamic stable morphology probably originates as the result of the interdependence of microsegregation and liquid cristallinity. From this point of view it is very fascinating to compare the microstructures originating in solution and in the bulk for such materials.

Practical applications in which copolymers are characterized by some degree of structural asymmetry have been suggested. For instance a flexible block may be chosen as it donates a flexural compliance, whereas the more rigid portion offers tensile strength. In addition to the mechanical properties, the orientational order and the electrical conductance of certain rigid blocks could be exploited in optical and electrical devices.

Comb type and linear block copolymer systems with enzymatically synthesized amylose are reported, which are outlined in the following.

Comb-type copolymers with amylase

The first comb like structures synthesized by enzymatic grafting *from* polymerization from a polymeric backbone were reported by Husemann et al.[53, 54] Acetobromo oligosaccharides were covalently bound to 6-trityl-2,3-dicarbanilyl-amylose chains and subsequently elongated by enzymatic polymerization with potato phosphorylase, the result being amylopectin-like structures with various degrees of branching. Pfannemüller et al. extended this work by grafting amylose chains onto starch molecules. The modified starches where studied by the uptake of iodine and by light scattering measurements of carbanilate derivates[55] and appeared to be star like in electron microscopy studies[56].

A full series of star-, network- and comb-like hybrid structures with oligosaccharides were synthesized by Pfannemüller et al. (see figure 4) and it was shown that the attached oligosaccharides can be extended via the enzymatic polymerization with potato phosphorylase.[57, 58, 28-30]

Another type of comb like amylose hybrids synthesized via enzymatic grafting with phosphorylase is based on polysiloxane backbones. To achieve these structures double bonds were incorporated to the reducing end of oligosaccharides which were then attached to poly(dimethylsiloxane-co-methylsiloxane) copolymers via hydrosililation[59, 60] or to silane monomers which were subsequently polymerized to polysiloxanes[61]. Various mono-, di-, tri and oligosaccharides were attached to siloxane backbones and their solution properties were studied with viscosimetry and static and dynamic light scattering.[62] The pendant oligosaccharide moieties could be extended with enzymatic grafting from polymerization[63, 64]

Kobayashi et al. succeeded in attaching maltopentaose to the *para* position of styrene and performed free radical polymerizations towards the homo-

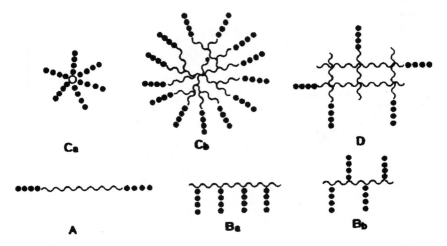

Figure 4. Maltotetraose hybrids with various carriers resulting in different chain architecture. A: poly(ethylene oxide); B_a and B_b: poly(acrylic acid), amylose, cellulose and other polysaccharides; C_a: cyclodextrin and multifunctional acids; C_b: amylopectin; D: crosslinked poly(acryl amide). (Reproduced with permission from reference 57. Copyright 1978 Wiley-VCH Verlag GmbH & Company KGaA.)

polymers[65, 66] as well as copolymers with acrylamide[65]. Kobayashi et al. also reported on the successful attachment of maltopentaose to poly(L-glutamic acid).[67] Kakuchi et al. showed that the saccharide modified styrene monomers could also be polymerized with TEMPO-mediated controlled radical polymerization.[68] In all cases the authors could successfully elongate the attached oligosaccharide structures with enzymatic polymerization, the product being comb type block copolymers with amylose.

Linear block copolymers with amylose

Various linear block copolymers of the AB, ABA and ABC type with enzymatically polymerized amylose blocks were reported. Ziegast and Pfannemüller converted the hydroxyl end groups of poly(ethylene oxide) into amino groups via tosylation and further reaction with 2-aminoalkylthiolate.[69] To the resulting mono- and di-amino functionalized poly(ethylene oxide) maltooligosaccharide lactones were attached and subsequently elongated to amylose via enzymatic polymerization.[70] Pfannemüller et al. performed a very detailed study on the solution properties of the synthesized A-B-A triblock copolymers as they can be considered model substances for "once broken rod" chains.[71] With static and dynamic light scattering they found that the flexible joint between the two rigid amylose blocks has no detectable effect on the common static and dynamic properties of the chain. With dielectric measurements it however became obvious that the directional properties of the electric dipoles of the broken rigid chains showed a different behavior to the non-broken rods (pure amylose). Akyoshi et al. also synthesized amylose-block-poly(ethylene glycol) block copolymers via enzymatic grafting from oligosaccharide terminated poly(ethylene oxide) and studied the solution properties of these amphiphilic block copolymers by static and dynamic light scattering.[72, 73]

It was also shown that the enzymatic polymerization of amylose could be started from oligosaccharide modified polymers that are not soluble in the medium of polymerization (aqueous buffers). Amylose-block-polystyrene block copolymers could be synthesized by attaching maltooligosaccharides to anionically synthesized amino terminated polystyrene and subsequent enzymatic elongation to amylose.[74, 75] Block copolymers with a wide range of molecular weights and copolymer composition were synthesized via this synthetic route. The solution properties of star type as well as crew cut micelles of these block copolymers were studied in water and THF and the according scaling laws were established.[76] In THF up to four different micellar species were detectable, some of them in the size range of vesicular structures, whereas the crew cut micelles in water were much more defined. Bosker et al. studied the interfacial behavior of amylose-block-polystyrene block copolymers at the air-water interface with the Langmuir-Blodgett technique.[77]

Recently two groups reported on controlled radical polymerizations started from maltooligosaccharides (ATRP[78] and TEMPO mediated radical polymerization[79]) which will certainly lead to new synthetic routes towards amylose containing block copolymers.

Even though the products are not block copolymer structures the work of Kadokawa et al. should be mentioned here. In a process the authors named "vine-twining polymerization" (after the way vine plant grow helically around a support rod) the enzymatic polymerization of amylose is performed in the presence of synthetic polymers in solution and the authors showed that the grown amylose chains incorporate the polymers into its helical cavity while polymerizing.[80-84]

Hybrid structures with amylopectin like structures

The results reviewed above clearly show that the combination of enzymatically polymerized amylose with surfaces of inorganic materials and synthetic polymers results in very interesting materials with superior properties. We are currently extending this concept by synthesizing hybrids with enzymatically synthesized amylopectin. This has several advantages including the better solubility of amylopectin versus amylose, the higher amount of functional group which will facilitate further modification etc.

The exact way of amylopectin biosynthesis in plants is still not known today. In our current research we are using a tandem reaction of two enzymes to synthesize "artificial" amylopectin or rather (hyper)branched amylose *in vitro*. One enzyme is responsible for building the linear (amylose) part while the other enzyme introduces the branches, phosphorylase and glycogen branching enzyme respectively.

Phosphorylase can be easily isolated from potatoes and, after purification, used to catalyze the polymerization of glucose-1-phosphate in order to obtain linear polysaccharide chains with α-(1→4) glycosidic linkages, as can be seen in figure 5.

The glycogen branching enzyme belongs to the transferase family and is able to transfer short, α-(1→4) linked, oligosaccharides from the non-reducing end of starch to an α-(1→6) position (see figure 5).

By combining the glycogen branching enzyme with glycogen phosphorylase it becomes possible to synthesize branched structures via an one-pot synthesis as phosphorylase will polymerize linear amylose and the glycogen branching enzyme will introduce the branching points which are again extended by phosphorylase.

iP   : inorganic phosphate
◯-P : glucose-1-phosphate
◯   : glucose residue

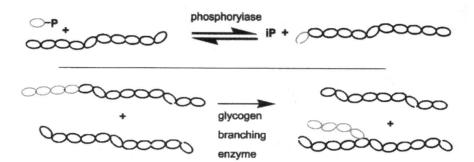

Figure 5. Schematic representation of the reactions catalyzed by glycogen phosphorylase (above) and glycogen branching enzyme (below)

Both described enzymes are isolated from natural sources. Phosphorylase is isolated from potatoes whereas the glycogen branching enzyme is produced by various bacterial sources. [85, 86] Depending on the source the properties of the products and the reaction conditions may differ.

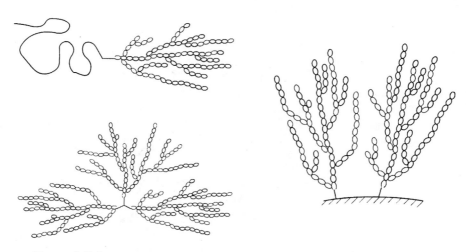

Figure 6. Schematic representation of hybrid structures with amylopectin

As shown above, hybrid structures bearing amylose blocks can be synthesized by covalent attachment of primer recognition units for phosphorylase and subsequent enzymatic *grafting from* polymerization. Following the same route we are currently synthesizing hybrid materials bearing (hyper)branched polysaccharide structures as shown in figure 6 with the described tandem reaction of two enzymes.

The branched structure, high amount of functional groups, biocompatibility of these structures make these architectures suitable for applications in the biomedical field and in the food industry.

Acknowledgment

We are indebted to M. Palomo Reixach, M.J.E.C. van der Maarel and L. Dijkhuizen from the Center for Carbohydrate Bioprocessing for their indispensable help with the branching enzymes.

References

1. Iwanow, L. *Ber. d. deutsch. bot. Ges.* **1902**, *20*, 366.
2. Zaleski, W. *Ber. d. deutsch. bot. Ges.* **1906**, *24*, 285.
3. Zaleski, W. *Ber. d. deutsch. bot. Ges.* **1911**, *29*, 146.
4. Suzuki, U.; Yoshimura, K.; Takaishi, M. *Tokyo Kagaku Kaishi* **1906**, *27*, 1330.
5. Bodńar, J. *Biochem. Z.* **1925**, *165*, 1.
6. Cori, G. T.; Cori, C. F. *J. Biol. Chem.* **1936**, *116*, 119.
7. Cori, G. T.; Cori, C. F. *J. Biol. Chem.* **1936**, *116*, 129.
8. Cori, C. F.; Colowick, S. P.; Cori, G. T. *J. Biol. Chem.* **1937**, *121*, 465.
9. Cori, C. F.; Cori, G. T. *Proc. Soc. Exp. Biol. Med.* **1937**, *36*, 119.
10. Kiessling, W. *Biochem. Z.* **1938**, *298*, 421.
11. Wolfrom, M. L.; Pletcher, D. E. *J. Am. Chem. Soc.;* **1941**, *63*, 1050.
12. Palm, D.; Klein, H. W.; Schinzel, R.; Buehner, M.; Helmreich, E. J. M. *Biochemistry* **1990**, *29*, 1099.
13. Schäfner, A.; Specht, H. *Naturwissenschaften* **1938**, *26*, 494.
14. Kiessling, W. *Naturwissenschaften* **1939**, *27*, 129.
15. Kiessling, W. *Biochem. Z.* **1939**, *302*, 50.
16. Cori, C. F.; Schmidt, G.; Cori, G. T. *Science* **1939**, *89*, 464.
17. Hanes, C. S. *Proc. Roy. Soc. B* **1940**, *128*, 421.
18. Hanes, C. S. *Proc. Roy. Soc. B* **1940**, *129*, 174.
19. Ostern, P.; Holmes, E. *Nature* **1939**, *144*, 34.
20. Cori, G. T.; Cori, C. F.; Schmidt, G. *J. Biol. Chem.* **1939**, *129*, 629.
21. Ostern, P.; Herbert, D.; Holmes, E. *Biochem. J.* **1939**, *33*, 1858.
22. Green, D. E.; Stumpf, P. K. *J. Biol. Chem.* **1942**, *142*, 355.
23. Weibull, C.; Tiselius, A. *Arkiv för Kemi Mineralogi Och Geologi* **1945**, *19*, 1.
24. Whelan, W. J.; Bailey, J. M. *Biochem. J.* **1954**, *58*, 560.
25. Pfannemüller, B.; Burchard, W. *Makromol. Chem.* **1969**, *121*, 1.
26. Pfannemüller, B. *Naturwissenschaften* **1975**, *62*, 231.
27. Suganuma, T.; Kitazono, J. I.; Yoshinaga, K.; Fujimoto, S.; Nagahama, T. *Carbohydr. Res.* **1991**, *217*, 213.
28. Emmerling, W. N.; Pfannemüller, B. *Makromol. Chem.* **1978**, *179*, 1627.
29. Emmerling, W. N.; Pfannemüller, B. *Stärke* **1981**, *33*, 202.
30. Emmerling, W. N.; Pfannemüller, B. *Makromol. Chem.* **1983**, *184*, 1441.
31. Ziegast, G.; Pfannemüller, B. *Makromol. Chem.* **1984**, *185*, 1855.
32. Pfannemüller, B. *Stärke* **1988**, *40*, 476.
33. Müller-Fahrnow, A.; Hilgenfeld, R.; Hesse, H.; Saenger, W.; Pfannemüller, B. *Carbohydr. Res.* **1988**, *176*, 165.
34. Pfannemüller, B.; Kühn, I. *Makromol. Chem.* **1988**, *189*, 2433.
35. Taravel, F. R.; Pfannemüller, B. *Makromol. Chem.* **1990**, *191*, 3097.

36. Tuzov, I.; Cramer, K.; Pfannemüller, B.; Kreutz, W.; Magonov, S. N. *Adv. Mater.* **1995**, *7*, 656.
37. Biermann, M.; Schmid, K.; Schulz, P. *Stärke* **1993**, *45*, 281.
38. von Rybinski, W.; Hill, K. *Angew. Chem.-Int. Edit.* **1998**, *37*, 1328.
39. Hill, K.; Rhode, O. *Fett-Lipid* **1999**, *101*, 25.
40. Ziegast, G.; Pfannemüller, B. *Carbohydr. Res.* **1987**, *160*, 185.
41. Niemann, C.; Nuck, R.; Pfannemüller, B.; Saenger, W. *Carbohydr. Res.* **1990**, *197*, 187.
42. Zhao, B.; Brittain, W. J. *Progr. Polym. Sci.* **2000**, *25*, 677.
43. Advincula, R.; Zhou, Q. G.; Park, M.; Wang, S. G.; Mays, J.; Sakellariou, G.; Pispas, S.; Hadjichristidis, N. *Langmuir* **2002**, *18*, 8672.
44. Edmondson, S.; Osborne, V. L.; Huck, W. T. S. *Chem. Soc. Rev.* **2004**, *33*, 14.
45. Loos, K.; von Braunmühl, V.; Stadler, R.; Landfester, K.; Spiess, H. W. *Macromol. Rapid Commun.* **1997**, *18*, 927.
46. Fiske, C. H.; Subbarow, Y. *J. Biol. Chem.* **1925**, *66*, 375.
47. Loos, K. *unpublished results*.
48. Breitinger, H.-G. *Tetrahedron Lett.* **2002**, *43*, 6127.
49. Stupp, S. I. *Curr. Opin. Colloid. In.* **1998**, *3*, 20.
50. Klok, H. A.; Lecommandoux, S. *Adv. Mater.* **2001**, *13*, 1217.
51. Lee, M.; Cho, B. K.; Zin, W. C. *Chem. Rev.* **2001**, *101*, 3869.
52. Loos, K.; Munöz-Guerra, S., In *Supramolecular Polymers*, 2nd ed.; Ciferri, A., Ed. CRC Press: Boca Raton, 2005; pp 393.
53. Husemann, E.; Reinhardt, M. *Makromol. Chem.* **1962**, *57*, 109.
54. Husemann, E.; Reinhardt, M. *Makromol. Chem.* **1962**, *57*, 129.
55. Burchard, W.; Kratz, I.; Pfannemüller, B. *Makromol. Chem.* **1971**, *150*, 63.
56. Bittiger, H.; Husemann, E.; Pfannemüller, B. *Stärke* **1971**, *23*, 113.
57. Andresz, H.; Richter, G. C.; Pfannemüller, B. *Makromol. Chem.* **1978**, *179*, 301.
58. Emmerling, W.; Pfannemüller, B. *Chem. Ztg.* **1978**, *102*, 233.
59. Jonas, G.; Stadler, R. *Makromol. Chem.-Rapid Commun.* **1991**, *12*, 625.
60. Jonas, G.; Stadler, R. *Acta Polym.* **1994**, *45*, 14.
61. Haupt, M.; Knaus, S.; Rohr, T.; Gruber, H. *J. Macromol. Sci. Pure* **2000**, *37*, 323.
62. Loos, K.; Jonas, G.; Stadler, R. *Macromol. Chem. Phys.* **2001**, *202*, 3210.
63. von Braunmühl, V.; Jonas, G.; Stadler, R. *Macromolecules* **1995**, *28*, 17.
64. von Braunmühl, V.; Stadler, R. *Macromol. Symp.* **1996**, *103*, 141.
65. Kobayashi, K.; Kamiya, S.; Enomoto, N. *Macromolecules* **1996**, *29*, 8670.
66. Wataoka, I.; Urakawa, H.; Kobayashi, K.; Akaike, T.; Schmidt, M.; Kajiwara, K. *Macromolecules* **1999**, *32*, 1816.
67. Kamiya, S.; Kobayashi, K. *Macromol. Chem. Phys.* **1998**, *199*, 1589.
68. Narumi, A.; Kawasaki, K.; Kaga, H.; Satoh, T.; Sugimoto, N.; Kakuchi, T. *Polym. Bull.* **2003**, *49*, 405.

69. Ziegast, G.; Pfannemüller, B. *Makromol. Chem.-Rapid Commun.* **1984**, *5*, 363.

70. Ziegast, G.; Pfannemüller, B. *Makromol. Chem.-Rapid Commun.* **1984**, *5*, 373.

71. Pfannemüller, B.; Schmidt, M.; Ziegast, G.; Matsuo, K. *Macromolecules* **1984**, *17*, 710.

72. Akiyoshi, K.; Kohara, M.; Ito, K.; Kitamura, S.; Sunamoto, J. *Macromol. Rapid Commun.* **1999**, *20*, 112.

73. Akiyoshi, K.; Maruichi, N.; Kohara, M.; Kitamura, S. *Biomacromolecules* **2002**, *3*, 280.

74. Loos, K.; Stadler, R. *Macromolecules* **1997**, *30*, 7641.

75. Loos, K.; Müller, A. H. E. *Biomacromolecules* **2002**, *3*, 368.

76. Loos, K.; Böker, A.; Zettl, H.; Zhang, A. F.; Krausch, G.; Müller, A. H. E. *Macromolecules* **2005**, *38*, 873.

77. Bosker, W. T. E.; Agoston, K.; Stuart, M. A. C.; Norde, W.; Timmermans, J. W.; Slaghek, T. M. *Macromolecules* **2003**, *36*, 1982.

78. Haddleton, D. M.; Ohno, K. *Biomacromolecules* **2000**, *1*, 152.

79. Narumi, A.; Miura, Y.; Otsuka, I.; Yamane, S.; Kitajyo, Y.; Satoh, T.; Hirao, A.; Kaneko, N.; Kaga, H.; Kakuchi, T. *J. Polym. Sci. Pol. Chem.* **2006**, *44*, 4864.

80. Kadokawa, J.-i.; Kaneko, Y.; Nakaya, A.; Tagaya, H. *Macromolecules* **2001**, *34*, 6536.

81. Kadokawa, J.-i.; Kaneko, Y.; Tagaya, H.; Chiba, K. *Chem. Commun.* **2001**, 449.

82. Kadokawa, J.-i.; Kaneko, Y.; Nagase, S.; Takahashi, T.; Tagaya, H. *Chem.-Eur. J.* **2002**, *8*, 3321.

83. Kadokawa, J.-i.; Nakaya, A.; Kaneko, Y.; Tagaya, H. *Macromol. Chem. Phys.* **2003**, *204*, 1451.

84. Kaneko, Y.; Kadokawa, J.-i. *Chem. Rec.* **2005**, *5*, 36.

85. Takata, H. T., T.; Okada, S.; Takagi, M.; Imanaka, T. *J. Bact.* **1996**, *178*, 1600.

86. van der Maarel, M. J. E. C. V., A.; Sanders, P.; Dijkhuizen, L. *Biocatal. Biotransfor.* **2003**, *21*, 199.

Chapter 25

Sugar Based Polymers: Overviews and Recent Advances of Vinyl Sugars

Yutaka Tokiwa[1] and Masaru Kitagawa[2]

[1]National Institute of Advanced Industrial Science and Technology (AIST), Central 6, 1-1-1 Higashi, Tsukuba, Ibaraki 305-8566, Japan
[2]Toyobo Co., Ltd., Research Center, Katata 2-1-1, Ohtsu, Shiga 520-0292, Japan

Sugar based polymers, which are obtained by polymerization of vinyl sugars, have recently received increased attention from two viewpoints. One is the development of environmentally friendly material from renewable resources. Another is physiologically active material that mimics carbohydrate on cell surface. This article provides an overview of known sugar based polymers and the recent advances of the poly(vinylalcohol) (PVA) with sugar pendants.

Introduction

Recently, the development of new functional materials from renewable resources such as various kinds of saccharides has received remarkable interest (1, 2, 3). Various types of synthetic polymer containing saccharides have been investigated (4, 5). There are two types of sugar-based polymers: (i) sugar-containing linear polymer; (ii) polymer with sugar pendant (Figure 1).

Currently, there is a great interest in the role of sugar *in vivo* as it plays an important part not only as an energy source but also in cell-cell recognition systems such as cell growth, cancer metastasis, bacterial and viral infections (6). Hence, the polymers with sugar pendant have recently been the focus of intensive research (4, 5). These polymers are useful tools in investigating glycobiological phenomena because of their glycocluster effects. The sugar branched polymers being environmentally friendly and having biological

a) Sugar containing linear polymers

b) Polymers with sugar pendant

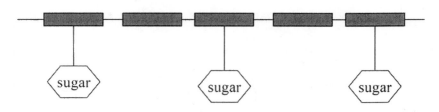

Figure 1. Sugar based synthetic polymers

recognition materials can be used to improve the antistatic property, dyeability, adhesion, printability and biocompatibility of widely used polymers.

Several types of the polymer with sugar pendant and having various synthetic polymer backbones such as vinyl polymer, peptide, etc are already reported (7, 8, 9). In this study, we discussed the polymerizable sugar derivatives including vinyl group and recent advances of PVA with sugar pendant.

Polymerizable Sugar Derivatives Involving Vinyl Group

To produce glycopolymers with sugar pendants, polymerizable sugar derivative monomers need to be developed, thus, several types of vinyl sugars have been investigated. Wulff et al. described the vinyl sugar consisting of three units, the bonding unit, spacer unit and sugar unit (Figure 2) (10). The sugar density, biodegradability and hydrophobicity are dependent on the type of vinyl sugar being used. Various kinds of polymers could be designed by modifying these three parts.

The bonding unit determines the biodegradation of the glycopolymer. Several types of bonding units have been reported such as acrylester, acrylamide, vinyl alcohol ester, vinyl ether, styrene and vinyl alkyl types. Among these bonding units, there have been numerous reports using acrylate and acrylamide sugars for a long time because of easy polymerizability and flexibility of the back-bone polymer (Figure 3).

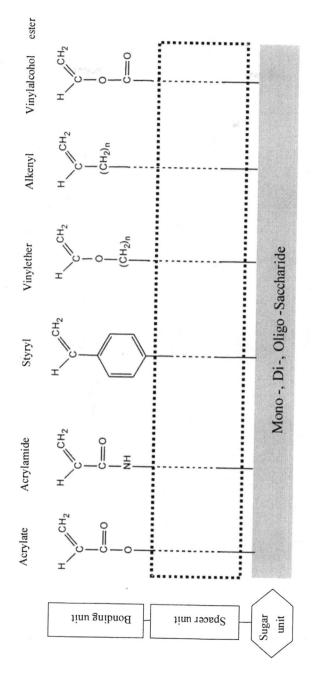

Figure 2. Various kinds of vinyl sugars

382

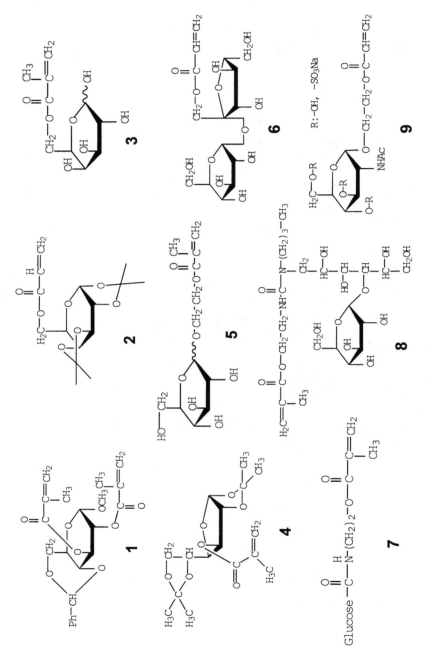

Figure 3. Acrylate sugars

From the 1940's to the 1960's, the acryloyl group was induced directly to the hydroxyl group of monosaccharide protected by isopropylidene or benzylidene such as methyl 2,3-di-*O*-acryloyl 4,6-benzylidene glucoside **1** (*11*), 1,2:3,4-di-*O*-isopropylidene- 6-*O*-acryloyl α-D-galactopyranose **2**, 6-*O*-methacryloyl D-galactoside **3** (*13*), 1,2; 5,6-diisopropylide 3-methacryloyl glucofuranose **4** (*14*). Martin et al. prepared the copolymers by free radical polymerization of 1,2:3,4-di-*O*-iso- propylidene- 6-*O*-acryloyl α-D-galacto-pyranose **2** for controlled release tablets as matrix-forming material of drug delivery systems (*12*). Recently, Yaacoub et al. reported the emulsion polymerization using the acrylate sugar **4** (*15, 16*), and further Ohno et al. examined nitroxide-mediated free radical polymerization using 1,2; 5,6-diisopropylide 3-*O*-acryloyl glucofuranose (*17*) and/or the atom transfer radical polymerization with an alkyl halide/copper complex using the acrylate sugar **4** (*18*). Ye et al. prepared the block copolymers using the acrylate sugar **4** and 1,1-dihydroperfluorooctyl methacrylate, and the deprotected copolymer formed aggregate structure in liquid CO_2 (*19*). These acrylates sugars have a polymerizable group at the hydroxyl group except hemiacetal hydroxyl group and are prepared via protected sugar derivatives blocked by benzylidene and/or isopropylidene group. Kitazawa et al. developed an efficient synthetic method for polymerizable sugar monomer having methacrylate at hemiacetal hydroxyl group, glucosyloxy ethylmethacrylate **5**, by use of heteropolyacid catalyst (*20*). Recently, Lowe et al. reported reversible addition-fragmentation chain transfer (RAFT) polymerization using this sugar monomer **5** which was already commercialized, and the homopolymerization displays the characteristics of a controlled living polymerization (*21*). Dordick reported a one-step synthesis of sugar acrylate, 1'-acryloyl sucrose **6**, using enzyme regioselectively, and the obtained sugar acrylate was polymerized to get polyacrylate with sugar pendant (*22*). Just recently, 2-(methacryloyloxy)ethyl carbamate glucose **7** was prepared by non-regioselective reaction of 2-(methacryloyloxy)ethyl isocyanate and glucose. The homopolymer of the monomer **7** showed excellent mechanical property and thermal stability as binders for fiber board (*23*). Klein et al. synthesized a new polymerizable sugar alcohol derivative, 2-[*N*'- (1-deoxymaltit-1-yl)-*N*'-butylureido] ethyl methacrylate **8** by reductive amination of disaccharide with alkylamine and its subsequent reaction with 2-isocyanatoethyl methacrylate (*24*). Grande et al. reported the synthesis of glycosaminoglycan mimetic polymers, polyacrylates with glucosamine pendants, at high yield using acryloxyethyl glucosamine derivatives **9** by cyanoxy-mediated free radical polymerization (*25*).

Different vinyl sugars having acrylamide group are shown in Figure 4. Lee et al. prepared the acrylamide derivatives containing thioglycosides **10** to elucidate the roles of sugars in biological reactions (*27*). Klein et al. developed a new synthetic method of 1-acrylamido-1-deoxymaltitol **11** by reductive aminonation and subsequent reaction with acrylilic anhydride (*26*). Bady et al.

384

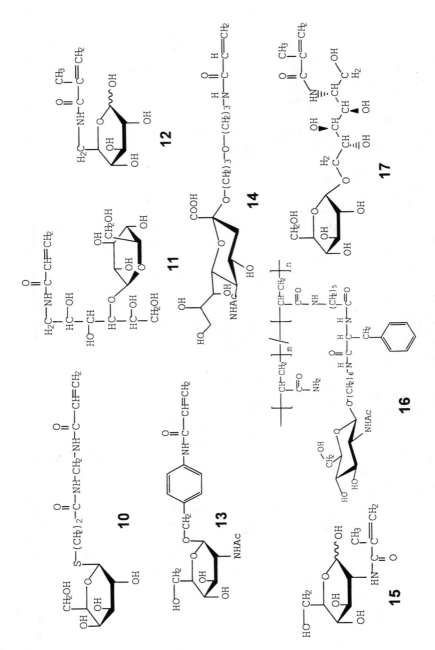

Figure 4. Acrylamide sugars.

reported the preparation of 6-deoxy-6-methacryloylamido D-glucose **12** using an easy method via key-intermediates, 6-azido derivatives (*28*). Carpino et al. prepared 1-*O*-(4-methacryloylaminophenyl) β-D-glucopyranoside **13** for the development of pharmacologically active polymers (*31*). Whitesides et al. reported an α-sialoside linked to acrylamide 5-acetamido-2-*O*-(*N*-acryloyl-8-amino-5-oxaoctyl)-2,6-anhydro-3,5-dideoxy D-galacto-α-nonulopyranosonic acid **14** which inhibit agglutination of erythrocytes by influenza A virus (*29*). Iwakura et al. synthesized *N*-methacryloyl D-glucosamine **15** by the one-step reaction of glucosamine hydrochloride with methacrylic anhydride (*30*). Nishimura et al. prepared the polyacrylamide having pendent sugar residues from the view point of pseudoglycoproteins related to the asparagine linked type glycoprotein, and that the water soluble glycopolymers having *N*-acetyl glucosamine branches **16** is used as a polymeric sugar acceptor substrate which is glycosylated with glycosyl transferase, to develop excellent methodology in oiligosaccharide synthesis (*32*). Poly(vinyl sugars) are usually used in the wet state. Hirata and Nakagawa reported the synthesis of methacrylamide sugar alcohol, D-glucopyranosyl-α(1-6)-2-deoxy-2-methacrylamido D-mannitol **17** and the hydration properties of the polymer was studied using differential scanning calorimetry (*33*).

The vinyl sugars having styrene group are shown in Figure 5. Chung reported the synthesis of polymerizable glucoside containing styrene group as an amphiphilic material, 3-(4-vinylphenyl)propyl β-D-glucoside **18** (*34*). Ohno et al. prepared *N*-p-vinylbenzyl-2,3,5,6-tetra-*O*-acetyl-4-*O*-(2,3,4,6-tetra-*O*-acetyl β-D-galacto pyranosyl) D-gluconamide **19** and nitroxide mediated free radical polymerization of the styryl sugar was achieved (*35*). Kobayashi et al. reported a variety of styryl sugars which the reducing end of sugars is reacted to vinylbenzylamine and was under intense study for interaction between hepatocytes and oligosaccharide containing polystyrene **22** for artificial liver (*36, 37*). Sato et al. reported layered thin films composed of lectin and poly[*N*-p-vinylbenzyl-O-α-D-glucipyranosyl1-(1-4)-D-gluconamide] **22** which is comercialized as sugar bearing polymer and the film was disintegrated upon the addition of sugars. They suggested that the film may be useful for constructing sensitive devices (*38*). Koyama et al. prepared p-vinylphenyl 3,4,6-tri-*O*-acetyl-2-amino-2-deoxy β-D-glucoside **23** by the glycosidation reaction of glycosyl halaide and vinyl phenol sodium salt and the obtained polymer showed a strong affinity for lectins (*39*). Wulff investigated on surface modified polymers and synthesized the styryl sugar with *N,N*-dimethylbarbituric acid **21** (*40*) as a plastic improvement material. To develop amphiphilic block copolymers bearing sugars, they investigated the controlled free radical polymerizations of the styryl sugars **20**, which prepared by Grignard reaction, mediated by 2, 2, 6, 6-tetramethyl piperidine-*N*-oxyl (TEMPO) (*41*). Loykulnant and Hirano synthesized p-[3-(1.2:5,6-di-*O*-isopropylidene α-D-gluco franose-3-oxyl)alkyl] styrene **24** by phase transfer catalyzed etherification and their anionic

386

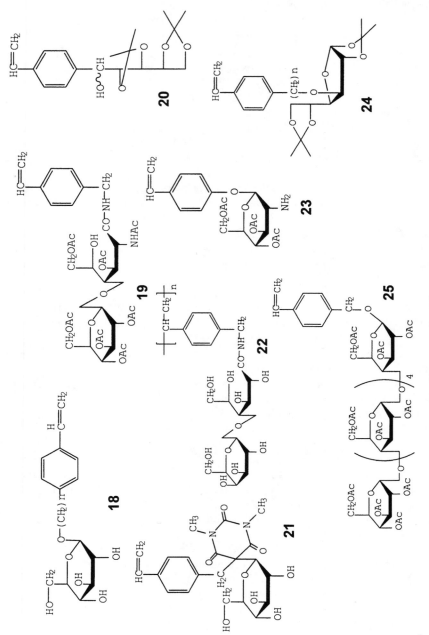

Figure 5. Styryl sugars

polymerizations proceeded in a living manner (*42*). Park et al. reported that a reducing glucose-carrying polymer, poly[3-*O*-(4'-vinylbenzyl β-D- glucose) interacted with erythrocytes carrying the glucose transporter on the cell membrane, and it was proposed that the polymers are useful for new type of cell recognition polymer (*43*). Narumi et al. also examined the TEMPO-mediated living radical polymerization of styryl sugar, 4-vinylbenzyl glucoside paracetate and 4-vinylbenzyl maltohexaoside peracetate **25**, and self-assembled block copolymers such as micelles were obtained (*44*).

The vinyl sugars having vinylether group are shown in Figure 6. Whistler et al. prepared 6-*O*-vinyl-1,2;3,4-di-*O*-isopropylidene D-galactopyranose **26** by bubbling acetylene with diisopropylidene galactose (*45*).

D'Agosto et al. reported the synthesis and the living cationic polymerization of 1,2: 3,4-di-*O*-isopropylidene-6-*O*-(2-vinyloxy ethyl) D-galactopyranose **27** (*46*) and oligonucleotide binding via reducing end of sugar in deprotected polymers were examined (*47*). Yamada et al. synthesized the vinyl ether having an isopropylidene protected glucose residue **28** and the emulsifying property of an ampiphilic block copolymer by the sequential living cationic polymerization was examined (*48*). Labeau et al. reported diisopropylidene glucofranoside vinyl ether **29** and amphiphilic diblock copolymers with styrene were prepared (*49*).

Figure 7 shows the different vinyl sugars having alkenyl group. Nishimura et al. have developed the efficient synthetic strategies for the preparation of polymerizable sugar monomers which have an n-pentenyl group at the reducing end 30, 31, 32 (*50, 51*) by the reaction of an oxazoline derivative with 4-penten-1-ol. Otey et al. prepared an allyl glucoside 33 by reacting glucose and allyl alcohol by a catalytic amount of H_2SO_4 (*52*). Blinkvsky and Dordick reported that β-galactosidase catalyzed the selective glycosidation of galactose in the 1-position with propargyl and allyl alcohol, and allyl β-D-galactoside 34 was obtained without the need for blocking and deblocking reactions (*53*). Grande et al. investigated the applicability of cyanoxyl-mediated free radical polymerization of the α-anomoer of pentenyl 2-acetamido-2-deoxy β-D-glucoside 30, n-undecenyl sugar 35 and their sulfated monomers to develop the biomimetic polymer against glucoaminoglycans (*25, 54*).

The vinyl sugars mentioned above were acrylate, acrylamide, styryl, vinylether and alkenyl sugars, and their polymer backbones after polymerization are not biodegradable.

PVA with Sugar Pendants

Syntheses of vinyl sugar ester

In 1977, Tokiwa and Suzuki found out that aliphatic polyesters were hydrolyzed by some lipases, excreted from microorganisms and esterase from

388

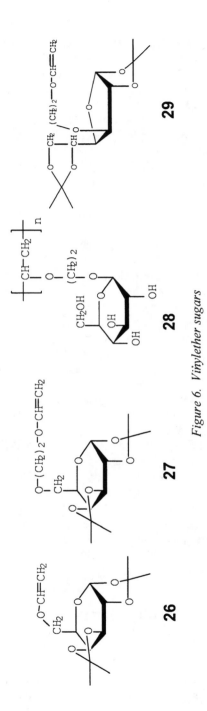

Figure 6. Vinylether sugars

389

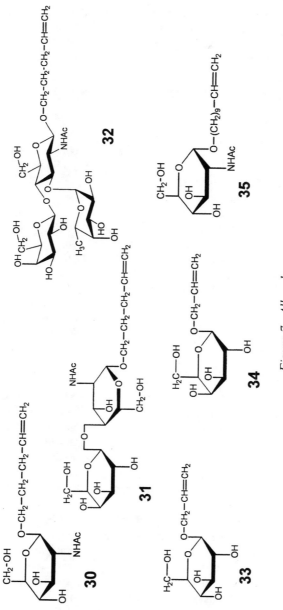

Figure 7. Alkenyl sugars

hog pancreas (55). Afterward, copolyamide-esters and copolyesters containing aromatic and aliphatic ester blocks were developed as biodegradable plastics (56, 57). To date, some biodegradable aliphatic polyesters are now commercially available. Furthermore, a lipase catalyzed polyester was synthesized by Okumura et al. in 1984 (58). Now the repetitive production and chemical recycling of polyesters using lipases is proposed as one of the effective methods for establishing green chemistry (59,60).

Concerning sugar containing polyesters (Figure 8), Patil et al. investigated enzymatic polymerization on sucrose with adipic acid derivative, bis(2,2,2-trifluoroethyladipate), to yield a polyester containing sugar adipate **36** (*61*), and reported biodegradation of the polymer. Uyama and Kobayashi carried out the enzymatic polymerization of adipic acid derivative, divinyl adipate, with a glycol using lipase to give the polyester containing glycol adipate **37** (*62*). Further, Park et al. reported the enzymatic synthesis of sucrose 6,6'-divinyl adipate and then the diester was employed as monomers in polycondensation with various diols to yield linear polyesters **38** (*63*).

This enzymatic polymerization proceeds with equimolar or excess molar of sugar against dicarboxylate derivatives as shown in Figure 9. Wang et al. reported that enolesters such as vinylacetate and isopropenyl acetate are effective for transesterification reaction (*64*). Enol group is good leaving group that changes into aldehyde or ketone but irreversible after elimination. The reaction rate of vinylester is 20-100 times higher than ethylester. We are moving ahead with a strategy to produce vinyl sugar using divinyl adipate by enzyme and further polymerized by radical initiator (*65*). Noting the polymerizability of vinyl ester, and the good leaving group for the transesterfication reaction, we designed a polymerizable sugar ester using sugar and divinyl dicarboxylate such as divinyl adipate (Figure 10).

Vinyl sugar ester might be obtained when excess molar of divinyl adipate against sugar was reacted. There is little known about PVA with sugar pendants, although the polymer backbone offers advantages in the biodegradability and biocompatibility. One of the reasons of few reports on this polymer is probably because of the difficulty in preparing the vinyl alcohol sugar ester monomers compared with acrylate sugar and others. This method using divinyl carboxylate and sugar could be useful in the preparation of PVA type synthetic polymers with sugar pendants.

Enzymatic modification of sugars offers a highly efficient process compared to conventional chemical synthesis using the blocking/deblocking process. But generally sugar is poorly soluble in organic solvents. One approach to overcome the disadvantages on the enzymatic synthesis of sugar ester is by modifying the sugars to increase the solubility in organic solvents. Therisod reported that monosaccharides with blocked C-6 hydroxyl groups were transesterified in acetone, tetrahydrofuran and methylene chloride by lipases (*66*). Martin also reported that 6-position of methyl glycoside was substituted with vinylacrylate by lipases (67). Ikeda reported that glucose complex with phenylboronic acid was increased in solubility against organic solvents and was acylated with vinylacrylate enzymatically (68). Another approach on the enzymatic esterification of sugar is using hydrophilic organic solvents which have good solubility against sugar. Dimethylformamide (DMF) has a good solubility against sugars compared with other organic solvents. There is little information, however, on its synthesis in DMF which is superior in solubility against sugar to other organic solvents. Only *Bacillus subtilis* protease (subtilisin) has been found to be active in DMF as well as in pyridine (69).

Enzymatic synthesis of vinyl sugar ester

Several commercially available lipases and proteases were screened for the synthesis of a sugar ester using glucose and activated fatty acid ester, 2,2,2-trichloro-ethylbutyrate in DMF (70). Although lipase type II from porcine pancreas and *Alcaligenes* lipase showed transesterification activity, most of the lipases used were inactive in micro aqueous DMF. However, some proteases gave high conversion activity. Alkaline proteases showed high glucose conversion activity compared with neutral proteases. Alkaline protease from *Streptomyces* sp. had the highest activity in DMF.

Moreover, transesterification of glucose with an excess molar of divinyl adipate by the protease in DMF, a good yield of 6-*O*-vinyladipoyl glucose was obtained (Figure 11). The reaction solvent is an important factor on the enzymatic synthesis. DMSO has higher solubility against sugar but it is also a denaturing solvent of enzyme protein .Wescott and Klibanov found that substrate specificity, enantioselectivity, prochiral selectivity, regioselectivity and chemoselectivity of enzymes were changed dramatically by the nature of the solvents (71).

Recently, Almarsson and Klibanov reported that the transesterification reactions of *N*-acetyl-L-phenylalanine-*O*-ethyl ester with 1-propanol catalyzed by protease from *Bacillus licheniformis* can be increased more than 100-fold by the addition of denaturing organic solvents such as DMSO and enhancement of conformational flexibility has been hypothesized (72). We studied the effect of DMSO on the enzymatic synthesis and the protease-catalyzed transesterifications between hexoses and divinyl adipate were examined (73). Esterification of different hexoses (D-glucose, D-mannose, D-galactose and α-methyl D-galactoside) with divinyl adipate in DMF was catalyzed by *Streptomyces* sp. alkaline protease to give corresponding 6-*O*-vinyl adipoyl sugars. When DMSO was added to the solvent, galactose was selectively esterified at only the C-2 position (Figure 12).

392

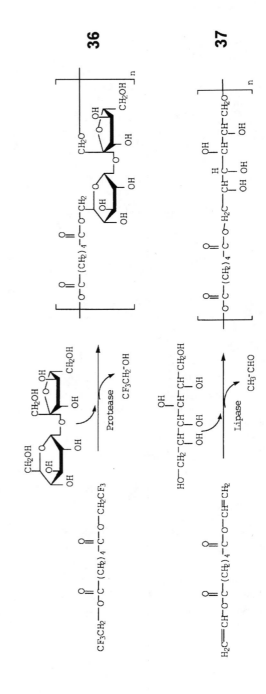

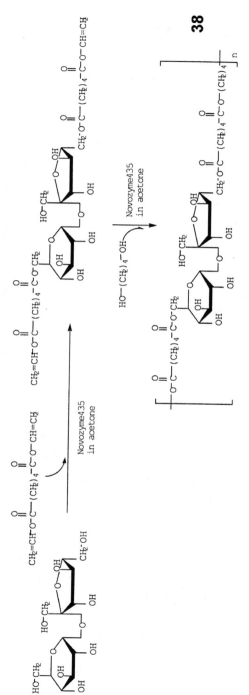

Figure 8. Enzymatic synthesis of sugar containing polyesters using adipic acid esters

394

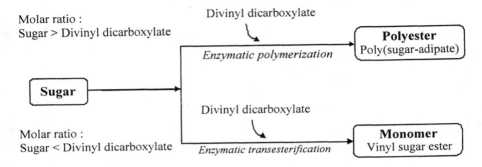

Figure 9. Enzymatic transesterification reaction of sugar with divinyl dicarboxylate

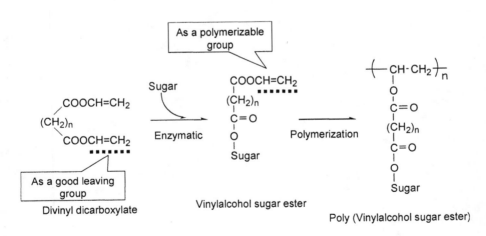

Figure 10. Strategy for PVA type sugar based polymer

395

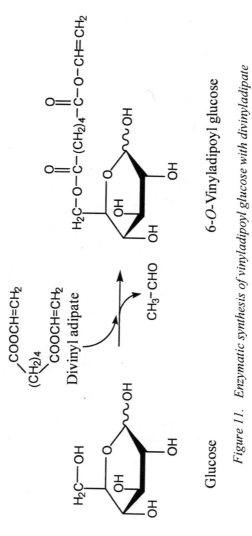

Glucose

6-*O*-Vinyladipoyl glucose

Figure 11. Enzymatic synthesis of vinyladipoyl glucose with divinyladipate

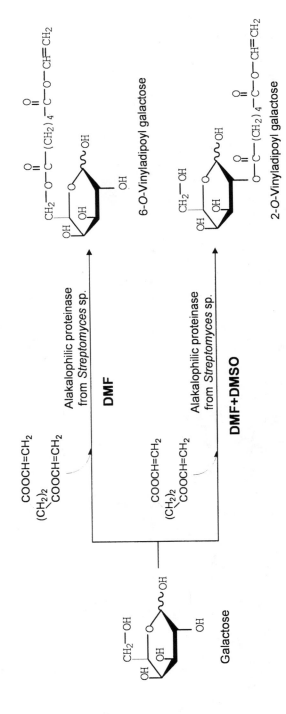

Figure 12. Regio-selective transesterification of galactose with divinyl adipate

Generally, primary hydroxyl group such as 6-position of hexopyranose is esterified easily by enzyme. It was suggested that substrate specificity of alkaline protease from *Streptomyces* sp. was changed in the presence of DMSO and secondary hydroxyl group at 2-position of galactose was preferentially modified.

Recently, Wu et al. reported the highly anomer-selective 6,6'-acylation from anomeric mixtures of several mono- and disaccharide with divinyl dicarboxylates ranging from 4 to 10 carbon atoms catalyzed by protease and α-anomeric vinyl sugars were produced (74). The alkaline protease recognizes the α-anomer preferentially in pyridine which the slow mutarotation α- and β-anomers of sugar was observed, and it means higher regio-selectivity of enzyme catalyst (Figure 13).

The water content is also an important factor on the enzymatic synthesis in hydrophilic organic solvents (75). The enzymatic activity in organic solvent depends on water content. Small amount of water is required to reach the maximum activity in hydrophobic solvents. However, hydrophilic organic solvents remove the essential water bound in the protein and inactivate enzyme activity (Figure 14 A) (76). Furthermore, optimum water content of the reaction in hydrophilic organic solvent is known to be higher than that in hydrophobic organic solvent. Kise et al. reported that in esterification of amino acids by α-chymotrypsin in alcohols, optimum water content is about 2% (77).

The transesterifications of divinyl adipate with glucose in DMF catalyzed by alkaline protease from *Bacillus subtilis* at various water contents were examined. The enzymatic reaction by the *Bacillus* protease was carried out in the presence of more than 2% water and maximum reaction rate was observed at a water content of 20%. It seems that the protease activity in DMF can be recovered by the addition of water (Figure 14 B).

Radical polymerization of vinyl sugar ester

Polymerization of 6-*O*-vinyladipoyl D-glucose was carried out in water and/or DMF by two types of radical initiators, azo- and redox initiators, which generate hydroxyl radicals via Fenton reaction (Figure 15). By polymerization using azo-initiators in DMF, linear polymers from vinyl compounds containing reducing sugar moieties were obtained. On the other hand, polymerization proceeded rapidly and a cross-linking polymer was obtained when Fenton reagent in water was used (78).

Redox catalyzed polymerizations using ferrous ion and peroxidative compound have been reported to be useful for grafting reaction of natural fiber such as chitosan (79). Ascorbic acid is also used instead of the ferrous ion (80). Ascorbic acid is reported to produce activated oxygen under mild condition, and recently it is used for linear polymer synthesis of sugar alcohol derivatives containing vinyl group (81). These catalysts, which promote the reaction under

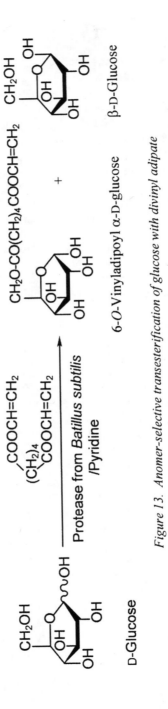

Figure 13. Anomer-selective transesterification of glucose with divinyl adipate

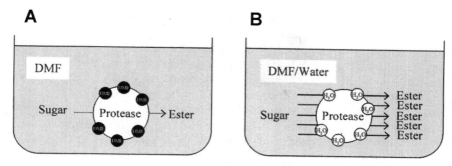

Figure 14. Water effect on the transesterification of sugar with divinyl adipate in hydrophilic organic solvent

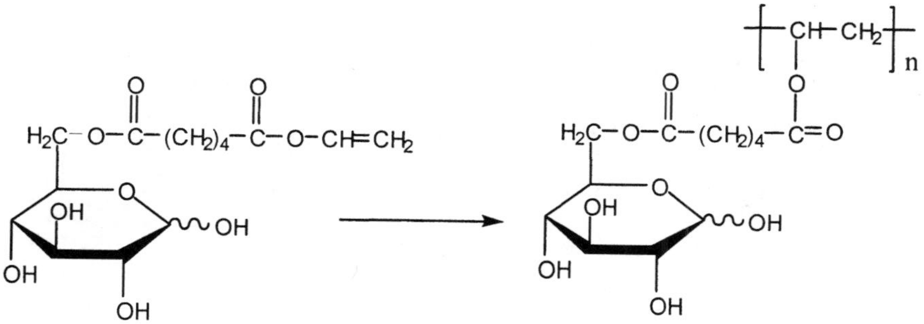

Figure 15. Polymerization of vinyladipoyl glucose

room temperature, atmospheric pressure, neutral pH and water as a solvent, have attracted considerable interest as an environmentally friendly method.

In the case of 6-O-vinyladipoyl D-glucose, polymerization occurred even under air atmosphere similar to under vacuum condition by ascorbic acid and hydrogen peroxide, but the produced polymer had lower molecular weight (82). Kashimura et al. studied the autoxidation of reducing sugars (83, 84). It is known that enediol derivatives react with hydrogen peroxide or oxygen molecules and produce hydroxyl radicals or superoxide, and further that the enediol moiety scavenges superoxide (85). The reducing sugar moiety of 6-O-vinyladipoyl D-glucose has an enediol structure. In the reaction condition, the polymerizable reducing sugar ester is also expected to react with superoxide, hydrogen peroxide and oxygen molecules, and scavenge and/or produce activated oxygen species. Hemiacetal hydroxyl group of reducing sugar ester seems to be involved in the complexed polymerization reaction under air atmosphere.

Along with rapid development of controlled radical polymerization, recently, synthesis of well-defined polymer with sugar pendant has been reported intensively. Albertin et al. reported the controlled radical polymerization of vinyl sugar ester using reversible addition-fragmentation chain transfer agent in protic media to get a polymer having a narrow-polydispersity (*86,87*). They investigated the free radical polymerization of 6-*O*-vinyladipoyl D-glucose in water and alcohol solutions, and thorough oxygen removal from the reaction mixture proved to be essential for the polymerization. They also suggested that it might be due to the presence of a reducing end in the sugar moiety. In the case of polymerization for vinyl sugars having a reducing sugar end, participation of hemiacetal hydroxyl group must be paid attention.

Biodegradation of Poly(vinylalcohol sugar ester)

Biodegradation of a PVA having sugar pendants was confirmed by a biochemical oxygen demand (BOD) tester using the oxygen consumption method (*88*). PVA is known as a biodegradable polymer by PVA assimilating microbes which are not widely distributed (*89*). The poly(vinylalcohol sugar ester) was degraded depending on the molecular weights. Polymer having a molecular weight (Mn) of 3600 was degraded (70-80 %) within 28 days in soil or in activated sludge without addition of PVA assimilating microbes.

Miura et al. synthesized PVA type glycoconjugate polymers by lipase-catalyzed transesterification of sugar alcohols such as maltitol and lactitol with divinyl dicarboxylates and subsequent radical polymerization to get PVA with sugar alcohol pendants **39** (Figure 16) (*90*). They found out that the biodegradabilities of these polymers were modest, but higher than that of PVA. Takasu et al. synthesized glucosamine or chitobiose-substituted PVA **40** by a

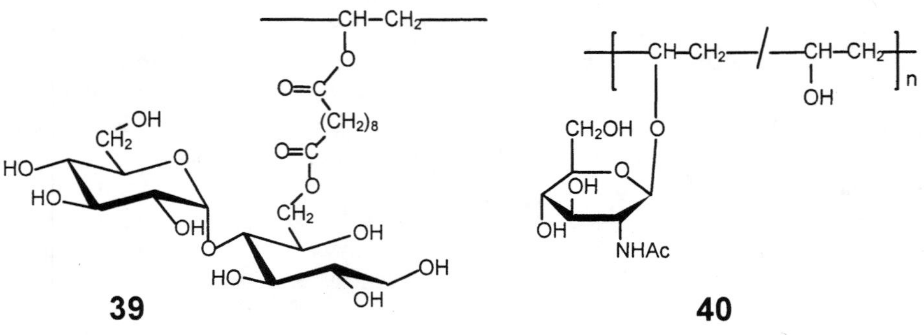

Figure 16. Biodegradative PVA sugar

chemical glycosidation of hydroxyl groups in PVA with sugar oxazolne and investigated their biodegradation (91). They demonstrated an acceleration of biodegradation of PVA by a partial glycosidation. Recognition of pendant sugar residues bonded to PVA was supposed to play an important role in the degradation mechanism.

PVA is the only C-C backbone polymer that is biodegradable under both aerobic and anaerobic conditions. However, the degradation rate under natural environmental conditions, if in the absence of PVA degrading microorganisms is too slow. The PVA with sugar pendants might be a new approach to improve the biodegradability of PVA.

Physiological application of poly(vinylalcohol sugar ester)

The site in sugar moiety which modified with polymerizable group is important to the characteristics of obtained polymers because it is known that oxygen radical species are generated by autoxidation of reducing sugars under physiological conditions (83). The reactions are important in biological systems such as inactivation of viruses and cleavage of DNA (84). Enolization of acyclic forms of reducing sugars may be a rate determining step in the reduction. Extensive investigations have been conducted on superoxide generation from monosaccharides. Enediol structure of reducing sugar moiety in poly(vinylalcohol sugar ester) also must be related to the reaction with oxygen species (Figure 17). Generation of superoxide from a polymer containing reducing sugar branches, poly(6-O-vinyladipoyl-D-glucose), was observed by the reduction of nitroblue tetrazolium, and the activity tended to increase with increasing molecular weight of the polymer (92).

Co-polymers of vinylalcohol sugar ester were shown in Figure 18. Wu et al. described the nano-capsule preparation using vinyl raffinose which is potential targeting drug-controlled delivery systems **41** (93). Further the 6-O-vinyladipoyl D-glucose was polymerized with vinyl fatty acid esters to produce amphiphilic polymers having branches of sugar and fatty acids **42** (94), and their surface tension lowering effect and biodegradability were observed. Thermosensitive polymers **43** (95) was synthesized using 6-O-vinyladipoyl D-glucose and N-isopropyl acrylamide, and the lower critical-solution temperature of copolymers measured with a light scattering photometer and a differential scanning calorimeter increased with increasing 6-O-vinyladipoyl D-glucose segment composition.

Miura et al. examined the enzymatic synthesis of 6-O-vinylsebacoyl trehalose and reported that the obtained Galacto-type trehalose polymer **44** (81) showed inhibition activity against Shiga toxin-1 based on a stereochemical structure similar to that of globosyl Gb2 disaccharide as shown in Figure 19.

402

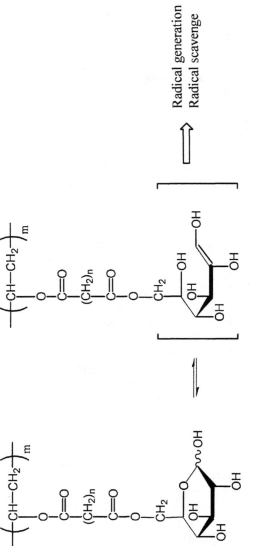

Figure 17. PVA type polyvinyl sugar having reactive reducing group

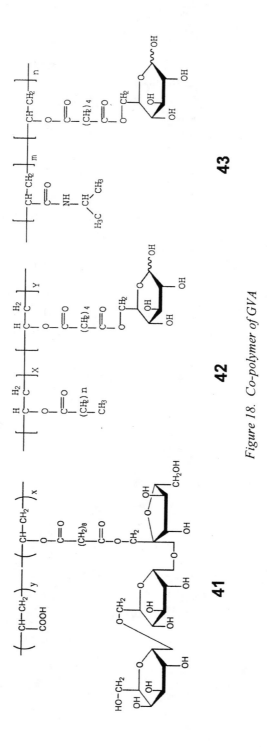

Figure 18. Co-polymer of GVA

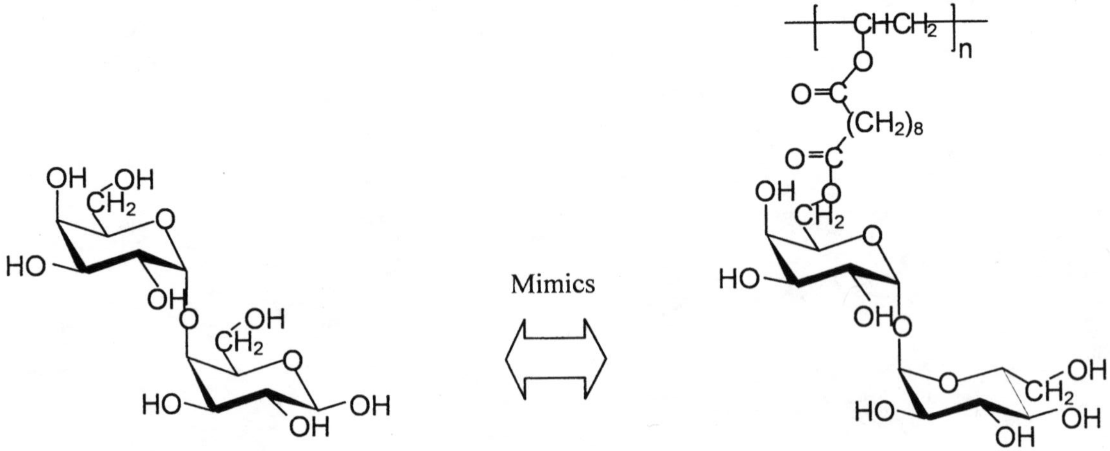

404

Galacto-type trehalose polymer

44

Mimics

Globobioside (Gb2)

Figure 19. Bio-functional PVA sugar polymer

Conclusion

Various synthetic methods for vinyl sugars have been developed. The chemical preparation of most vinyl sugars using monosaccharide need for protected sugars such as 1,2:4,6-diisopropylidene glucofuranoside, 1,2:3,4-diisopropylidene galactoside and sugar acetate derivatives. Further, for modification at hemiacetal hydroxyl group, various reactions were developed such as glycosidation reaction, reductive amination and Grignard reaction etc.

Biological safety, including biocompatibility, biodegradability, and bioabsorbability and non-cytotoxity is important in developing generally useful materials. Among the various polymers with sugar pendants described above, the PVA backbone is known to have the advantage of this biological safety. The PVA sugar polymers are also expected to be used as biomedical materials such as polymeric drugs, drug delivery systems, cell culture substrates, artificial tissues, etc.

On the other hand, highly selective enzymatic catalysis has been used to modify sugars containing multiple hydroxyl groups in microaqueous media. Enzymes are non-harmful catalysts which catalyze under mild conditions. Recent advances in enzymatic catalysts would be useful tools not only for polyester synthesis but also for vinyl monomer synthesis (96).

We have investigated the enzyme-catalyzed modification of sugar with divinyl dicarboxylate. The sugar branched-polymers consist of three parts, i.e. main chain, spacer arm and sugar moiety. The sugar density, biodegradability and hydrophobicity can be changed by modification of the main chain. By modification of these three parts, we could design various types of polymers with only two step reactions (enzymatic synthesis of sugar ester monomer and its chemical polymerization) as shown in Table 1. Further studies need to be done to reduce the production cost of vinylalcohol sugar esters and to find out the various applications of these vinyl sugars.

Acknowledgments

Our experiments described in this article were carried out at the National Institute of Industrial Science and Technology (AIST). Thanks are due to our colleagues: Dr. Tsuchii, Dr. Y. Hiraguri, Dr. T. Raku, Dr. F. Hong from AIST, Dr. C. Panpaka from Mahidol University, Thailand Dr. Vaidya from National Chemical Laboratory, India, Dr. M. Takagi from San Paulo University, Brazil, Dr. S. Shibatani from Toyobo Co. Ltd. and T. Yokochi from Mie University, Japan. The authors would like also to express their gratitude to Mr. T. Totani of Konan Chemical Co., Ltd., for his guidance and encouragement. The authors are grateful to Prof. Dr. N. Kashimura of Mie University, Japan for helpful discussion and to Ms. B. Calabia for helping us in the revision of the manuscript.

Table 1. **Vinyl sugars synthesized using various sugars and divinyl
dicarboxylate by enzymatic catalyst**

	Vinyl sugars	Ref
Mono-saccharide	6-*O*-Vinyladipoyl D-glucose	*65*
	6-*O*-Vinylsebacoyl D-glucose	*97*
	6-*O*-Vinyladipoyl D-mannose	*73*
	6-*O*-Vinyladipoyl D-galactose	*73*
	2-*O*-Vinyladipoyl D-galactose	*73*
	6-*O*-Vinyladipoyl D-allose	*98*
	Methyl 6-*O*-vinyladipoyl D-glucoside	*73*
	Methyl 6-*O*-vinyladipoyl D-mannoside	*73*
	Methyl 6-*O*-vinyladipoyl D-galactoside	*73*
	2-*O*-Vinyladipoyl D-fucose	*99*
	2-*O*-Vinyladipoyl L-fucose	*99*
	4-*O*-Vinyladipoyl L-rhamnose	*99*
	5-*O*-Vinyladipoyl D-arabinose	*100*
	5-*O*-Vinyladipoyl L-arabinose	*100*
Di-saccharide	6-*O*-Vinylsuccinyl maltose	*102*
	6-*O*-Vinyladipoyl maltose	*101*
	6-*O*-Vinylsebacoyl maltose	*102*
	6-*O*-Vinyladipoyl trehalose	*101*
	6-*O*-Vinylsebacoyl trehalose	*102*
	6,6'-*O*-di-Vinyladipoyl sucrose	*63*
	6-*O*-Vinylsebacoyl galactosyl-(1-1)-glucoside	*81*
	6'-*O*-Vinylsuccinyl lactose	*102*
	6'-*O*-Vinyladipoyl lactose	*102*
	6'-*O*-Vinylsebacoyl lactose	*102*
Tri-saccharide	1-*O*-Vinyldecanedioyl raffinose	*93*
Sugar alcohol	6-*O*-Vinyladipoyl maltitol	*103*
	6-*O*-Vinylsebacoyl maltitol	*103*
	6-*O*-Vinyladipoyl lactitol	*103*
	6-*O*-Vinylsebacoyl lactitol	*103*

References

1. Carbohydrates as Organic Raw Materials; Editor Lichtenthaler, F. W.; VCH, Weinheim, **1991**.
2. Carbohydrates as Organic Raw Materials II; Editor Descotes, G.; John Wiley & Son Ltd., **1993**.
3. Carbohydrates as Organic Raw Materials III; Editor Bekkum, H.; Roper, H.; Voragen, F.; VCH, Weinheim, **1996**.
4. Miyata, T.; Nakamae, K. *Trend Polym. Sci.* **1997**, *5*, 198-206.
5. Dordick, J. S.; Linhardt, R. J.; Rethwisch, D. G. *CHEMTECH* **1994**, 33-39.
6. Neoglycoconjugates Part A: Synthesis: Methods in Enzymamology; **1994**, vol. 242, pp3-326.
7. Tsutsumiuchi, K.; Aoi, K.; Okada, M. *Macromol. Rapid Commun.* **1995**, *16*, 749-755.
8. Manning, D. D.; Hu, X.; Beck, P.; Kiessling, L. L. *J. Am. Chem. Soc.* **1997**, *119*, 3161-3162.
9. Charych, D.; Nagy, J. O.; Spevak, W.; Bednarski, M. D. *Science*, **1993**, *261*, 585-588.
10. Wulff, G.; Schmid, J.; Venhoff, T. *Macromol. Chem. Phys.* **1996**, *197*, 259-274.
11. Haworth, W. N.; Gregory, H.; Wiggins, L. F. *J. Chem. Soc.* **1946**, 489-491.
12. Garcia-Martin, M. G. *Polymer*, **2000**, *41*, 821-826.
13. Black, W. A. P.; Colquhoun, J. A.; Dewar, E. T. *Makromol. Chem.* **1968**, *117*, 210-214.
14. Kimura, S.; Imoto, M. *Makromol. Chem.* **1961**, *50*, 155-160.
15. Al-Bagoury, M.; Yaacoub, E. J. *J. Appl. Polym. Sci.* **2003**, *90*, 2091-2102.
16. Koch, U.; Yaacoub, E. J. *J. Polym. Sci. Part A: Polym. Chem.* **2003**, *41*, 788-803.
17. Ohno, K.; Izu, Y.; Ymamoto, S.; Miyamoto, T.; Fukuda, T. *Macromol. Chem. Phys.* **1999**, *200*, 1619-1625.
18. Ohno, K.;Tsuji, Y.; Fukuda, T. *J. Polym. Sci. Part A: Polym. Chem.* **1998**, *36*, 2473-2481.
19. Ye, W.; Wells, S.; Desimone, J. M. *J. Polym. Sci. A: Polym. Chem.* **2001**, *39*, 3841-3849.
20. Kitazawa, T.; Okumura, M.; Kinomura, K.; Sakakibara, T. *Chem. Lett.* **1990**, *19*, 1733-1736.
21. Lowe, A. B.; Sumerlin, B. S.; McCormick, C. L. *Polymer*, **2003**, *44*, 6761-6765.
22. Patil, D. R.; Dordick, J. S.; Rethwisch, D. G. *Macromolecules* **1991**, *24*, 3462-3463.
23. Sugimoto, H.; Nakanishi, E.; Miura, T.; Inomata, K.; Ogawa, A.; Hirata, S. *Seni Gakkaishi* (Japanese), **2006**, *62*, 130-134.

408

24. Klein, J.; Kunz, M.; Kowalczyk, J. *Makromol. Chem.* **1990**, *191*, 517-528.
25. Grande, D.; Baskaran, S.; Chaikof, E. L. *Macromolecules*, **2001**, *34*, 1640-1646.
26. Klein, J.; Begli, A. H. *Makromol. Chem.* **1989**, *190*, 2527-2534.
27. Lee, R. T.; Cascio, S.; Lee, Y. C. *Anal. Biochem.* **1979**, *95*, 260-269.
28. Bady, B.; Boullanger, P.; Domard, A.; Cros, P.; Delair, T.; Pichot, C. *Macromol. Chem. Phys.* **1996**, *197*, 3711-3728.
29. Lees, W. J.; Spaltenstein, A.; Kingery-Wood, J. E.; Whitesides, G. M. *J. Med. Chem.* **1994**, *37*, 3419-3433.
30. Iwakura, Y.; Imai, Y.; Yagi, K. *J. Polym. Sci. Part A-1*, **1968**, *6*, 1625-1632.
31. Carpino, L. A.; Ringdorf, H.; Ritter, H. *Makromol. Chem.* **1976**, *177*, 1631-1635.
32. Yamada, K.; Fujita, E.; Nishimura, S. *Carbohydr. Res.* **1998**, *305*, 443-461.
33. Hirata, Y.; Nakagawa, T. *Bull. Chem. Soc. Jpn.* **2000**, *73*, 1905-1912.
34. Chung, D. C.; Kostelnik, R. J.; Elias, H. G. *Makromol. Chem.* **1977**, *178*, 691-700.
35. Ohno, K.; Tsuji, Y.; Miyamoto, T.; Fukuda, T. Goto, M.; Kobayashi, K.; Akaike, T. *Macromolecules*, **1998**, *31*, 1064-1069.
36. Kobayashi, K.; Sumitomo, H.; Ina, Y. *Polym. J.* **1985**, *17*, 567-575.
37. Kobayashi, K.; Tsuchida, A.; Usui, T.; Akaike, T. *Macromolecules* **1997**, *30*, 2016-2020.
38. Sato, K.; Kodama, D.; Anzai, J. *Anal. Sci.* **2005**, *21*, 1375-1378.
39. Koyama, Y.; Yoshida, A.; Kurita, K. *Polym. J.* **1986**, *18*, 479-485.
40. Wulff. G.; Zhu, L.; Schmidt, H. *Macromolecules* **1997**, *30*, 4533-4539.
41. Chen, Y.; Wulff, G. *Macromol. Chem. Phys.* **2001**, *202*, 3426-3431.
42. Loykulnant, S.; Hirao, A. *Macromolecules* **2000**, *33*, 4757-4764.
43. Park, K. H.; Takei, R.; Goto, M.; Maruyama, A.; Kobayashi, A.; Kobayashi, K.; Akaike, T. *J. Biochem.* **1997**, *121*, 997-1001.
44. Narumi, A.; Matsuda, T.; Kaga, H.; Satoh, T.; Kakuchi, T. *Polym. J.* **2001**, *33*, 939-945.
45. Whistler, R. L.; Panzer, H. P.; Goatley, J. L. *J. Org. Chem.* **1962**, *27*, 2961-2962.
46. D'Agosto, F.; Charreyre, M. Delolme, F.; Dessalces, G.; Cramail, H.; Deffieux, A.; T.; Pichot, C. *Macromolecules* **2002**, *35*, 7911-7918.
47. D'Agosto, F.; Charreyre, M. T.; Pichot, C.; Mandrand, B. *Macromol. Chem. Phys.* **2002**, *203*, 146-154.
48. Yamada, K.; Minoda, M.; Fukuda, T.; Miyamoto, T. *J. Polym. Sci. A: Polym. Chem.* **2001**, *39*, 459-467.
49. Labeau, M. P.; Cramail, H.; Deffieux, A. *Macromol. Chem. Phys.* **1998**, *199*, 335-342.
50. Nishimura, S.; Matsuoka, K.; Furuike, T.; Ishii, S.; Kurita, K.; Nishimura, K. M. *Macromolecules* **1991**, *24*, 4236-4241.

51. Nishimura, S.; Matsuoka, K.; Furuike, T.; Nishi, N.; Tokura, S.; Nagami, K.; Murayama, S.; Kurita, K. *Macromolecules* **1994**, *27*, 157-163.
52. Otey, F. H.; Westhoff, R. P.; Mehltretter, C. L. *Ind. Eng. Chem. Prod. Res. Develop.* **1972**, *11*, 70-73.
53. Blinkovsky, A.; Dordick, J. *Tetrahedron: Asym.* **1993**, *4*, 1221-1228.
54. Grande, D.; Baskaran, S.; Baskaran, C.; Gnanou, Y.; Chaikof, E. L. *Macromolecules*, **2000**, *33*, 1123-1125.
55. Tokiwa, Y.; Suzuki, T. *Nature*, **1977**, *270*, 76-78.
56. Tokiwa, Y.; Suzuki, T.; Ando, T. *J. Appl. Polym. Sci.* **1979**, *24*, 1710-1711.
57. Tokiwa, Y.; Suzuki, T. *J. Appl. Polym. Sci.* **1981**, *26*, 441-448.
58. Okumura, S.; Iwai, M.; Tominaga, Y. *Agric. Biol. Chem.* **1984**, *48*, 2805-2808.
59. Kobayashi, S.; Uyama, H.; Takamoto, T. *Biomacromolecules* **2000**, *1*, 3-5.
60. Matsumura, S. *Macromol. Biosci.* **2002**, *2*, 105-126.
61. Patil, D. R.; Rethwisch, D. G.; Dordick, J. S. *Biotechnol. Bioeng.* **1991**, *37*, 639-646.
62. Uyama, H.; Kobayashi, S. *Chem. Lett.* **1994**, *23*, 1687-1690.
63. Park, O. J.; Kim, D. Y.; Dordick, J. S. *Biotechnol. Bioeng.* **2000**, *70*, 208-216.
64. Wang, Y. F.; Lalonde, J. J.; Momongan, M.; Bergbreiter, D. E.; Wong, C. H. *J. Am. Chem. Soc.* **1988**, *110*, 7200-7205.
65. Kitagawa, M.; Tokiwa, Y. *J. Carbohydr. Chem.* **1998**, *17*, 893-899.
66. Therisod, M.; Klibanov, A. M. *J. Am. Chem. Soc.* **1987**, *109*, 3977-3981.
67. Martin, . D.; Ampofo, S. A.; Linhardt, R. J.; Dordick, J. S. *Macromolecules* **1992**, *25*, 7081-7085.
68. Ikeda, I.; Klibanov, A. M. *Biotechnol. Bioeng.* **1993**, *42*, 788-791.
69. Riva, S.; Chopineau, J.; Kieboom, A. P. G.; Klibanov, A. M. *J. Am. Chem. Soc.* **1988**, *110*, 584-589.
70. Shibatani, S.; Kitagawa, M.; Tokiwa, Y. *Biotechnol. Lett.*, **1997**, *19*, 511-514.
71. Wescott, C. R.; Klibanov, A. M. *Biochim. Biophys. Acta* **1994**, *1206*, 1-9.
72. Almarsson, O.; Klibanov, A. M. *Biotechnol. Bioeng.* **1996**, *49*, 87-92.
73. Kitagawa, M.; Fan, H.; Raku, T.; Shibatani, S.; Maekawa, Y.; Hiraguri, Y.; Kurane, R.; Tokiwa Y. *Biotechnol. Lett.* **1999**, *21*, 355-359.
74. Wu, Q.; Lu, D.; Cai, Y.; Xue, X.; Chen, Z.; Lin, X. *Biotechnol. Lett.* **2001**, *23*, 1981-1985.
75. Garman, L. A. S.; Dordick, J. S. *Biotechnol. Bioeng.* **1992**, 39, 392-397.
76. Kitagawa, M.; Raku, T.; Fan, H.; Shimakawa, H.; Tokiwa, Y. *Macromol.Biosci.* **2002**, *2*, 233-237.
77. Kise, H.; Shirato, H.; Noritomi, H. *Bull. Chem. Soc. Jpn.* **1987**, *60*, 3613-3618.
78. Kitagawa, K.; Takegami, S.; Tokiwa, Y. *Macromol. Rapid Commun.* **1998**, *19*, 155-158.

410

79. Lagos, A.; Reyes, . *J. Polym. Sci. A: Polym. Chem.* **1988**, *26*, 985-991.
80. Larpent, .; Tadros, T. F. *Colloid Polym. Sci.* **1991**, *269*, 1171-1183.
81. Miura, Y.; Wada, N.; Nishida, Y.; Mori, H.; Kobayashi, K. *J. Polym. Sci. A: Polym. Chem.* **2004**, *42*, 4598-4606.
82. Kitagawa, M.; Tokiwa, Y. *Carbohydr. Poly.* **2006**, *64*, 218-223.
83. Kashimura, N.; Morita, J.; Komano, T. *Carbohydr. Res.* **1979**, *70*, C3-C7.
84. Morita, J.; Kashimura, N.; Komano, T. *Agric. Biol. Chem.* **1980**, *44*, 2971-2978.
85. Isbell, H. S. *Carbohydr. Res.* **1976**, *49*, C1- C4.
86. Albertin, L.; Kohlert, C.; Stenzel, M.; Foster, L. J. R.; Davis, T. P. *Biomacromol.* **2004**, *5*, 254-260.
87. Albertin, L.; Stenzel, M.; Barner-Kowollik,; Foster, L. J. R.; Davis, T. P. *Polymer* **2005**, *46*, 2831-2835.
88. Tokiwa, Y.; Fan, H.; Hiraguri, Y.; Kurane, R.; Kitagawa, M.; Shibatani, S.; Maekawa, Y. *Macromolecules* **2000**, *33*, 1636-1639.
89. Suzuki, T. *Agric. Biol. Chem.* **1976**, *40*, 497-504.
90. Miura, Y.; Ikeda, T.; Wada, N.; Sato, H.; Kobayashi, K. *Green Chem.* **2003**, *5*, 610-614.
91. Takasu, A.; Takada, M.; Itou, H.; Hirabayashi, T.; Kinoshita, T. *Biomacromol.* **2004**, *5*, 1029-1037.
92. Kitagawa, M.; Tokiwa, Y. *Chem. Lett.* **1998**, 281-282.
93. Wu, Q.; Chen, Z. C.; Lu, D. S.; Lin, X. F. *Macromol. Biosci.* **2006**, *6*, 78-83.
94. Tokiwa, Y.; Raku, T.; Kitagawa, M.; Kurane, R. *Clean Prod. Proc.* **2000**, *2*, 108-111.
95. Raku, T.; Tokiwa, Y. *J. Appl. Polym. Sci.* **2001**, *80*, 384-387.
96. Gross, R. A.; Kumar, A.; Kalra, B. *Chem. Rev.* **2001**, *101*, 2097-2124
97. Kitagawa, M.; Tokiwa, Y. *Biotechnol. Lett.* **1998**, *20*, 627-630.
98. Tokiwa, Y.; Fan, H.; Raku, T.; Kitagawa, M. *Polym. Prep.* **2000**, *41*, 1818.
99. Raku, T.; Tokiwa, Y. *Macromol. Biosci.* **2003**, *3*, 151-156.
100. Tokiwa, Y.; Kitagawa, M.; Fan, H.; Raku, T.; Hiraguri, Y.; Shibatani, S.; Kurane, R. *Biotechnol. Lett.* **1999**, *13*, 173-176.
101. Kitagawa, M.; Chalermisrachai, P.; Fan, H.; Tokiwa, Y. *Macromol. Symp.* **1999**, *144*, 247-256.
102. Wu, Q.; Lu, D.; Xiao, Y.; Yao, S.; Lin, X. *Chem. Lett.* **2004**, *33*, 94-95.
103. Miura, Y.; Ikeda, T.; Kobayashi, K. *Biomacromol.* **2003**, *4*, 410-415.

Silicone-Containing Materials

Chapter 26

Synthetic Peptides Derived from the Diatom *Cylindrotheca fusiformis*: Kinetics of Silica Formation and Morphological Characterization

Patrick W. Whitlock[1,4], Siddharth V. Patwardhan[2], Morley O. Stone[3], and Stephen J. Clarson[1,*]

[1]Department of Chemical and Materials Engineering, University of Cincinnati, Cincinnati, OH 45221–0012
[2]School of Biomedical and Natural Sciences, Nottingham Trent University, Clifton Lane, Nottingham NG11 8NS, United Kingdom
[3]Air Force Research Laboratory, Wright-Patterson Air Force Base, Dayton, OH 45433–7702
[4]Current address: Drexel University, College of Medicine, Philadelphia, PA 19129
*Corresponding author: Stephen.Clarson@UC.Edu

It is well known that a variety of organisms are able to produce intricately patterned species-specific siliceous materials and that they are able to do so with great fidelity. In order to understand the role(s) that biomolecules play in the formation of nano- and micro-structured biosilica, organic material associated with biosilica has been isolated, identified and studied. To date, the main focus has been a variety of investigations of R5 (in both its modified and unmodified forms) and its ability to form silica *in vitro*. The R5 peptide is a nineteen amino acid sequence which corresponds with silaffin-1A$_1$ of the diatom Cylindrotheca fusiformis. In order to study the role of biomolecules in (bio)silicification, we have chosen to systematically study the silica forming ability of three peptides derived from the diatom C. fusiformis – namely, R1, R2, and R5. The R1 peptide is a thirty three amino acid sequence which corresponds to silaffin-1B from C. fusiformis and the R2 peptide is a twenty two amino acid sequence which corresponds to silaffin-1A$_2$ from C. fusiformis. To our knowledge, this is the first study of silica formation involving the R1 and R2 peptides. In this study, we have compared and contrasted silica formation in the presence of R1 and R2 with that of R5 as a function of silica precursor concentration, pH, and temperature of the reaction medium. The key findings presented herein are discussed in the context of possible implications for *in vivo* biosilicification. We also hypothesize how these peptides and their individual kinetics and associated silica morphologies might produce species-specific nano- and micro-structured biosilica *in vivo*.

Introduction

Intra- and extra-cellular biosilica[1] formation has been observed in various living systems. Marine diatoms and sponges are excellent natural examples of nanostructured biosilica.[2-4] It is well known that these organisms are able to produce intricately patterned siliceous materials within their cell walls. Marine organisms perform this task with such great fidelity that these structures are used by biologists for taxonomy.[5-8] In the case of marine and freshwater diatoms, the ornate hierarchical structures of biosilica are intriguing not only to biologists but also to materials chemists. Furthermore, the controlled 'shaping' of biosilica spicules as observed in sponges is equally interesting. Biosilicas and their formation are of immense importance due to the ornate structures produced and the sophistication achieved *in vivo* unmatched by current, corresponding *in vitro* synthetic methods. It has been postulated that various chemical conditions (such as presence of biomolecules, ionic species and pH)[9-12] and physical parameters (e.g. diffusion and phase separation)[13, 14] play important roles in biosilicification. However, the exact mechanisms remain unclear. Several recent investigations have begun to elucidate the mechanism(s) of biosilica morphogenesis.[9, 10, 12]

In order to understand the roles that biomolecules play in the fabrication of nano- and micro-structured biosilica, organic material associated with biosilica was identified, isolated and studied. In addition to some earlier studies,[15-17] the following recent findings are of particular importance. Several proteins isolated from plants,[9] diatom species (silaffins)[12, 18, 19] and sponges (silicateins)[10, 20] have previously been shown to facilitate the *in vitro* polymerization of silica from a silicon catecholate complex,[9, 21, 22] tetramethoxysilane (TMOS)[12] and tetra-ethoxysilane (TEOS),[23] respectively. Bioextracts from plants *Equisetum talmateia* and *Equisetum arvense* have been isolated from biosilica.[9] These bioextracts, when present in condensing silicic acid systems, produced unusual crystalline silica at ambient temperature and *ca.* pH 7.[9, 21, 22] When biosilica from sponge spicules was selectively dissolved, entrapped silicatein proteins were recovered.[10] Silicatein proteins have been proposed to catalyse siloxane bond formation in silica, silicone and silsesquixane synthesis.[23] Similarly, various biomolecules were extracted from the diatom *Cylindrotheca fusiformis*.[12, 18, 19] Among the extracts from *C. fusiformis*, silaffin proteins and polyamines are of particular interest due to their ability to form silica *in vitro* using tetramethoxysilane as the precursor.[12, 24] All the aforementioned biomolecules isolated from their respective biological systems, based on their *in vitro* activity, were proposed to be responsible for biosilicification in their corresponding systems.

In order to further elucidate the mechanism of biosilica formation and the role of biomolecules in biosilicification, various model studies have been undertaken[25-35] and the roles of these organic additives hypothesized.[36] In particular, a nineteen amino acid peptide (R5 peptide) derived from the *sil1* gene of diatom *C. fusiformis* was investigated for its ability to synthesize silica *in vitro*.[12, 37-40] We, as well as others, are interested in the sil1p protein of *C. fusiformis* for obvious reasons. More specifically, a low cost, low impact biomimetic synthesis avoids harsh conditions at which normal silica synthesis takes place. More importantly, biomimetic silicification holds the promise that specific, desired morphologies may be replicated for use as components in future nano-applications. We have previously shown that the R5 peptide sequence from *C. fusiformis* could be used to produce novel silica nanostructures via manipulation of the physical reaction environment.[40] We were also able to incorporate the R5 sequence to produce a hybrid organic/inorganic ordered nanostructure of silica spheres in a polymer hologram created by two photon photopolymerization.[38] These investigations have reinforced the technological importance of bioinspired studies in developing novel applications as well as gaining an understanding of silica production *in vivo*.

It is noted that Kroger *et al.* have proposed that *p*R5 (a peptide of R5 repeat units) does not possess silica formation activity below pH 7 when compared with the activity of silaffin-1A (a mixture of -1A$_1$ and -1A$_2$) *in vitro*.[12] One possible explanation for this observation, which was questioned subsequently,[41] was the lack of any post-translational modifications in *p*R5. However, the observed ability of the R5 sequence to produce a consistent morphology[37] led us to investigate how novel manipulation of the reaction environment might allow further control over product formation and morphology. Thus in order to understand the role of R5 and other biomolecules in (bio)silicification, we have chosen to systematically study the silica precipitating activity of three relevant peptides under different conditions. The peptides of interest include R5, a nineteen amino acid peptide which corresponds to silaffin-1A$_1$ of the diatom *C. fusiformis*; R2 a twenty two amino acid peptide which corresponds to silaffin-

$1A_2$ and R1 a thirty three amino acid peptide which is analogous to silaffin-1B (see Table 1).[12] To our knowledge, this is the first study of silica formation involving the R1 and R2 peptides.

Table 1. Sequence and molecular weight of peptides used.

	Sequence	MW[a]
R5	SSKKSGSYSGSKGSKRRIL	2013
R2	SSKKSGSYSGYSTKKSGSRRIL	2364
R1	SSKKSGSYYSYGTKKSGSYSGYST KKSASRRIL	3601

[a] MW is in g mol⁻¹.

The knowledge gained from this work is important because it may provide additional information on the process of (bio)silification using three peptides derived from those found in the diatom *C. fusiformis*. Recent site–directed mutagenesis studies of the R5 peptide and its mutants have demonstrated that the C–terminus RRIL motif may be critical in the formation of active silica precipitating assemblies[42] although it was proposed that this C-terminus RRIL sequence may be proteolytically removed *in vivo*.[43] This recent study on the role of R5 mutants in silicification has prompted us to further investigate silica precipitating peptide sequences derived from *C. fusiformis*. In this study we report how R1, R2 and R5 peptides control silica formation *in vitro*. Specifically, the effect of solution pH ranging from 2 to 10 was studied. Our decision to study the effects of pH on *in vitro* silicification was largely a result of the current debate on the actual physiological pH in biological systems depositing biosilica.[11, 41] It is known that diatoms can produce biosilica in external environments ranging from acidic to basic (pH as high as 12).[44] Similarly, it is known that diatoms proliferate in aqueous medium over a wide range of temperatures[44] and thus we also present relevant data describing the effect of temperature on peptide mediated silica formation.

Experimental

Materials

Tetramethoxysilane (TMOS) and ammonium molybdate were obtained from Aldrich. All the peptides were custom synthesized by New England Peptide Inc. Sodium phosphate and citric acid required for the preparation of buffer, ammonium hydroxide, hydrochloric acid and HPLC grade water were all obtained from Fisher.

Synthetic Procedures and Characterisation

R5-mediated silica synthesis

Three polycationic peptide sequences derived from the *Sil1p* protein of *C. fusiformis*[45] were exposed to varying reaction parameters in an effort to observe the influence of these parameters on reaction kinetics and morphology. The peptides of interest include R5, a 19 amino acid peptide which corresponds with silaffin-1A$_1$; R2 a 22 amino acid peptide, which corresponds with silaffin-1A$_2$ and R1 a 33 amino acid peptide which is analogous to silaffin-1B (see Table 1).

Citrate – phosphate buffer was prepared freshly by mixing 500 mM sodium phosphate with 500 mM citric acid resulting in titration to appropriate pH.[46] A stock solution of 20 mg/ml R5 peptide solution was prepared in citrate – phosphate buffer of appropriate pH for each experiment. Immediately prior to each experiment, a fresh solution of 1M tetramethoxysilane (TMOS) was prepared via addition to 1 mM HCl. (The pre-hydrolysis method adopted for TMOS was the one described by Kroger, Sumper and co-workers[45]).

Each silica precipitation study was carried out in a microfuge tube at a constant temperature. Each reaction was permitted to occur for five minutes, washed three times using HPLC – grade water, and centrifuged at 14,000 rpm for five minutes in order to remove any free silicic acid. The amount of silica precipitated was then determined as described below (see '*Analysis of kinetics*').

In order to determine the effect of substrate (TMOS) concentration upon the relative reaction kinetics of R5 – precipitated silica and to examine product morphology, a constant 40 µL volume of R5 solution was added to five separate stirred quartz cuvettes containing a Teflon stir bar, an appropriate amount of buffer, and a specified amount of the hydrolyzed TMOS solution. The final TMOS concentrations thus studied were: 0.017, 0.033, 0.067, 0.133 and 0.267 M. Temperature was maintained at 5 °C unless otherwise specified.

In addition to varying the substrate (TMOS), we were also interested in the effects of temperature on particle formation kinetics and product morphology. The particle growth was studied using turbidity measurements as described below. It is noted that, in the absence of any peptide, only gelation, and no particle formation was observed, which was also reflected in the weak absorbance (Figure S1). A typical reaction mixture containing 40 µL of TMOS (0.067 M) pre-hydrolyzed in 1 mM HCl and 520 µL citrate – phosphate buffer at pH 7.55 was prepared in a stirred quartz cuvette. Immediately after addition of these reactants to the cuvette, 40 µL of the 20 mg/ml R5 (final concentration of

0.615 mM) peptide solution was added and the reaction was permitted to take place at the following desired temperatures: 5, 15, 25, 35 and 45°C. The sample at 45°C immediately became opaque making it unable to characterize. At a specified time, the reactions were terminated by pipetting the product into a 1.5 mL centrifuge tube and centrifugation was carried out at 5,000 g for 10 minutes. The supernatant was removed and 1.5 ml of deionized water was added. The material was gently pipetted to resuspend and then centrifuged again. This washing step was repeated for a total of three washes. The product was then resuspended and transferred to a new sample vial, diluted with 5ml of deionized water, and labeled for storage until analysis using electron microscopy.

R1- and R2-mediated silica synthesis

Here we sought to investigate the effect of other peptide sequences derived from the *Sil1p* protein on silica formation. 10 µl of TMOS (in 1 mM HCl) was added to a pre-mixed solution of 80 µl citrate – phosphate buffer containing 10 ul of 20 mg/ml of the peptide of interest. The final concentration of TMOS was maintained at 0.067 M. Stock solutions of both the R2 and the R1 peptides were prepared in citrate – phosphate buffer pH 7.55 to a final concentration of 20 mg/ml. The resulting concentrations used for each peptide are as follows: R1 (2.12 x10^{-7} moles), R2 (3.34 x10^{-7} moles) and R5 (3.69 x10^{-7} moles). Each reaction was permitted to take place for five minutes at 25°C. The entire contents of the tube were washed with 100 µl of HPLC – grade water using vigorous pipetting. This solution was then centrifuged at 14,000g for five minutes and washed with three 100 µL volumes of HPLC – grade water. The resulting precipitate was then reconstituted in 100 µl of fresh HPLC – grade water and carefully and completely redistributed in a quartz cuvette. The solutions were then scanned at 290 nm using UV/Vis spectroscopy as an indicator of pH-dependent silica formation at pH 2, 4, 6, 7, 8, and 10 for R1, R2 and R5 peptides.

Analysis of kinetics

Spectrophotometric measurement of apparent absorbance was carried out at wavelength of 290 nm for the duration of the reaction. After it had been determined to be the most reliable peak for monitoring the absorbance due to silica formation (unpublished results). Measurement of turbidity at this wavelength was performed in order to rapidly determine and compare the relative growth kinetics of the three peptides of interest. All reactions were permitted to occur at specified temperature and reactant concentrations in a stirred quartz cuvette using a PTFE stir bar. All reactions were conducted at ambient pressure. All the reactants were constantly stirred in order to avoid any aggregation that might affect the data. Absorbance was recorded and plotted versus time. The data resulted in sigmoidal curves and in order to compare the rates of silica formation and the amount of silica formed after a given time, the data was fitted using a sigmoidal function. Curve fitting was performed using the following sigmoidal curve equation $A = A_2 + (A_1-A_2)/(1+e^{[t-t0]/dt})$, where A is the absorbance at a given time, A_1 and A_2 are initial and final absorbance respectively, t_0 is point of inflection and dt is a constant. The mean R^2 of all fits was 0.922 indicating a good fit. Two parameters were used for the comparison between peptide mediated silica formation – growth rate (min^{-1}) and final absorbance (A_2). The growth rates were the slopes of tangents (i.e. 1st derivatives) at the inflection point. Statistical analysis was carried out on triplicates of selected samples. The estimated errors in the growth rate and A_2 were ~11% and ~6.5% respectively. These values are included in all plots below. In order to validate this method for particle growth study, data was obtained from a set of standard silica samples with varying silica particle concentration. The data revealed a correlation between silica concentration and absorbance at 290 nm wavelength (non-linear dependence; data not shown). It, thus, suggests that the analytical method used herein could be able to provide qualitative information, if not quantitative, from comparison between samples.

In the TMOS concentration dependent silicification study, an assay was performed to determine the amount of silica precipitated. In order to measure the amount of silica precipitated, 10 μl of 1M NaOH was added to the washed samples in clean microfuge tube and incubated at 55°C for 20 minutes. This incubation allows all precipitated silica to break down to monomeric orthosilicic acid. Silicic acid concentration was then determined via the β–molybdosilicate method by measuring the yellow molybdosilicate complex formed as described by Iler.[47] The reaction velocities were calculated to study the enzymatic effect of the peptides under consideration.

Electron microscopy

Field Emission Scanning Electron Microscopy (FE-SEM) was performed in order to study the product morphologies of the samples. The aforementioned silica samples, which were suspended in water, were centrifuged (r. p. m. = 13400) and washed with ethanol. They were then placed on SEM sample holders and dried. The dried silica samples were sputter coated with gold and were then analyzed under the electron microscope.

Results and Discussion

1. Silicification Mediated by R1, R2 and R5 Peptides

A comparative study of silica formation using three peptides – R1, R2 and R5 – was carried out. As presented in Figure 1, the maximum silica formation was observed in the presence of the R5 peptide. The turbidity analysis provides evidence that the R1 peptide is capable of the most rapid rate of silica precipitation, even when using a molar concentration of peptide less than that of R2 and R5. Furthermore, the R1 peptide was able to synthesize fairly distinct particulate silica (*ca.* 200 nm in size, see Figure 2a, b) when compared with the silica prepared using the R5 peptide (see below, Figure 7, 8). These particles were fused with each other and found to form network-like structures. The representative elemental mapping of silica formed using the R1 peptide is shown in Figure 2c. Clearly, the particles were made up of silicon and oxygen. Aluminum seen in the elemental analysis corresponds to the sample holder. It is belived that the R5 peptide is occluded in the silica structures. Indeed the silica digestion followed by SDS-PAGE analysis has shown that the peptide is tightly associated with silica particles.[40] In the case of silica formed with the R2 peptide, particles of sizes *ca.* 170 nm, although not distinct, were observed by SEM and are shown in Figure 2d. R2-mediated silica particles were found to be porous (see Figure 2d). The corresponding EDS spectrum of silica formation in the presence of R2 peptide is presented in Figure 2e.

Although extensive experimentation is required to explain the behaviour of these peptides in silica formation, in order to obtain some understanding, we have estimated peptide properties by performing molecular modelling.[48, 49] Several parameters were calculated and, in particular, we have focused on the peptide length, pI, number of cationically charged residues and predicted secondary conformation. All the peptides exhibited high pI (>10) suggesting that they will be positively charged under our experimental conditions and be able to flocculate silica particles. R1, being a considerably longer peptide (33 residues) could bridge several particles and hence may be able to show the fastest silica precipitation rates. When the silica precipitation rates in the presence of the R2 and R5 peptides are compared, R5 shows faster precipitation. Although the R2 and R5 peptides have a comparable number of residues (22 and 19 respectively), because R5 contains more basic residues (~31%), its charge density becomes higher than that of R2 and this enables the R5 peptide to precipitate silica particles more effectively.

2. Effect of pH on R1-, R2- and R5-mediated silicification

Thus far, all of the preceeding reactions have been carried out at *ca.* pH 7. In order to understand the effect of pH on the bevavior of the three peptides,

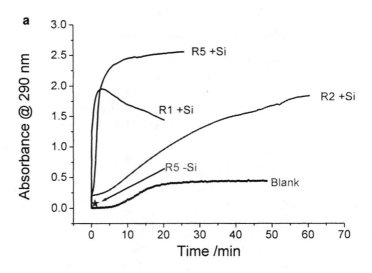

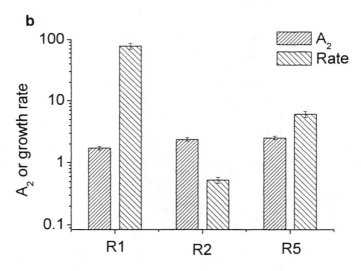

Figure 1. (a) Turbidity measurements of silica formation with and without the presence of R1, R2 and R5 peptides (+Si). For comparison, absorbance of R5 in the absence of silica (-Si) is also shown. (b) Plot of growth rate and A_2 as functions of peptides used. Error bars are from statistical analysis.

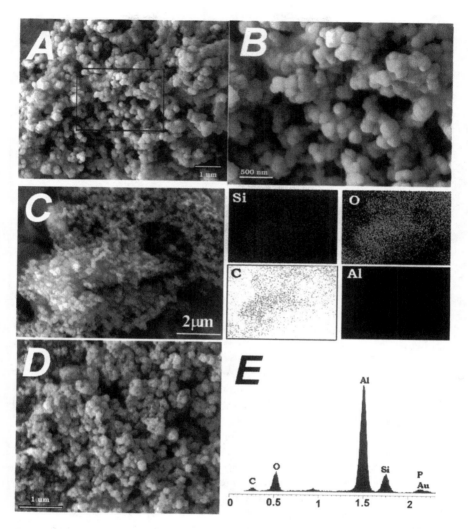

Figure 2. (a), (b) SEM micrographs of silica formed using R1. Highlighted area in (a) is presented at higher magnification in (b). (c) Elemental mapping of silica formed by using R1 peptide. (d) SEM micrographs of silica formed using R2 peptide. (e) Typical EDS of silica formed in the presence of R2 peptide. Scale bar = 1 μm (a), (d); 500 nm (b) and 2μm (c). (See page 2 of color insert.)

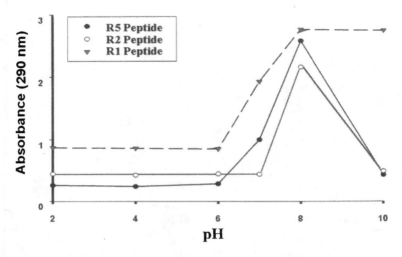

Figure 3. Turbidity measurements of silica formation using R1, R2 and R5 at varied pH.

silica formation was peformed at pH ranging from 2 to 10 in the presence of each of these peptides. Turbidity analysis was carried out to monitor the course of each reaction and is presented in Figure 3. R5 peptide shows almost no silica formation up to pH 6 which is consistent with previous work.[45] However, the amount of silica formation increased with pH in the range pH 6-8 and then decreased at pH 10. Although the reason for such sudden rise and drop in the activity is unclear, this behaviour may demonstrate the effects that the R5 peptide exhibits during (bio)silicification. A similar behavior was observed in the R2-mediated silica formation. The amount of silica formed in the presence of the R2 peptide was found to increase from pH 7 to pH 8 with a decrease in activity similar to that of R5 observed at pH 10. However, in the case of R1-mediated silica formation, an increase in the amount of silica formed was observed from pH 6 to pH 8, which then remained contant at pH 10. All the peptides showed maximum precipitation activity in the range pH 6-8, while R1 additionally possessed unusually high silica precipitation activity even at pH 10; a condition under which silica becomes soluble. This result is very interesting. However, further work is necessary in order to explain such unusual behaviour. It is interesting to note that the secondary structure predictions carried out suggested that the R1 peptide may possess ~18% helical structure.[49] However, the R2 and R5 peptides were proposed to be random coils or extended chains (>95%) and this has been confirmed in the case of the R5 peptide using Circular Dichroism spectroscopy.[50] Further work directed towards understanding the secondary conformation of peptides/proteins in mineral formation and details can be found elsewhere.[34, 51]

3. Effect of Temperature on R5-Mediated Silicification

Biosilicification observed in diatoms occurs over a wide range of temperature. In fact, diatoms are found in waters from freezing temperature up to 40°C.[35] In order to understand the catalytic/scaffolding mechanism by which biomolecules facilitate silica formation, we studied the effect of temperature on the peptide mediated silica formation. The silica formation using the R5 peptide was studied from 5° to 45°C. The maximum turbidity was observed in the case of silica synthesized at 5° and 15°C using R5 peptide, implying a greater amount of particle formation (Figure 4). Conversely, less precipitate was observed at 25° and 35°, as recorded by turbidity measurements (see Figure 4). A comparison of the growth rates and A_2 is presented in Figure 4b. In addition to the higher rates of silica formation observed at lower temperatures, a greater amount of silica formation was also observed at 5°C and 15°C using the R5 peptide. From the data presented for R5 mediated silicification as a function of temperature, it can be concluded that the process of particle formation is faster at lower temperatures (5°C and 15°C). The exact reason for such unusual behaviour is unclear and further investigations are being undertaken. However, it should be noted that silicification is a series of complex equilibria and that the effect on both forward and reverse reactions needs to be considered. Thus, if forward reactions are highly favoured, it could be that higher temperatures favour the reversion reaction. The conformation of the peptide molecules will also be affected by temperature and hence their ability to perform as catalyst/structure directing agent/template. Indeed, molecular dynamic simulation of the R5 peptide has shown some indication of the change in peptide conformation as a function of temperature (unpublished) and future work is being undertaken.

4. Effect of Substrate (TMOS) Concentration on R5-Mediated Silicification

In order to determine if any enzymatic activity is observed for the R5 peptide in (bio)silicification, we performed silica formation in the presence of the R5 peptide over a range of TMOS concentrations. The silica formation was studied using both the β-molybdosilicate method and turbidity analysis. The data obtained using the former technique was used to calculate reaction velocities. Figure 5 depicts the effects of TMOS concentration on the reaction rate. Figure 5a presents the data as a Michaelis – Menton plot, while Figure 5b is plotted in Lineweaver – Burke form. Saturation kinetics normally associated with enzyme kinetics were not observed in the case of R5-mediated silicification. The plot in Lineweaver – Burke form (Figure 5b) displays neither a traditional linear form, nor a sigmoidal form commonly found in a cooperative binding scheme.[52] The R5 peptide therefore was observed to operate in a manner inconsistent with

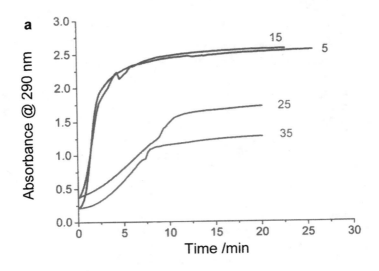

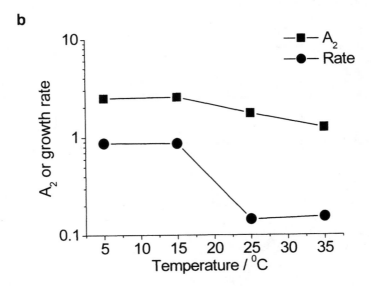

Figure 4. Effect of temperature on R5 mediated silica formation. (a) Turbidity measurements of silica formation using R5 at different temperatures. (b) Plots of growth rate (min⁻¹) and A₂ as functions of temperature. R5 concentration = 6.15 x 10⁻⁴ M and TMOS concentration = 0.067 M. Error bars ≤ symbol size.

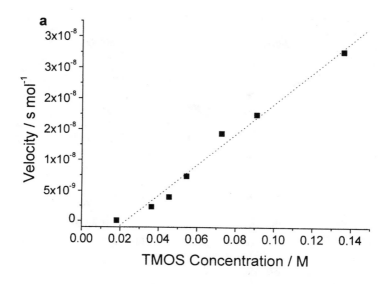

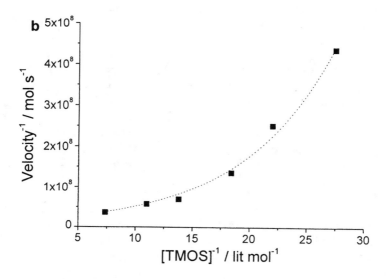

*Figure 5. Silica precipitating activity of R5 determined using the β –
silicomolybdate method. Data plotted as a (a) Michaelis – Menton plot, (b)
Lineweaver – Burke plot.*

traditional enzyme catalysis. This could mean that either the R5 peptide does not possess enzymatic activity or that it may have an unique catalytic effect as found recently.[51] Alternatively, the entrapment of the R5 peptide in the silica aggregates as they form is a possible explanation for such an unusual behaviour seen in Figure 5a,b.

Figure 6 shows a comparison between the amount of silica precipitated and the rates of silicification as monitored by turbidity measurements at 290 nm. A maximum amount of silica formation has been observed at a TMOS concentration of 0.067 M. As expected, the rate of silicification was found to be concentration dependent. The rate of silica formation was found to increase with increasing TMOS concentration, which was expected. It should be noted that all the concentration dependent studies were carried out at 5°C.

5. Morphology of Silica in R5-Mediated Silicification

The SEM data of the silica structures formed at different temperatures and their corresponding EDS analyses are shown in Figure 7. The formation of particulate structures was observed (see Figure 7a and b). At 5°C, the silica particles formed (sizes *ca.* 430 nm) were porous and were observed to be 'loose' aggregates of tiny (<20 nm) particles (Figure 7a). Silica prepared using the R5 peptide at 15°C was found to be more 'compact' with morphologies different from that seen for the silica particles synthesized at 5°C. It appears that the structures formed at 15°C were composed of particles of sizes *ca.* ≤150 nm (Figure 7b). Samples prepared at 45°C gelled immediately and hence were not analyzed. A correlation between product morphology and temperature of its formation could not be determined. From EDS data (Figure 7c) silica formation can be confirmed from the sharp Si and O signatures.

The SEM analysis of concentration dependence confirms the formation of particulate morphologies (Figure 8). The particles thus seen were composed of tiny (<50 nm) porous particles as clearly observed in Figure 8b and c. The apparent particle sizes for TMOS concentrations of 0.017, 0.033, 0.067, 0.133 and 0.267 M were respectively ~100, 300, 400, 50 and 140 nm (Figure 8a-e respectively). Silica synthesized at TMOS concentration of 0.017M was highly fused as shown in Figure 8a. The corresponding elemental analysis confirmed the formation of silica (Figure 8f). The digestion silica synthesised in the presence of the R5 peptide followed by SDS-PAGE analysis has shown that the peptide is tightly associated with silica particles.[40]

Conclusions

The data presented for the three different peptide sequences and their role(s) in silicification indicate that each of the peptides, has a distinct way of

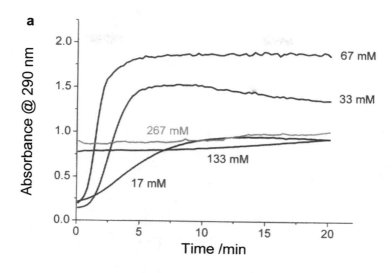

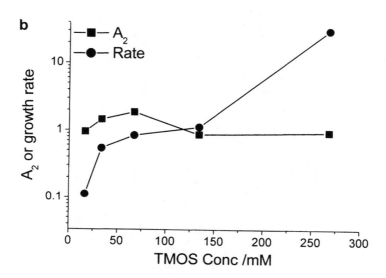

Figure 6. (a) Turbidity measurements of silica formation using R5 at different TMOS concentrations.(b) Plots of growth rate and A₂ as functions of TMOS concentration. Error bars ≤ symbol size.

428

Figure 7. SEM micrographs of silica formed using R5 at 5°C (a) and at 15°C (b). (c) Typical EDS of R5 mediated silica. The R5 concentration used was 6.15 x 10⁻⁴ M and that of TMOS was 0.067 M. Scale bars = 2.55μm.

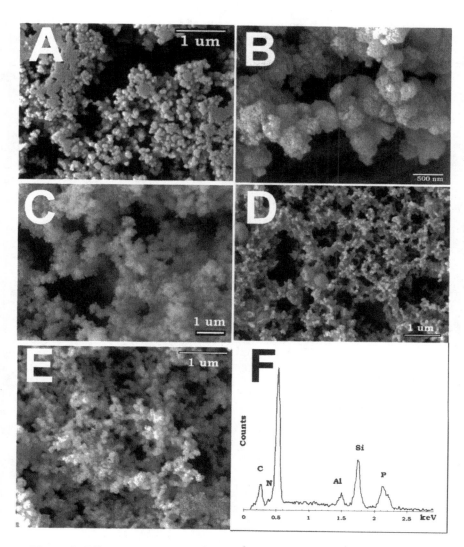

Figure 8. Effect of TMOS concentration on the morphology of R5 mediated silica formation. SEM micrographs of silica formed using R5 at 5° C using TMOS concentrations of 0.017M (a); 0.033M (b); 0.067M (c); 0.133M (d); and 0.267M (e). (f) Typical EDS of silica from sample shown in (b). Scale bars = 1μm in all micrographs except 500 nm in (b).

interacting with growing silica and controlling silicification. Saturation kinetics normally associated with enzyme kinetics were not observed in the case of R5-mediated silicification. Neither was a traditional cooperative binding scheme followed and this observation constitutes one of the most interesting aspects of the work. The morphologies of particles synthesized *in vitro* differ from each other depending on the peptide used. Peculiar kinetics were also observed for each peptide. When rate of precipitation was considered, the following trend was observed: R1>>R5>R2. At the same time, the R1 peptide was able to facilitate the formation of distinct and 'condensed' particles. On the other hand, the R5 peptide produced loose aggregates of tiny (<50 nm) particles. The aggregates synthesized using R2 and R5 peptide were also found to be porous.

It becomes clear that each of the peptides considered herein possess an unique role in silicification. As these peptides were directly derived from the diatom *C. fusiformis*, one can postulate that similar regulating roles of analogous biomolecules may exist *in vivo*. The subtle kinetic and morphological properties inherent to these peptide sequences could be postulated to be partly responsible for the formation of temporally coordinated building blocks, each with its own unique morphological contribution, which aid the formation of the ornate structures produced *in vivo*. However, it should be kept in mind that there are certainly many other parameters at work in biosilicification *in vivo*. Further investigations of the roles of various biomolecules, including genetically designed proteins, in (bio)silicification are currently being undertaken.

Acknowledgements

SVP and SJC thank Professor Kristi L. Kiick, Professor Carole C. Perry, Professor Nils Kröger and Professor Michael Brook for helpful discussions.

References

1. Biosilica should be understood as silica synthesized *in vivo* and should not be confused with silica produced *in vitro*, even though one is using biomolecules such as peptides and proteins to facilitate the reaction(s). Similarly, biomineralization or biosilicification should be used for describing the biosynthesis of minerals that occurs *in vivo*.
2. Bendz, G.; Lindqvist, I., The Biochemistry of Silicon and Related Problems. Plenum: New York, 1977.
3. Simpson, T. L.; Volcani, B. E., Silicon and Siliceous Structures in Biological Systems. Springer-Verlag: New York, 1981.
4. Evered, D.; O'Connor, M., Silicon Biochemistry. Wiley: New York, 1986.

5. Sullivan, C. W.; Volcani, B. E., In Silicon and Siliceous Structures in Biological Systems, Simpson, T. L.; Volcani, B. E., Eds. Springer-Verlag: New York, 1981; pp Chap. 2, p. 15.

6. Haeckel, E., Art Forms in Nature. Dover: New York, 1974.

7. Simpson, T. L.; Volcani, B. E., In Silicon and Siliceous Structures in Biologic Systems, Simpson, T. L.; Volcani, B. E., Eds. Springer-Verlag: New York, 1981; pp Chap. 1, p. 1.

8. Hildebrand, M.; Wetherbee, R., Components and control of silicification in diatoms. Prog. Mol. Subcell. Biol. **2003**, *33*, 11-57.

9. Harrison, C. C., Evidence for intramineral macromolecules containing protein from plant silicas. Phytochemistry **1996**, *41*, 37-42.

10. Shimizu, K.; Cha, J.; Stucky, G. D.; Morse, D. E., Silicatein alpha: Cathepsin L-like protein in sponge biosilica. Proc. Natl. Acad. Sci. USA **1998**, *95*, (11), 6234-6238.

11. Vrieling, E. G.; Gieskes, W. W. C.; Beelen, T. P. M., J. Phycol. **1999**, *35*, 548.

12. Sumper, M.; Kroger, N., Silica formation in diatoms: the function of long-chain polyamines and silaffins. J. Mater. Chem. **2004**, *14*, (14), 2059 - 2065.

13. Gordon, R.; Drum, R. W., The chemical basis of diatom morphogenesis. Int. Rev. Cytol. **1994**, *150*, 243.

14. Sumper, M., A phase separation model for the nanopatterning of diatom biosilica. Science **2002**, *295*, (5564), 2430-2433.

15. Hecky, R. E.; Mopper, K.; Kilham, P.; Degens, E. T., The amino acid and sugar composition of diatom cell-walls. Mar. Biol. **1973**, *19*, 323-331.

16. Swift, D. M.; Wheeler, A. P., Evidence of a Biosilica Organic Matrix in a Fresh-Water Diatom and Its Role in Regulation of Silica Polymerization. Abstracts of Papers of the American Chemical Society **1990**, *200*, 181-COLL.

17. Swift, D. M.; Wheeler, A. P., Evidence of an organic matrix from diatom biosilica. J. Phycol. **1992**, *28*, (2), 202-209.

18. Poulsen, N.; Kroger, N., Silica Morphogenesis by Alternative Processing of Silaffins in the Diatom Thalassiosira pseudonana. J. Biol. Chem. **2004**, *279*, (41), 42993-42999.

19. Wenzl, S.; Deutzmann, R.; Hett, R.; Hochmuth, E.; Sumper, M., Quaternary Ammonium Groups in Silica-Associated Proteins. Angew. Chem. Int. Ed. **2004**, *43*, 5933-5936.

20. Krasko, A.; Lorenz, B.; Batel, R.; Schröder, H. C.; Müller, I. M.; Müller, W. E. G., Expression of silicatein and collagen genes in the marine sponge Suberites domuncula is controlled by silicate and myotrophin. Eur. J. Biochem. **2000**, *267*, (15), 4878-4887.

21. Perry, C. C.; Keeling-Tucker, T., Crystalline silica prepared at room temperature from aqueous solution in the presence of intrasilica bioextracts. Chem. Commun. **1998**, 2587-2588.

22. Perry, C. C.; Keeling-Tucker, T., Model studies of colloidal silica precipitation using biosilica extracts from Equisetum telmateia. Colloid Polym. Sci. **2003**, *281*, (7), 652-664.

23. Morse, D. E., Silicon biotechnology: harnessing biological silica production to construct new materials. Trends Biotechnol. **1999**, *17*, (6), 230-232.

24. Kroger, N.; Deutzmann, R.; Bergsdorf, C.; Sumper, M., Species-specific polyamines from diatoms control silica morphology. Proc. Natl. Acad. Sci. USA **2000**, *97*, (26), 14133-14138.

25. Patwardhan, S. V.; Clarson, S. J., Silicification and Biosilicification Part 3. Silicon Chem. **2002**, *1*, (3), 207-214.

26. Patwardhan, S. V.; Mukherjee, N.; Steinitz-Kannan, M.; Clarson, S. J., Bioinspired Synthesis of New Silica Structures. Chem. Commun. **2003**, *10*, 1122-1123.

27. Patwardhan, S. V.; Clarson, S. J., Silicification and biosilicification - Part 7: Poly-L-arginine mediated bioinspired synthesis of silica. J. Inorg. Organomet. Polym. **2003**, *13*, (4), 193-203.

28. Coradin, T.; Durupthy, O.; Livage, J., Interactions of amino-containing peptides with sodium silicate and colloidal silica: A biomimetic approach of silicification. Langmuir **2002**, *18*, 2331-2336.

29. Sudheendra, L.; Raju, A. R., Peptide-induced formation of silica from tetraethylorthosilic ate at near-neutral pH. Mater. Res. Bull. **2002**, *37*, 151.

30. Cha, J. N.; Stucky, G. D.; Morse, D. E.; Deming, T. J., Biomimetic synthesis of ordered silica structures mediated by block copolypeptides. Nature **2000**, *403*, (6767), 289-292.

31. Belton, D.; Paine, G.; Patwardhan, S. V.; Perry, C. C., Towards an understanding of (bio)silicification: the role of amino acids and lysine oligomers in silicification. J. Mater. Chem. **2004**, *14*, (14), 2231-2241.

32. Belton, D.; Patwardhan, S. V.; Perry, C. C., Putrescine homologues control silica morphogenesis by electrostatic interactions and the hydrophobic effect. Chem. Commun. **2005**, (27), 3475-7.

33. Belton, D.; Patwardhan, S. V.; Perry, C. C., Spermine, spermidine and their analogues generate tailored silicas. J. Mater. Chem. **2005**, *15*, (43), 4629-4638.

34. Patwardhan, S. V.; Maheshwari, R.; Mukherjee, N.; Kiick, K. L.; Clarson, S. J., Conformation and Assembly of Polypeptide Scaffolds in Templating the Synthesis of Silica: An Example of a Polylysine Macromolecular "Switch". Biomacromolecules **2006**, *7*, (2), 491-7.

35. Annenkov, V. V.; Patwardhan, S. V.; Belton, D.; Danilovtseva, E. N.; Perry, C. C., Step-by-step Full Synthesis of Propylamines Derived from Diatom Silaffins and their Activity in Silicification. Chem. Commun. **2006**, 1521-1523.

36. Patwardhan, S. V.; Clarson, S. J.; Perry, C. C., On the role(s) of additives in bioinspired silicification. Chem. Commun. **2005**, *9*, 1113-1121.

37. Whitlock, P. W.; Naik, R. R.; Brott, L. L.; Clarson, S. J.; Tomlin, D. W.; Stone, M. O., Polymer Preprints **2001**, *42*, 252.

38. Brott, L. L.; Pikas, D. J.; Naik, R. R.; Kirkpatrick, S. M.; Tomlin, D. W.; Whitlock, P. W.; Clarson, S. J.; Stone, M. O., Ultrafast holographic nanopatterning of biocatalytically formed silica. Nature **2001**, *413*, 291.

39. Patwardhan, S. V., Silicification and Biosilicification: Role of Macromolecules in Bioinspired Silica Synthesis. Ph.D. Dissertation, University of Cincinnati: 2003.

40. Naik, R. R.; Whitlock, P. W.; Rodriguez, F.; Brott, L. L.; Glawe, D. D.; Clarson, S. J.; Stone, M. O., Controlled formation of biosilica structures in vitro. Chem. Commun. **2003**, 238-239.

41. Patwardhan, S. V.; Clarson, S. J., Silicification and biosilicification: Part 5 - An investigation of the silica structures formed at weakly acidic pH and neutral pH as facilitated by cationically charged macromolecules. Mat. Sci. Eng. C. **2003**, *23*, (4), 495-499.

42. Knecht, M. R.; Wright, D. W., Functional analysis of the biomimetic silica precipitating activity of the R5 peptide from Cylindrotheca fusiformis. Chem. Commun. **2003**, (24), 3038-3039.

43. Kroger, N.; Deutzmann, R.; Sumper, M., Silica-precipitating peptides from diatoms - The chemical structure of silaffin-1A from Cylindrotheca fusiformis. J. Biol. Chem. **2001**, *276*, (28), 26066-26070.

44. Steinitz-Kannan, M., private communication.

45. Kroger, N.; Deutzmann, R.; Sumper, M., Polycationic peptides from diatom biosilica that direct silica nanosphere formation. Science **1999**, *286*, (5442), 1129-1132.

46. Chang, J. S.; Kong, Z. L.; Hwang, D. F.; Chang, K. L. B., Chitosan-Catalyzed Aggregation during the Biomimetic Synthesis of Silica Nanoparticles. Chem. Mater. **2005**, ASAP, cm052161d.

47. Kaufman, P. B.; Dayanandan, P.; Takeoka, Y.; Bigelow, W. C.; Jones, J. D.; Iler, R. K., Silicon in shoots of higher plants. In Silicon and siliceous structures in biological systems, Simpson, T. L.; Volcani, B. E., Eds. Springer-Verlag: New York, 1981.

48. ProtParam tool. In 2006.

49. NPS@ : HNN secondary structure prediction. In 2006.

50. Farmer, R.; Patwardhan, S. V.; Kiick, K. L., unpublished results.

51. Patwardhan, S. V.; Shiba, K.; Schroder, H. C.; Muller, W. E. G.; Clarson, S. J.; Perry, C. C., The Role of Bioinspired Peptide and Recombinant Proteins in Silica Polymerisation. In The Science and Technology of Silicones and Silicone-Modified Materials, Clarson, S. J.; Fitzgerald, J. J.; Owen, M. J.; Smith, S. D.; Van Dyke, M. A. American Chemical Society Symposium Series 964, Washington DC, 2007, pp. 328-347.

52. Dixon, M.; Webb, E. C., Enzymes. 3rd Ed.; Academic Press: New York, 1979.

Chapter 27

Enzymatic Dihydroxylation of Aryl Silanes

Wyatt C. Smith[1], Gregory M. Whited[1], Thomas H. Lane[2],
Karl Sanford[1], and Joseph C. McAuliffe[1]

[1]Danisco Genencor, 925 Page Mill Road, Palo Alto, CA, 94304
[2]Dow Corning Corporation, 2200 W. Salzburg Road, Midland, MI, 48686

Aromatic dioxygenases were used to oxidize aryl silanes to
chiral (1S,2S)-3-sila-cyclohexa-3,5-diene-1,2-cis-diols with
excellent stereospecificity (>95% ee). Bioconversions were
conducted using a whole cell biocatalyst, *E. coli* JM109
expressing cloned dioxygenases, as well as by the DDT-
degrading organism *Ralstonia eutropha* A5. Treatment of silyl
cis-dihydrodiols with the enzyme cis-diol dehydrogenase gave
the corresponding silyl catechols. The silane-functional cis-
diols were also chemically converted into a range of
derivatives by modification of the hydroxyl, silyl or olefinic
functions. These silicon containing chiral cis-diols and
catechols represent a novel class of compounds having
potential application in the synthesis of fine chemicals, silicon-
based pharmaceuticals, polymers and optical materials.

Enzymatic dioxygenation of aromatic compounds to *cis*-dihydrodiols was first described by Gibson and coworkers in 1968, who suggested that the conversion of benzene to catechol by the microorganism *Pseudomonas putida* proceeded through the intermediate 3,5-cyclohexadiene *cis*-1,2-diol (1). The reaction was extensively studied in the following years, and has been the subject of several comprehensive reviews (2,3,4). A number of related enzymes are known to perform this transformation, including toluene dioxygenase (EC 1.14.12.11), naphthalene dioxygenase (EC 1.14.12.12), and other aromatic oxygenases, which together act upon a broad range of aromatic, heterocyclic and polycyclic substrates. Substituted aromatic compounds are converted to chiral products, depicted in Scheme 1.

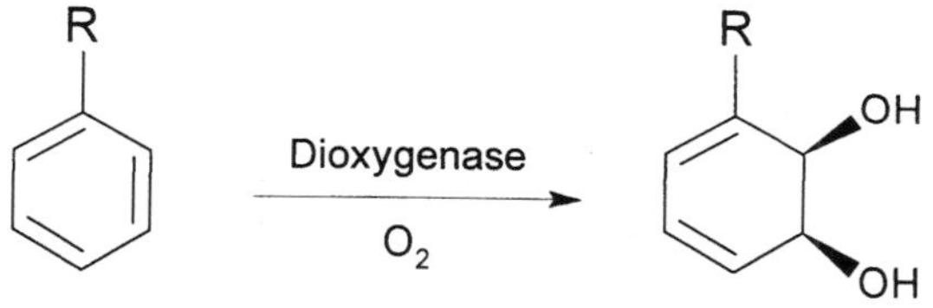

Scheme 1. Dioxygenase-catalyzed production of a cis-dihydrodiol.

The *cis*-dihydroxylation reaction catalyzed by these dioxygenases is typically highly enantioselective (often >98% ee) and, as a result, has proven particularly useful as a source of chiral synthetic intermediates (2,4). Chiral *cis*-dihydrodiols have been made available commercially and a practical laboratory procedure for the oxidation of chlorobenzene to (*1S, 2S*)-3-chlorocyclohexa-3,5-diene-1,2-*cis* diol by a mutant strain of *Pseudomonas putida* has been published (6). Transformation with whole cells can be achieved either by mutant strains that lack the second enzyme in the aromatic catabolic pathway, *cis*-dihydrodiol dehydrogenase (E.C. 1.3.1.19), or by recombinant strains expressing the cloned dioxygenase. This biocatalytic process is scalable, and has been used to synthesize polymer precursors such as 3-hydroxyphenylacetylene, an intermediate in the production of acetylene-terminated resins (7). A synthesis of polyphenylene was developed by ICI whereby the product of enzymatic benzene dioxygenation, *cis*-cyclohexa-3,5-diene-1,2-diol, was acetylated and polymerized as shown in Scheme 2 (8).

Other industrial applications of *cis*-dihydrodiols have also been pursued. Merck has considered using the diol derived from indene in the production of the HIV-protease inhibitor indinavir (9). A bioprocess for indigo production was developed by Genencor where the keystep involved the dioxygenation of indole using toluene dioxygenase, summarized in Scheme 3 (10).

Although microbial oxidation of a large number of substituted aromatic compounds to *cis*-dihydrodiols (*cis*-diols) has been previously demonstrated,

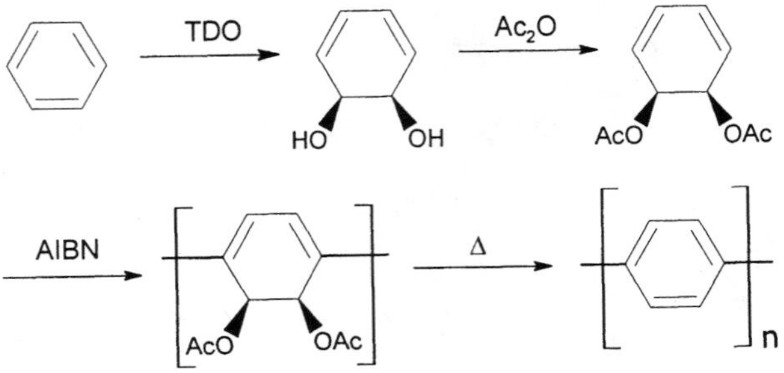

Scheme 2. Polyphenylene synthesis from benzene cis-dihydrodiol.

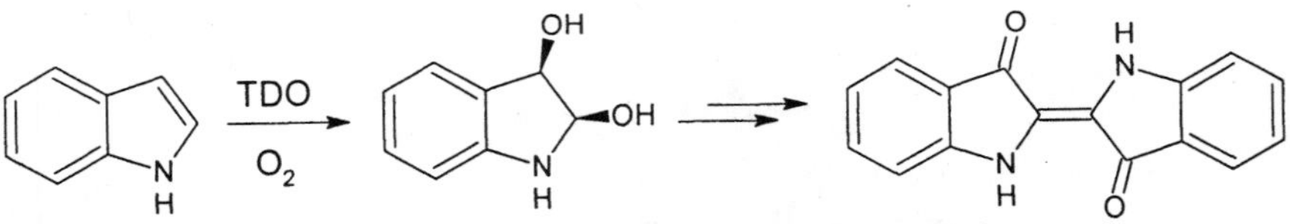

Indole   Indole *cis*-dihydrodiol   Indigo

Scheme 3. Toluene dioxygenase mediates the key step in the conversion of indole to indigo in an engineered E. coli strain.

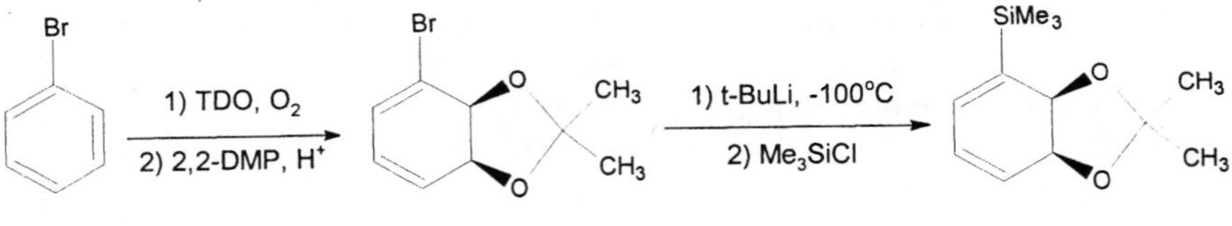

Scheme 4. Conversion of bromobenzene into a trimethylsilyl cyclohexadiene-1,2-cis-diol acetonide.

there was no example in the literature of a similar transformation of an aromatic compound bearing a silicon substituent (i.e. an aryl silane) prior to our investigation (11). A previous attempt to obtain silyl-functional *cis*-diols by enzymatic dioxygenation of trimethylsilylbenzene by Ley and coworkers was not successful (12), although the acetonide of trimethylsilylbenzene *cis*-diol was synthesized by lithiation of a bromoacetonide derived from bromobenzene, followed by addition of TMS chloride as shown in Scheme 4.

We sought to examine the enzymatic dioxygenation of aryl silanes using a number of different aromatic dioxygenases in order to determine if such transformations were possible and to define the substrate-specificity profile. We were also motivated by the rich chemistry of silicon-based materials, which includes the hydrosilylation of alkenes and ketones, the addition of electrophiles to vinyl and allyl silanes, and palladium catalyzed cross-coupling of vinyl silanes with aryl halides (13). As a result, silyl functional *cis*-diols have potential as chiral intermediates for drug development, as polymer precursors/modifiers and as elements in non-linear optical materials.

Enzymatic Dioxygenation of Aryl Silanes

To explore the enzymatic dioxygenation of aryl silane substrates, a small library of 26 silanes representing several structural classes was obtained from commercial sources. Compounds were selected based upon our desire to produce products that could be further elaborated to polymers, or that would contain chiral silicon atoms. Examples included silanes bearing alkoxy, vinyl and hydrido substituents, in addition to a number of diphenylsilanes. Several whole cell biocatalysts were selected for the study, summarized in Table 1.

Aryl silanes were initially screened against whole cells of *E. coli* JM109 (pDTG601), and to some extent, *S. yanoikuyae* B8/36. Each aryl silane was subjected to small-scale whole cell transformation by using resuspended cells in the presence of 0.1% glucose and 1 mg/mL substrate. Substrates were typically added directly to the transformation broth and incubated for 3-4 hours before extraction with dichloromethane and subsequent analysis by thin layer chromatography (TLC), GC/MS, and ^{1}H NMR (olefinic signals in the 6-7 ppm region being diagnostic for *cis*-dihydrodiol products) (11).

Controls to assess the stability of the aryl silane substrate under the conditions of the transformation were run using dioxygenase-free *E. coli* cells (no plasmid). Since dioxygenase activity was somewhat variable, a positive control was run in parallel with every transformation experiment using the natural substrate of the particular dioxygenase (toluene, naphthalene, or biphenyl). In most cases where products were undetectable, shake flask transformations were repeated using cells grown in a 14L fermentor, as cells grown in the fermentor under carefully controlled conditions provide cultures of extremely high cell density and enzyme activity. However, in no case was a *cis*-

438

Table I. Aryl dioxygenase expressing strains used to study the dioxygenation of aryl silanes.

Strain	Characteristics	Ref.
E. coli JM109 (pDTG601)	E. coli JM109 host containing the TDO genes (IPTG-inducible) (todC12BA) from Pseudomonas putida in expression plasmid pKK223-3 (Amp^r)	(5)
E. coli JM109 (pDTG602)	E. coli JM109 host containing the (+)-cis-(1S, 2R)-dihydroxy-3-methylcyclohexa-3,5-diene dehydrogenase genes (todD) from Pseudomonas putida F1 (IPTG-inducible) in expression plasmid pKK223-3 (Amp^r)	(5)
E. coli JM109 (pDTG141)	E. coli JM109 host containing the NDO genes (IPTG-inducible) (nahAaAbAcAd) from Pseudomonas sp9816 in plasmid pKK223-3 (Amp^r)	(14)
Sphingomonas yanoikuyae B8/36	S. yanoikuyae strain lacking cis-diol dehydrogenase activity, derived from B1 (wild strain) by chemical mutagenesis. Has a biphenyl-inducible biphenyl dioxygenase (BPDO) pathway.	(15)
Ralstonia eutropha A5	Wild strain containing biphenyl (BP)-inducible dioxygenases. Can utilize (BP) or 4-chlorobiphenyl as sole carbon source	(16)

SOURCE: U.S. patent 7179932.

diol product detected where there was none before. Of the enzymes tested, only toluene dioxygenase was found to accept aryl silane substrates. Complete disappearance of the substrate could sometimes be observed within 30 minutes (as determined by GC/MS analysis), but an accurate mass balance was somewhat difficult to determine due to the high vapor pressure of many of the substrates. Thus, these shake flask experiments were useful qualitative screens for dioxygenase substrates. In most cases these transformations produced relatively pure products following extraction and removal of solvents, for example the cis-diol derived from dimethylphenylvinylsilane (Figure 1). The successful bioconversions of aryl silanes to cis-diols are summarized in Table 2.

The conversion of dimethylvinylsilane 1a to the cis-diol 2a is noteworthy given the presence of an oxidizable vinylsilane function. Similarly, the sensitive hydrosilane function of dimethylphenylsilane 1b survived both exposure to aqueous conditions, in addition to enzyme mediated oxidation. The cis-diol products listed in Table 2 were moderately stable, provided that exposure to acidic conditions was minimized, discussed in further detail later in this chapter. The product of benzyl silane (1d) dioxygenation was the exception however, with the cis-diol (2d) undergoing a facile conversion to only the ortho-phenol at

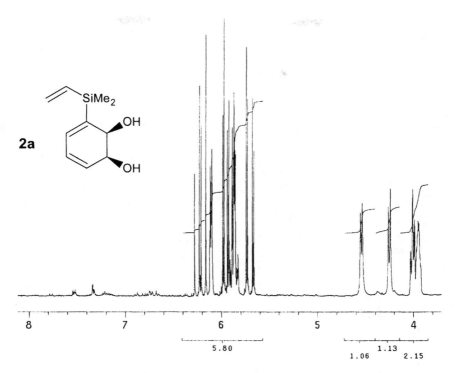

Figure 1. A 300MHz ^1H NMR spectrum of crude dimethylphenylvinylsilane cis-dihydrodiol, obtained in d6-DMSO.

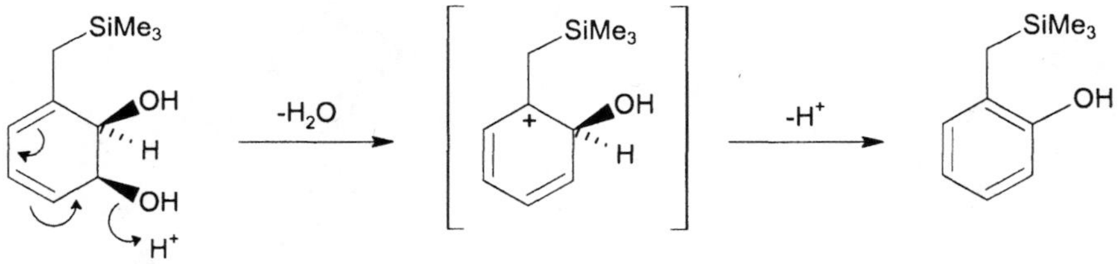

Scheme 5. Postulated mechanism for the conversion of benzylsilane cis-diol to an ortho-substituted phenol through a silicon stabilized carbocation.

440

Table II. Successful conversions of aryl silanes to *cis*-dihydrodiols using toluene dioxygenase (TDO).

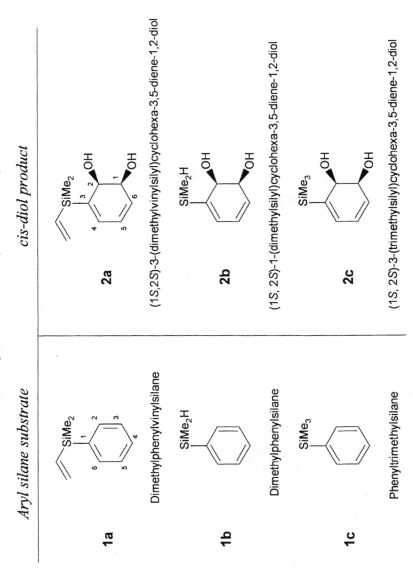

Aryl silane substrate	*cis-diol product*

1a Dimethylphenylvinylsilane — **2a** (1S,2S)-3-(dimethylvinylsilyl)cyclohexa-3,5-diene-1,2-diol

1b Dimethylphenylsilane — **2b** (1S, 2S)-1-(dimethylsilyl)cyclohexa-3,5-diene-1,2-diol

1c Phenyltrimethylsilane — **2c** (1S, 2S)-3-(trimethylsilyl)cyclohexa-3,5-diene-1,2-diol

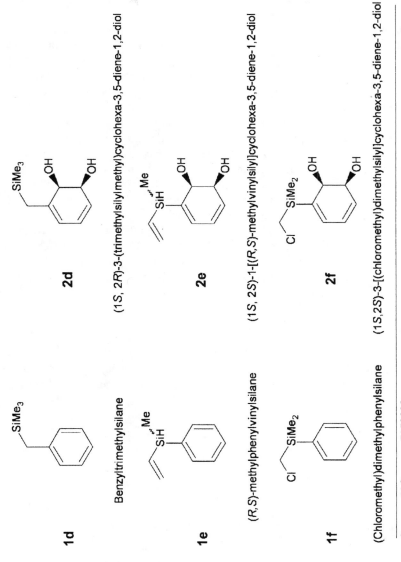

1d

Benzyltrimethylsilane

2d

(1S, 2R)-3-(trimethylsilylmethyl)cyclohexa-3,5-diene-1,2-diol

1e

(R,S)-methylphenylvinylsilane

2e

(1S, 2S)-1-[(R,S)-methylvinylsilyl]cyclohexa-3,5-diene-1,2-diol

1f

(Chloromethyl)dimethylphenylsilane

2f

(1S,2S)-3-[(chloromethyl)dimethylsilyl]cyclohexa-3,5-diene-1,2-diol

SOURCE: U.S. patent 7179932.

room temperature. This phenomenon was thought to be a consequence of the silicon β-effect, which would be expected to enhance the formation of the *ortho*-phenol through the mechanism depicted in Scheme 5.

Dihydroxylation of the racemic arylsilane, (*R,S*)-methylphenylvinylsilane (**1e**) with TDO gave a mixture of *cis*-diol diastereoisomers bearing a chiral silicon atom. The *cis*-dihydrodiol product was converted to the acetonide as described below and the product separated on a chiral cyclodextrin GC/MS column. The result revealed a 1:1 mixture of the *cis*-diol diastereoisomers, indicating that the enzyme did not discriminate between the two enantiomers. A similar observation was obtained by Bui and coworkers who found that TDO-mediated oxidation of a series of aromatic compounds with remote chiral centers did not result in any degree of kinetic resolution (17). Overall, the number of successful conversions mediated by toluene dioxygenase was restricted to unsubstituted phenyl silanes with small substituents around the silicon atom. Diphenylsilanes were not substrates, nor were silanes with two or more hydrosilane functions such as methylphenylsilane. Efforts to convert aryl silanes with more than one leaving group to silane *cis*-diols were also unsuccessful, for example, neither trialkoxyphenyl silanes nor dialkoxymethylphenyl silanes were converted to the corresponding silane *cis*-diols. It is possible that these enzymes are inhibited by such substrates, or that they are too bulky to be accommodated within the deeply buried active sites of the enzymes. An alternate screening approach of these recalcitrant substrates against an additional aryl dioxygenase expressed by *Ralstonia eutropha* A5 is described later in this chapter.

Large scale bioconversion

Two substrates were chosen for large scale bioconversion (i.e. >10 g scale), dimethylphenylvinylsilane (DMPVS) **1a** and dimethylphenylsilane (DMPS) **1b**.

Dimethylphenylvinylsilane (**1a**)   Dimethylphenylsilane (**1b**)

The dioxygenation was performed by the addition of the neat silanes to a 14L fermentor containing *E. coli* JM109 (pDTG601) expressing toluene dioxygenase. The tolerance of the cells to possible toxic effects of the aryl silanes was assessed by online monitoring of the respiratory quotient (RQ), an indicator of the aerobic respiration rate (18). RQ values of 0.9 or greater were maintained by adjusting the addition rate of the aryl silane substrates. The extent of conversion was monitored by either GC/MS or ^{1}H NMR. Samples of whole fermentor broth (10 mL) were taken at regular intervals and extracted with 1 mL of deuterochloroform (CDCl$_3$). The ^{1}H NMR spectrum of the extract allowed the extent of chemical conversion to be determined.

The aryl silane substrate was added to the fermentors until such time as levels of aryl silane began to increase (typically 25 to 75 g in total). At this point the fermentor broth was harvested, centrifuged to remove cells, and extracted three times with 1 L of ethyl acetate. Upon evaporation of the solvent, crude silane *cis*-diols 2a and 2b were obtained in yields of 40% and 64% respectively, with only minimal amounts of the aryl silane starting materials. It was important to include a bicarbonate wash of the organic extract in order to prevent the acid-catalyzed conversion to silyl phenols through dehydration. The crude *cis*-diols could be further purified by column chromatography over silica gel pretreated with a small amount of triethylamine. The enantiomeric excess (% ee) and absolute configuration of purified diols *cis*-(*1S,2S*)-3-(dimethylvinylsilyl)-cyclohexa-3,5-diene-1,2-diol (2a) and *cis*-(*1S,2S*)-3-(dimethylsilyl)cyclohexa-3,5-diene-1,2-diol (2b) was shown to be greater than 98% ee as determined by the ^{1}H NMR method of Resnick *et al.* (19). Both *cis*-dihydrodiols 2a and b were stable at –20°C for at least several months, although at room temperature they slowly converted to a mixture of *ortho*- and *meta*-phenols, compounds 3 a,b and 4 a,b respectively, as the result of an acid-catalyzed dehydration.

3a R^1 = OH, R^2 = H
3b R^1 = H, R^2 = OH
3c R^1 = R^2 = OH

4a R^1 = OH, R^2 = H
4b R^1 = H, R^2 = OH
4c R^1 = R^2 = OH

Catechol biosynthesis

In organisms that are able to fully degrade toluene, the second step in the catabolic pathway involves the dehydrogenation of *cis*-diols to the corresponding catechols by *cis*-diol dehydrogenase (E.C. 1.3.1.19), as shown in Scheme 6.

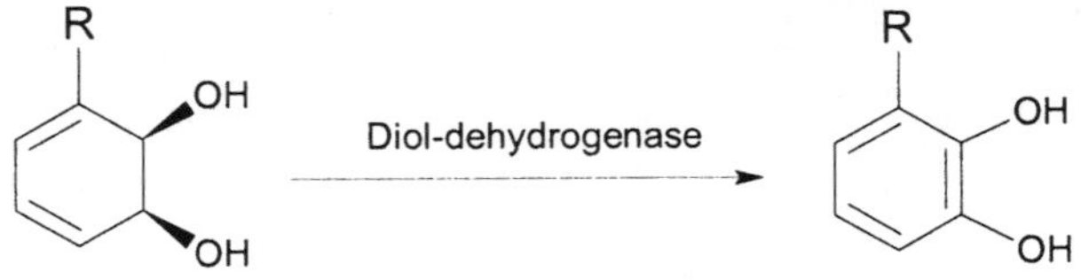

Scheme 6. Diol dehydrogenase mediated production of a catechol.

Biocatalytic production of catechols using *E. coli* (pDTG602) was demonstrated in a shake flask for *cis*-diols **2a** and **2b** resulting in the silyl functional catechols **3c** and **4c**, respectively. Following extraction of the aqueous phase with ethyl acetate, the products were analyzed by GC/MS (for **3c** [M$^+$] m/z 194; **4c** [M$^+$] m/z 168) and TLC, the latter by visualization with Gibbs reagent which provides a color test diagnostic for catechols (UV-active bands turned dark brown immediately after treatment of the plate with reagent). The yield of the conversions was low however, and optimization will be needed in order to provide larger amounts of these novel compounds.

Synthetic Derivatives of silyl *cis*-dihydrodiols

In order to prevent the degradation of the somewhat labile *cis*-diols upon storage, they were converted to their acetonide derivatives by treatment with 2,2-dimethoxypropane (2,2-DMP) and Amberlite 120-H$^+$ resin, followed by filtration and removal of solvents. Care needed to be taken in order to minimize losses of the somewhat volatile silanes during the concentration step (Table 3).

Diels-Alder adducts of *cis*-diol acetonides

Arene *cis*-diols can undergo cycloaddition reactions, acting as both diene and dienophile to produce homoadducts (12). Heteroadducts have also been synthesized, for example recent work by Banwell and coworkers exploited the reaction of toluene *cis*-diol with cyclopentenone as the initial step in their synthesis of complicatic acid and related triquinanes (20). We found that upon prolonged standing at room temperature in concentrated form, the acetonides **5a** and **5b** were slowly converted to the Diels-Alder adducts depicted in Figure 2.

Additional derivatives

Additional synthetic derivatives of the silyl *cis*-diols and the corresponding acetonides were prepared in order to demonstrate the versatility of these

Table III. Acetonide derivatives of *cis*-diols.

cis-diol substrate	*Acetonide derivative*
2a	5a
2b	5b
2c	5c
2d	5d
2e	5e
2f	5f

SOURCE: U.S. patent 7179932.

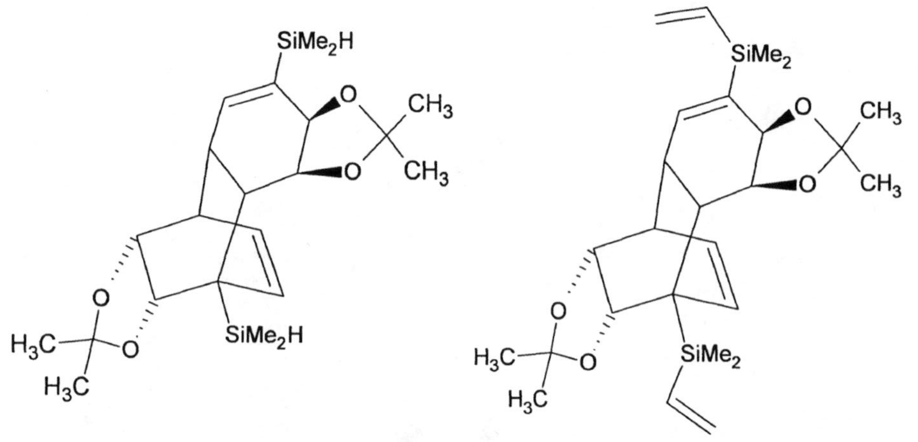

*Figure 2. Diels-Alder cycloadduct dimers of dimethylphenylsilane cis-diol acetonide (**5a**) and dimethylvinylphenylsilane cis-diol acetonide (**5b**).*

compounds. As mentioned above, dehydration to form the phenol is relatively facile (standing at room temperature, or slight acid), and so preservation of the chiral *cis*-diol moiety in subsequent synthetic steps is important. In addition to acetonides, the *cis*-diol group was protected through standard synthetic techniques including acylation, phenylboronation, or conversion to the dimethyl *tert*-butylsilyl ether. Further transformations of DMPS *cis*-diol **2b** and its acetonide derivative **5b**, both by modification of the ring or simple transformation of the hydrosilane function are summarized in Schemes 7 and 8.

Epoxidation of a *cis*-diol acetonide bearing a hydrosilane functionality

The chemical epoxidation of DMPS *cis*-dihydrodiol was performed using 2 equivalents 3-chloroperbenzoic acid (m-CPBA) in dichloromethane and gave a mixture of 4 compounds (Scheme 8). Surprisingly, the hydrosilane functionality was robust enough to allow the isolation two epoxidized products. Analysis of this mixture by ^{1}H NMR gave the spectrum depicted in Figure 3. The formation of these highly functionalized, chiral silanes in just 3 steps from dimethyl-phenylsilane demonstrates the potential of this methodology.

Naphthalene Dioxygenase Regioselectivity

Naphthalene dioxygenase (NDO) is known to perform a number of other reactions in addition to *cis*-dihydroxylations, including monohydroxylations, desaturations, *O*-and *N*-dealkylations and sulfoxidations (3). In our hands, NDO was unable to convert any aryl silanes into a *cis*-diol product, including (1-naphthyl)trimethylsilane. This lack of reactivity was thought to be related more to the steric bulk around the silicon atom, as opposed to an electronic effect. In contrast, following an incubation of 30 minutes in a shake flask with cells expressing NDO, dimethylphenylsilane (DMPS) was converted to a sole new product, identified as dimethylphenylsilanol. This was an interesting result given the fact that the analogous reaction with TDO produces only the *cis*-diol (**2b**). Figure 4 shows the results of GC/MS analysis of extracts from the two reactions. Note that the silane *cis*-diol derived from DMPS **2b** is converted to the corresponding phenol(s) upon injection into the GC/MS.

A control transformation with *E. coli* JM109 (no dioxygenase) cells was also run in parallel with both the NDO and TDO reactions and produced negligible amounts of dimethylphenylsilanol. In another control transformation, cumene, which can be considered the carbon analogue of DMPS, was converted completely into cumyl alcohol (1,1-dimethylbenzyl alcohol) by NDO, while TDO produces exclusively to the *cis*-diol Together, these results, support the hypothesis that silanol production was the result of an enzyme-mediated process.

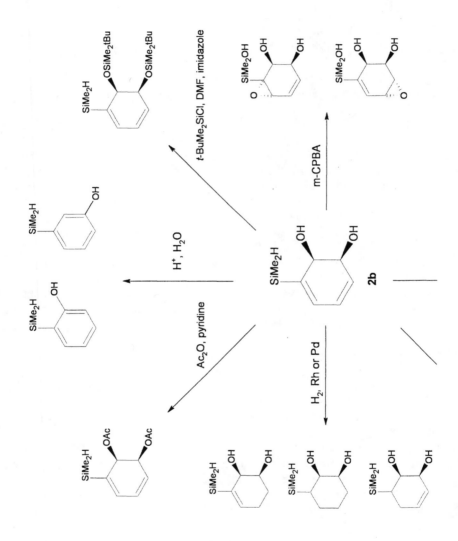

449

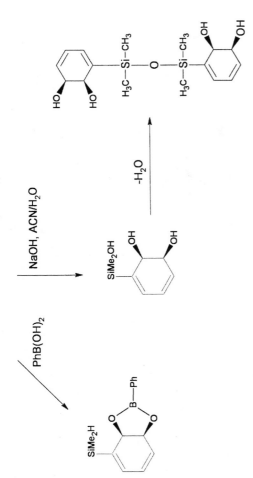

Scheme 7. Derivatives of dimethylphenylsilyl cis-diol **2b**.

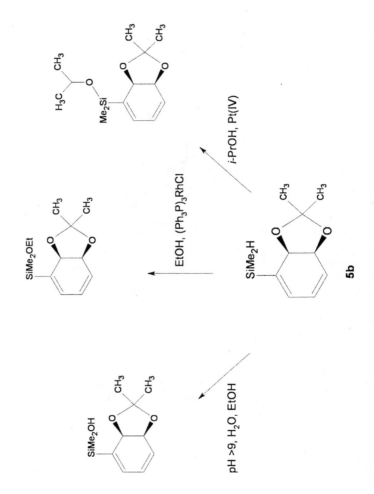

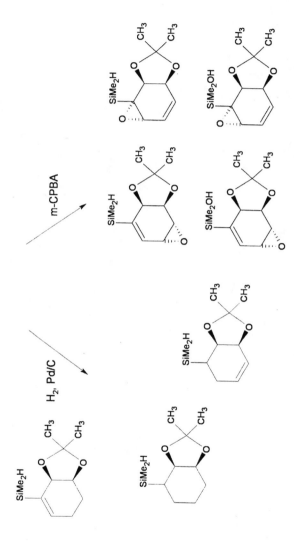

Scheme 8. Derivatives of dimethylsilylphenyl cis-diol acetonide **5b**

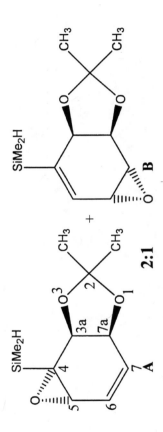

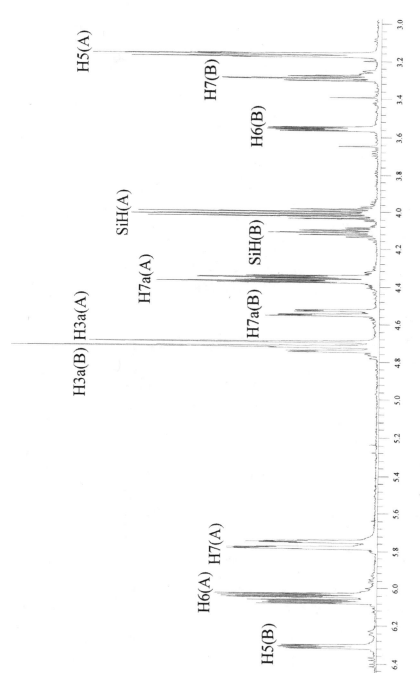

Figure 3. ^1H *NMR spectrum of a mixture of hydrosilane-functional epoxides.*

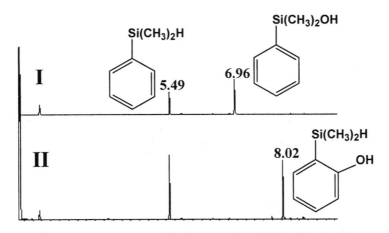

I. JM109(pDTG141) (NDO), 0.3 gDMPS/L, 30 min.
II. JM109(pDTG601) (TDO), 0.3 gDMPS/L, 30 min.

Figure 4. Distinct products could be detected by NDO and TDO transformations by GC/MS, corresponding to the silanol and cis-diol of DMPS respectively.

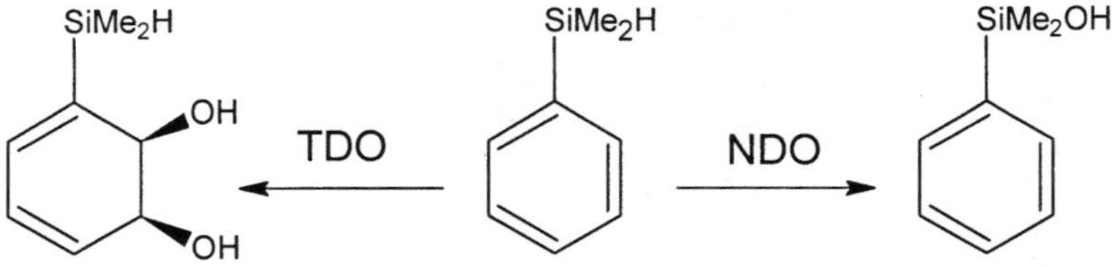

Scheme 9. Contrasting oxidation products of DMPS with TDO and NDO.

Oxidation versus Hydrolysis of Dimethylphenylsilane

We were interested in determining whether the production of the silanol from DMPS was the result of a monooxidation at a silicon atom, or the result of enzyme mediated hydrolysis, as both mechanisms would give rise to the same product. In the former case, however, water would be produced, as opposed to hydrogen gas (Scheme 10). Several attempts to settle this question through $^{18}O_2$ labeling did not result in enough of the respective products to answer this question in an unambiguous fashion, possibly due to subsequent exchange of the ^{18}O-silanol group with water.

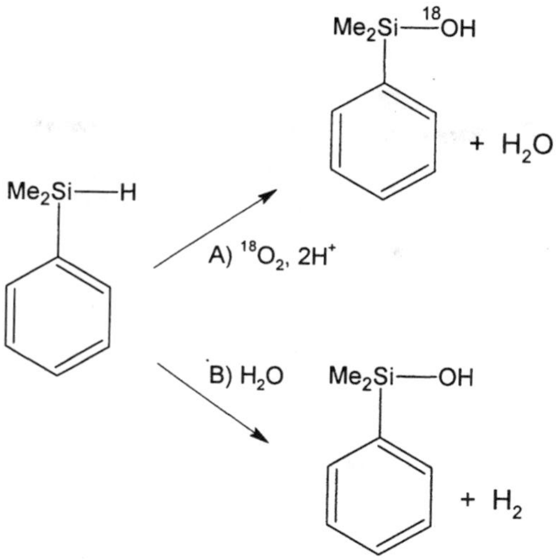

Scheme 10. Possible mechanisms for conversion of DMPS to the silanol by NDO. B) Oxidation by molecular oxygen; B) Enzyme-assisted hydrolysis of the hydrosilane function.

Aryl silanes as sole carbon source

We still wished to demonstrate the stereoselective oxidation of a prochiral diphenylsilane to a product containing a stereogenic silicon atom, and thus an alternative screening approach was employed. Certain bacterial strains possess the ability to utilize aromatic compounds as the sole source of carbon and energy. Oxidative metabolism of these compounds begins with the *cis*-dihydroxylation of the aromatic ring catalyzed by dioxygenase, followed by formation of the catechol by *cis*-diol dehydrogenase. *Ralstonia eutropha* A5 is a wild type strain isolated from PCB-contaminated river sediment for its ability to use 4-chlorobiphenyl as sole carbon source (21). The organism contains a biphenyl-inducible BPDO system that was also found to degrade the environmental contaminant 1,1,1,-trichloro-2,2-bis(4-chlorophenyl)ethane (DDT) (Figure 5), a substrate that no other known aromatic dioxygenase is able to transform (6). Thus, given its unusual substrate tolerance, this strain was used to screen the library of 26 aryl silane substrates, including those that were not transformed by TDO, NDO or the *Sphingomonas* BPO.

To employ a more rapid screen, *R. eutropha* A5 was grown on solid MSB in the presence of each member of our aryl silane library, the appearance of colonies indicating that the organism can use that substrate as sole carbon

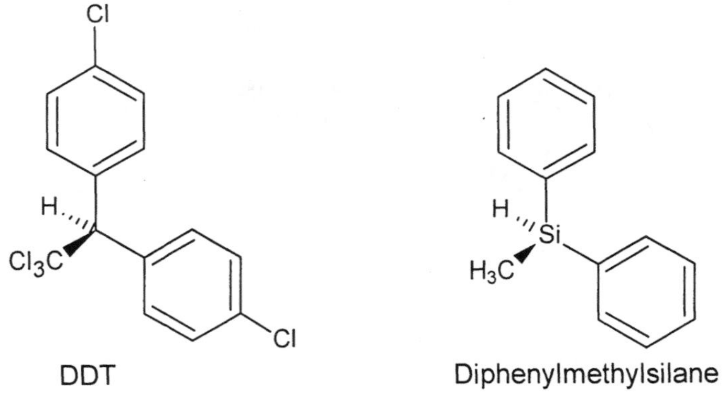

Figure 5. Comparison of the structures of DDT (1,1,1-trichloro-2,2-bis(4-chlorophenyl)ethane) and diphenylmethylsilane.

source. These data are summarized in Table IV with qualitative ratings of +, ++, or +++ to indicate the extent of growth, both the rate and the density. Interestingly, many of the compounds that were not substrates for NDO or TDO are apparently utilized by A5, and growth is accompanied by a color that may indicate formation of a catechol. The fact that aryl silanes that contain two or more reactive functions at silicon (i.e. methylphenylsilane, trimethoxy-phenylsilane) are transformed by A5 is a significant observation, since the *cis*-diols of these compounds would be amenable to subsequent chemical modification (i.e. polymerization). Because of its wider substrate tolerance, cloning and isolation of the A5 dioxygenase may be considered in the future.

Conclusion

We have demonstrated the efficient and stereoselective enzymatic dihydroxylation of several aryl silanes to the corresponding *cis*-diols. Of the aryl silanes screened, 6 of the 26 substrates were converted by toluene dioxygenase to a *cis*-diol product, with dimethylphenylsilane (DMPS) proving to be a particularly good substrate. In contrast to TDO, naphthalene dioxygenase (NDO) did not produce a *cis*-diol from any of the substrates tested, however DMPS was converted by NDO to the corresponding dimethylphenylsilanol. In addition, silyl catechols were produced upon treatment of silyl *cis*-diols with cells expressing the enzyme *cis*-diol dehydrogenase. The silyl *cis*-diols could be readily converted to silyl phenols upon acid treatment, or to the stable acetonide derivatives. Both the DMPS *cis*-diol and the corresponding acetonide were converted into a range of derivatives by chemical modification of the diol, olefinic or silane functions.

Table IV. Growth of *R. eutropha* A5 on solid MSB with various aryl silane and
aromatic substrates as sole carbon source.

Substrate	Growth
Biphenyl	+++
Cumene	++
Naphthalene	
Phenylsilane	+
Methylphenylsilane	+, brown
Dimethylphenylsilane	++, brown
Diphenylsilane	++, yellow
Diphenylmethylsilane	+++, brown
Benzyltrimethylsilane	-
Phenyldimethylvinylsilane	+
Phenyltrimethylsilane	+
Phenyltrimethoxysilane	+++, yellow
Phenyltriethoxysilane	+++, yellow
Diphenyldifluorosilane	+
Phenylmethylsilane	++
p-Tolyltrimethylsilane	+
Chloromethyldimethylphenylsilane	+
(Dimethylphenylsilyl)acetylene	-
Vinylphenylmethylsilane	++, brown
Phenylmethyldimethoxysilane	+
1-Chloro-4-trimethylsilylbenzene	+
1,3-Divinyl-1,3-diphenyl-1,3-dimethyldisiloxane	++, brown
(1-Naphthyl) trimethylsilane	+
4-Iodophenyltrimethylsilane	-
Phenylethynyltrimethylsilane	-
Diphenyldifluorosilane	+
1,4-bis(dimethylsilyl)benzene	-
(Dimethylphenylsilyl)methanol	-
1,3-Diphenyl-1,1,3,3-tetramethyldisiloxane	-

458

The aryl silane library was also screened for their ability to function as the sole carbon source for *Ralstonia eutropha* A5, a wild type strain expressing a biphenyl dioxygenase (BPO) enzyme. A number of silanes were observed to support growth, including diphenylsilanes and trialkoxysilanes. Overall the study indicated the feasibility of the enzymatic conversion of arylsilanes to a novel series of silane *cis*-dihydrodiols and catechols. Such compounds may find application as chiral polymer precursors, intermediates for natural product synthesis and other useful materials.

References

1. Gibson, D. T.; Koch, J. R.;Kallio, R. E. *Biochemistry*, **1968**, *7*, 2653-2662.
2. Boyd, D. R.; Sheldrake, G. N. *Nat. Prod. Rep.*, **1998**, *15*, 309-324.
3. Resnick, S. M.; Lee, K.; Gibson, D. T. *J. Indust. Microbiol.* **1996**, *17*, 348-457.
4. Hudlicky, T.; Gonzalez, D.; Gibson, D. T. *Aldrichimica Acta* **1999**, *32*, 35-62.
5. Zylstra, G. L.; Gibson, D. T. *J. Biol. Chem.* **1989**, *264*, 14940-14946.
6. Hudlicky, T.; Stabile, M. R.; Gibson, D. T.; Whited, G. M. In *Organic Syntheses*, *76*, 77.
7. Williams, M. G.; Olson, P. E., Tautvydas, K. J., Bitner, R. M., Mader, R. A., Wackett, L. P. *Appl. Microbiol. Biotechnol.*, **1990**, *34*, 316-321.
8. Ballard, D. G. H.; Courtis, A.; Shirley, I. M.; Taylor, S. C. *J. Chem. Soc., Chem. Commun.* **1983**, 954-955.
9. Davies, I. W.; Senanayake, C. H.; Castonguay, L., Larsen, R. D., Verhoven, T. R.; Reider, P. J. *Tetrahedron Lett.*, 1995, *36*, 7619-7622.
10. Berry, A.; Dodge, T. C.; Pepsin, M.; Weyler, W. *J. Ind. Microbiol. Biotechnol.* **2002**, *28*, 127-133.
11. McAuliffe, J. C.; Whited, G. M.; Smith, W.C. *US patent*, US 7179932, **2007**.
12. Ley, S., Redgrave, A. J., Taylor, S. C., Ahmed, S., and Ribbons, D. Synlett, 1991, 741-742.
13. Brook, M. A., *Silicon in Organic, Organometallic and Polymer Chemistry* Wiley, New York, **2000**.
14. Suen, W. C.; Gibson, D. T. *Gene* **1994**, *143*, 67-71.
15. Kim, E.; Zylstra, G. J. *J. Ind. Microbiol. Biotechnol.* **1999**, *23*, 294-302.
16. Nadeau, L. J.; Sayler, G. S.; Spain, J. C. *Arch. Microbiol.* **1998**, *171*, 44-49.
17. Bui, V. P.; Vidar Hansen, T.; Stenstrom, Y.; Hudlicky, T.; Ribbons, D.W. *New J. Chem.*, **2001**, *25*, 116-124.
18. Shuler M. L.; Kargi, F. *Bioprocess Engineering: Basic Concepts*, 2nd edn., Prentice-Hall, Upper Saddle River, **2002**, p 211.

19. Resnick, S. M.; Torok, D. S.; Gibson, D. T. *J. Am. Chem. Soc.* **1995**, *60*, 3546-3549.

20. Austin, K. A. B.; Banwell, M. G.; Harfoot, G. J.; Willis, A. C. *Tetrahedron Lett.*, **2006**, *47*, 7381-7384.

21. Shields, M. S.; Hooper, S. W.; Sayler, G. S. *J. Bacter.* **1985**, *163*, 882-889.

Indexes

Author Index

Subject Index

A

Acrylamide, biocatalytic route to, 13

Acrylamide sugars, polymers, 381*f*, 383, 384*f*

Acrylate sugars, polymers, 380, 381*f*, 382*f*

Adipic acid, cutinase-catalyzed polycondensation with alcohols, 264–265

Adipic acid/1,8-octanediol condensation polymerization immobilized CALB (*Candida antarctica* lipase B) on polyacrylic resin, 161, 162*f*
immobilized CALB on polystyrene resin, 173, 174*f*

Alcohols, cutinase-catalyzed polycondensation with diacids, 264–265

Alditols, lipase-catalyzed polymerization
hyperbranched polymers (HBPs), 276
influence of substrate chain length, 279–280
lipase regioselectivity, 277
lowering reaction temperature, 279
molecular weight vs. time, 278–279, 280*f*
potential to form high molecular weight polyol-polyesters, 281
reaction stages, 277–278
reactivity of secondary hydroxyl groups and chain growth, 280–281
regulation of vacuum, 277–279
structures of alditols in study, 278
time-course study of molecular weight vs. polyol structure, 280*f*

Alkenyl sugars, polymers, 381*f*, 387, 389*f*

Alkylsilanes, self-assembled monolayers on silicon or glass, 182

Amination reactions. *See* Soybean oil (SBO)

Amino acids
aromatic interactions using phenylalanine (Phe) analogues, 23–24
imparting structural stability to peptides, 23–24
maltose formation from starch hydrolysis by immobilized, 62, 63*f*
meniscus displacement vs. time curves for polymerization of 2-hydroxyethyl methacrylate (HEMA) onto α-amylase, 62, 63*f*
non-natural, incorporation into peptides and polypeptides, 24
percentage release from untreated and treated α-amylase/PVA blend, 57*f*
poly(HEMA) (PHEMA) and, interactions, 64–65
stabilizing α-helical peptides, 23
structures of natural and non-natural, 25*f*
See also Bioartificial materials; Peptides

Amylopectin
α-glucan, 363, 364*f*
hybrid structures with, like structures, 373–375
schematic of hybrid structures with, 375*f*
See also Amylose

Immobilized porcine pancreas
lipase (IPPL); Poly(acrylic
acid)/poly(vinyl alcohol)
(PAA/PVA) hydrogel fibers;
Poly(butylene succinate)
Lipids, novel biomaterials, 3

M

Macroinitiation. *See* Chemoenzymatic
synthesis
Maltose, percentage release from
starch in bioartificial film, 70, 71*f*
Maltotetraose hybrids
varying architecture, 371, 372*f*
See also Alditols, lipase-catalyzed
polymerization
Mechanism
Candida antarctica lipase B
(CALB), 233–234, 235
cellulase catalysis, 326*f*, 327*f*
chitinase catalysis, 328*f*, 329*f*
glycoside hydrolases, 325, 330
kinetic model for peptide synthesis,
298, 299
postulated, for conversion of
benzylsilane *cis*-diol to *o*-
substituted phenol, 438, 439,
442
Messenger ribonucleic acid (mRNA),
peptide aptamers, 197, 199*f*
Methyl esters
enzyme-catalyzed formation, 83
soybean oil (SBO), 80, 81*f*
See also Soybean oil (SBO)
Methyl methacrylate (MMA). *See*
Polyacrylic resins with
immobilized enzyme
Methyl *p*-nitrophenyl *n*-
hexylphosphate (MNPHP), lipase
activity, 159, 161, 168, 170
Molecular weight study. *See* Alditols,
lipase-catalyzed polymerization;
Poly(butylene succinate)

Monomers, biocatalytic routes, 13
Morphology, biosilica formation, 426,
428*f*, 429*f*
Mullin effect, phenomenon, 100

N

Nanoreactors
laccase and diblock copolymer,
5
See also Oxidation of steroids
Naphthalene dioxygenase,
regioselectivity, 447, 454*f*
Natural polymers
enzymatic hydrolysis and
degradation, 12–13
interaction between, and synthetic
polymers, 52–53
See also Bioartificial materials
Nature's polymers
tropoelastin and elastin, 38–39,
42
See also Elastin
Non-natural oligomers
biocatalysts with nonnatural
components, 205–206
combinatorial bioengineering, 203–
206
oligonucleotide aptamers with
nonnatural components, 203–
204
peptide aptamers with nonnatural
components, 204–205
Norovirus, glycohydrogels entrapping,
345, 346*f*, 355–359
Novozym 435
Candida antarctica lipase B
(CALB) on acrylic resin, 231
See also ω-Substituted lactones
Nuclear magnetic resonance (NMR),
experiments on modified peptides,
27, 29*f*
Nylon polymers. *See* Polyamides and
derivatives